Isaac Elishakoff
Multifaceted Uncertainty Quantification

Also of interest

Hierarchical Composite Materials.
Materials, Manufacturing, Engineering
Kaushik Kumar, J. Paulo Davim (Eds.), 2019
ISBN 978-3-11-054400-8, e-ISBN 978-3-11-054510-4

Freischneiden in der Festigkeitslehre
Philipp Steibler, 2017
ISBN 978-3-11-048118-1, e-ISBN 978-3-11-048123-5

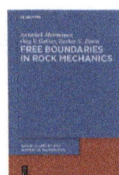

Free Boundaries in Rock Mechanics
Anvarbek Meirmanov, Oleg V. Galtsev, Reshat N. Zimin, 2017
ISBN 978-3-11-054490-9, e-ISBN 978-3-11-054616-3

Optimal Structural Design.
Contact Problems and High-Speed Penetration
Nikolay V. Banichuk, Svetlana Yu. Ivanova, 2017
ISBN 978-3-11-053080-3, e-ISBN 978-3-11-053118-3

Isaac Elishakoff

Multifaceted Uncertainty Quantification

—

DE GRUYTER

Author
Prof. Dr. Isaac Elishakoff
elishako@fau.edu
Department of Ocean and Mechanical Engineering
Department of Mathematical Sciences
Florida Atlantic University
777 Glades Road
Boca Raton, FL 33431-0991
United States of America

ISBN 978-3-11-135421-7
e-ISBN (PDF) 978-3-11-135423-1
e-ISBN (EPUB) 978-3-11-135473-6

Library of Congress Control Number: 2024906254

Bibliographic information published by the Deutsche Nationalbibliothek
The Deutsche Nationalbibliothek lists this publication in the Deutsche Nationalbibliografie;
detailed bibliographic data are available on the internet at http://dnb.dnb.de.

© 2024 Walter de Gruyter GmbH, Berlin/Boston
Typesetting: Integra Software Services Pvt. Ltd.

www.degruyter.com

Dedicated to my wife, Dr. Esther Elisha, who introduces much needed certainty into this world.

Preface

There are known knowns. There are things we know we know. We also know there are known unknowns. That is to say, we know there are some things we do not know. But there are also unknown unknowns, the ones we don't know we don't know. Donald Rumsfeld, *Known and Unknown: A Memoir*, 2011

The problem with experts is that they do not know what they do not know.

Nassim Nicholas Taleb, *The Black Swan: The Impact of the Highly Improbable*, 2007
Living systems in general hate uncertainty.
Beau Lotto, *Deviate*, 2017

Uncertainty quantification and propagation are the subjects of intense development nowadays. It felt instructive to stop and reflect on their numerous manifestations. Sources of uncertainty are numerous. They include parameter uncertainty; parametric variability; structural uncertainty; algorithmic uncertainty; experimental uncertainty; interpolation uncertainty. Last but not least, there are so-called aleatoric and epistemic uncertainties. There are two types of problems in uncertainty propagation: These are forward and inverse problems. Wikipedia (2020) states that

> The targets of uncertainty propagation analysis can be:
> - To evaluate low-order moments of the outputs, i.e., mean and variance.
> - To evaluate the reliability of the outputs. This is especially useful in reliability engineering where outputs of a system are usually closely related to the performance of the system.
> - To assess the complete probability distribution of the outputs. This is useful in the scenario of utility optimization where the complete distribution is used to calculate the utility.

And, following the immortal Shakespeare, from his *Hamlet* (1.5.167-8), we can say:

> There are more things in heaven and earth, Horatio,
> Than are dreamt of in your philosophy.

We, just like Hamlet or Horatio, can debate the uncertainty quantification and propagation issues, described by the above Wikipedia. Chaudhuri *et al.* (2015) write: ". . .uncertainties must be reduced to an acceptable level and this requires their quantification."

In her definitive article, Paté-Cornell (1996) identifies six levels of treatment:

Level 0: Hazard detection and failure modes' identification. Level 0 simply involves the detection of a potential hazard or of the different ways in which a system can fail, without attempting to assess the risk in any quantitative way. This approach is sufficient, in theory, to support strict zero-risk policies, or to make risk management decisions when the costs are low and the decision is clear. . .

Level 1: "worst-case" approach Level 1 is the "worst-case" approach. It does not involve any notion of probability. It is based on the accumulation of worst-case assumptions and yields, in theory, the maximum loss level. It is reasonable if the worst loss ("what do ! risk?") is sufficient to support the decision. . .

https://doi.org/10.1515/9783111354231-202

VIII —— Preface

Level 2: quasi-worst cases and plausible upper bounds Level 2 involves "plausible upper bounds" (or "quasi-worst cases"). This analysis represents an attempt to obtain an evaluation of the worst possible conditions that can be "reasonably" expected (1) when there is some uncertainty as to what the worst case might be or (2) when the worst case is so unlikely (or infrequent) that it is meaningless. Examples of these approaches include the Maximum Credible Earthquake ("largest" magnitude and "minimum" distance from site) used in some building codes, 34 or the Maximum Probable Flood used by the US Army Corps of Engineers in the construction and management of

Level 3: best estimates and central values Level 3 relies on a "best estimate" and/or on a central value (e.g., the mean, the median or the mode) of the outcome (e.g., loss) distribution, generally through "best estimates" of the different variables. It is currently one of the directions in which legislators and government agencies seem to be heading (in addition to other types of information including the effects of uncertainties) for a more realistic assessment of health risks than through the plausible upper bounds alone. . .

Level 4: probabilistic risk assessment, single risk curve Level 4 relies on the probabilistic risk analysis (PRA) process described briefly in the previous section and also found in the literature under the names of quantitative risk assessment (QRA) or probabilistic safety assessment (PSA). . .

Level 5: probabilistic risk analysis, multiple risk curves Level 5 allows the display of uncertainties about fundamental hypotheses by a family of curves. This can be done in several ways including a statistical treatment (Bayesian inference) of existing data. Another approach is to ask a group of experts to provide an assessment of the risk based on their preferred model, and on their evaluation of the distribution of parameter values given this model. . .

Helton and Burmaster (1996) stress that uncertainties are needed for (a) risk assessment in the federal government: managing the process; (b) environmental standards for the management and disposal of spent nuclear fuel, high-level, and transuranic waste; (c) safety goals for the operation of nuclear power plants.

This book, therefore, discusses various approaches to uncertainty analysis. These are, naturally, the random vibration (Chapter 2), reliability (Chapter 3), stochastic linearization (Chapter 5), and Monte Carlo method (Chapter 6). Chapter 6 deals with what might go wrong with probabilistic approaches, whereas Chapter 8 discusses possible limitations of probabilistic methods. Chapter 7 is devoted to fuzzy-sets bases approaches, whereas Chapter 10 reviews anti-optimization or worst-case analysis methodology, but sine this methodology could yield extremely conservative designs the optimization of anti-optimized solutions is called for.

Chapter 11 shows that the uncertain-but-bounded variables approach and stochastic methodology are fully compatible, although, at first sight, they might appear to be contradictory. Chapter 4 deals with numerical aspect, namely the finite element method in either stochastic, fuzzy, or anti-optimization viewpoints. Chapter 12 concludes with the comforting assertion that each of the above, competing methodologies are valid.

Author will be indebted to hear from critically minded reader by electronic mail elishako@fau.edu, by fax 561-297-3825, or by regular mail (the latter was extensively used by the founding fathers of probability, Pascal and Fermat, in their famous correspondence).

About the author

Prof. Isaac Elishakoff serves as a Distinguished Research Professor at the Florida Atlantic University, U.S.A. for 35 years. Prior to that he was a Full Professor at the Technion—Israel Institute of Technology, Israel, where he taught for 18 years. He got his Master's and Ph.D. degrees from Moscow Power Engineering Institute and State Research University, in Moscow, Russia, with Academician V.V. Bolotin as his advisor.

He is the author or co-author of 18 books, and editor or co-editor of another 14 books in addition to over 600 scientific papers.

He occupied several visiting positions, serving as a Distinguished Research Professor at the Technion-I.I.T., Israel; a Theodore von Kármán Fellow at the University of Aachen, Federal Republic of Germany; S. P. Timoshenko Scholar at Stanford University, U.S.A; Eminent Scholar at the Beijing University of Aeronautics and Astronautics, P. R. C.; Fellow of the Royal Engineering Academy, U.K. and the W. T. Koiter Chair Professor at the Delft University of Technology, The Netherlands, among many other honorific appointments in U.S.A., Italy, Japan, and China.

He is the recipient of Batsheva de Rotschild Prize in Israel; Worchester Warner Medal of American Society of Mechanical Engineers, U.S.A. and Blaise Pascal Medal from European Academy of Sciences in Engineering, of which he is a Fellow.

Three-volume book titled *Modern Problems in Structural Mechanics* (N. Challamel, J. Kaplunov, and I. Takewaki, eds.), London : ISTE-Wiley, was dedicated to him.

https://doi.org/10.1515/9783111354231-203

Contents

Chapter 1
Introduction: What this book is all about?

> As far as the laws of mathematics refer to reality, they are not certain; and as far as they are certain, they do not refer to reality (Albert Einstein).

According to Queen Beatrix of the Netherlands, reigning between the years 1980 and 2013, "If one thing today is certain, it is a feeling of uncertainty – a premonition that the future cannot be a simple extension of the present", as she stated in her speech of allegiance to the Dutch constitution, on April 30, 1980.

Some authors *ab initio* identify uncertainty with randomness. For example, Bucher's (2009) excellent book is tellingly titled *Computational Analysis of Randomness in Structural Mechanics.* The definitive book by Grigoriu (2002) about stochastic calculus states that its main objective "is the solution of stochastic problems, that is, determination of the probability law, moments, and/or other probabilistic properties of the state of a physical, economic, or social system. It is assumed that the operators and the inputs defining a stochastic problem are specified. We do not discuss the mathematical formulation of a stochastic problem, that is, the selection of the functional form of the equation and the probability laws of its random coefficients and input for the stochastic problem."

Likewise, there is a plenitude of books on structural reliability. These include books by Madsen, Krenk, and Lind (2006), Lemaire, Chateauneuf, and Mitteau (2009), Verma, Ajit, and Karanki (2010), Nowak and Collins (2013), Leira (2013), Der Kiureghian (2022), and others.

It can be safely stated that the stochastic community is doing a great job of educating funding agencies as well as the public in general that uncertainty should be accounted for. Unfortunately, most funding agencies or students of uncertainty do not know yet that probabilistic methodology is not the only one. Kosko (1994) speaks about probabilistic "monopoly" on uncertainty: "Fuzzy theory challenges the probability monopoly. Probabilists have attacked it with gusto to keep their monopoly status, to have, as Jaynes (1979) and Lindley (1987) want, the only uncertainty theory in the unit interval [0, 1]. But the fuzzy math is sound. Its worldview of shades of gray has a deep intuitive ring. And the new fuzzy products have come into their own in the marketplace. The probability monopoly is over." Likewise, Rudolf Kalman (2004) speaks, with an obvious but expressive exaggeration, in the humble opinion of this author, about the so-called probabilistic "quasi-religion" (Kalman, 1994).

In the short paper by Elishakoff (1990) the idea was propagated that uncertainty comprises three various analytically treatable items, as depicted in Fig. 1.1.

https://doi.org/10.1515/9783111354231-001

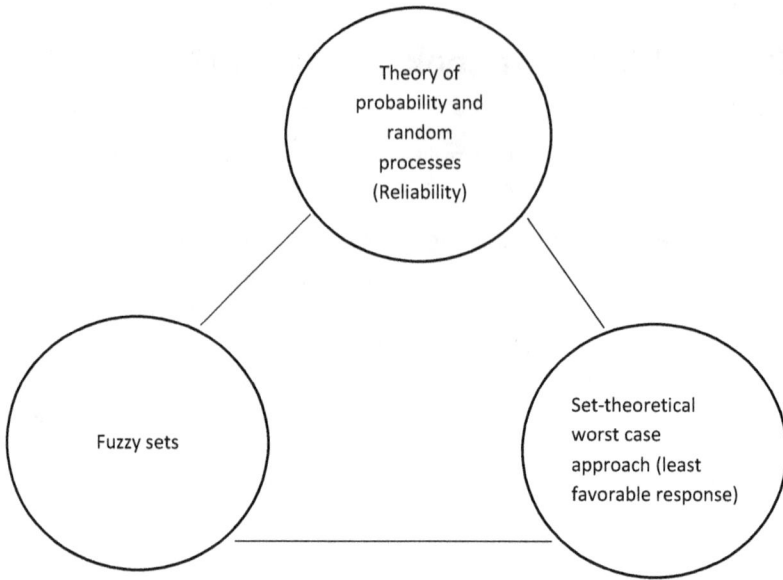

Fig. 1.1: Uncertainty triangle.

The components of uncertainty are (a) the theory of probability and random processes, leading to the determination of probabilistic characteristics or reliability, (b) fuzzy sets-based methodologies, and (c) approaches dealing with unknown-but-bounded variables. The latter approach includes interval methodology, ellipsoidal calculus, or super-ellipsoidal calculus.

Chapter 2
Topics in random vibration of structures

I have been searching for decades for certainty. (Jordan B. Peterson, 2021)

Uncertainties appear everywhere in the model. When using a mathematical model, careful attention must be given to the uncertainties in the model. (Feynman, 1988)

In this chapter some topics on random vibration of structures are reviewed. For this occasion, three topics are chosen: (a) effect of cross-correlations in linear vibration of continuous structures, (b) effect of choice of deterministic theory describing the dynamic behavior of the structure, (c) new versions of stochastic linearization of nonlinear continuous structures. The chapter attempts to partially answer a familiar but still nagging question: "Now that I'm here, where am I ?"

2.1 Introduction

Accounts on recent developments in random vibrations were written by Vanmarcke (1979), Crandall and Zhu (1986) and Lin *et al.* (1986). Since then, almost a decade has passed, and it appears that it is time for another review. Yet, due to the huge amount of literature in this field, it is almost impossible to mention everything that deserves to be mentioned. Inevitably, in such circumstances, personal perspectives pre-dominate the discussion. This review is no different: as such, it may contain personal biases. Nevertheless, this writer hopes that the points made will be well taken. (To recall from Bill Cosby, *"I do not know the key to success, but the key to failure is trying to please everybody."*)

We sadly observe that some researchers who dealt with single degree-of-freedom nonlinear systems in 1968, in many cases, have not moved on to two-degree of freedom nonlinear systems. To take the liberty of quoting one of the reviewers of this chapter, ". . . obviously, the classes of problems being considered in one and two DoF has become much more complex; more is being required of the analysis, and the level of the discussion has risen." It is understood that the study of a single DoF system is an important instructive step. However, this step must lead, it is the firm belief of the present writer, to the multi-degree-of-freedom approximation of complex, continuous structures. Here is another area of paramount importance, as noted to us by one of the reviewers, to say in his/her words:

many workers will adopt simple equations of motion without even showing that the equation has any connection to a reasonable physical model. While this is also true in the world outside random vibrations, it is especially tempting for the stochastic analyst to pick an equation of motion for which there is a theory. Thus, the great dependence on 'white noise,' on Duffing oscillators, on the Fokker-Planck equations. This is wrong and it hurts the credibility of the discipline.

https://doi.org/10.1515/9783111354231-002

Some researchers might maintain that even in multi-degree-of-freedom systems only several numbers of important modes "decide" the behavior of the system, and hence the single, or overall, two-degree-of-freedom system approximation may be justified. This statement is only partially true. Indeed, qualitative explanation of the imperfection sensitivity of shells in either a static or a dynamic setting can be given based on the single-degree-of-freedom system, say on Koiter's model (1972) or the Budiansky-Hutchinson model (1964). However, in order to obtain deterministic results comparable with those in experiments, or in order to perform accurate reliability predictions, one may need to include 15 or more modes, as shown by Elishakoff and Arbocz (1985).

This illustrates that although low-order approximations may sometimes explain the physical phenomenon, accurate quantitative results generally require many more modes.

2.2 Statistical energy analysis versus random vibrations

For realistic complex structures engineers do not appear to utilize the achievements of the modern random vibration theory but rely on the statistical energy analysis (SEA), as was indicated in the title of his lecture at Florida Atlantic University by Professor B.L. Clarkson (1994). This method in essence does not look in the space-wise averages of the responses.

The following justifications are given by those who use SEA. As Mehta maintains, ". . . Statistical theory does not predict the detailed level sequence [resonant frequencies] of any one nucleus . . . there is a reasonable expectation, through no rigorous mathematical proof, that a system under observation will be described correctly by an ensemble average . . . if this particular [system] turns out to be far removed from the ensemble average, it will show that [the system] . . . possessed specified properties of which we are not aware. This then, will prompt us to try to discover nature and origin of these properties."

Lyon (1975) mentions that ". . . the resonance frequencies and mode shapes of these models show great sensitivity to small details of geometry and construction. In addition, the computer programs used to evaluate the mode shapes and frequencies are known to be rather inaccurate for the higher order modes, even for rather idealized systems."

Finally, Smith and Lyon (1976) stress that ". . . In the classical approach to a vibratory problem one usually asks, 'What is the dynamic displacement of a particular point at a particular instant?' Now in many practical problems this is a most unreasonable question. As with the question 'What is the present population of China?', no reasonable effort can yield an answer. Even if an answer were forthcoming, from that ideal computer that the analysts dream of, it would not be useful because *particular* points and *particular* instants are not really of concern, and the collection of data for *all* points and for *all* instants would be overwhelming. To get a useful answer, some

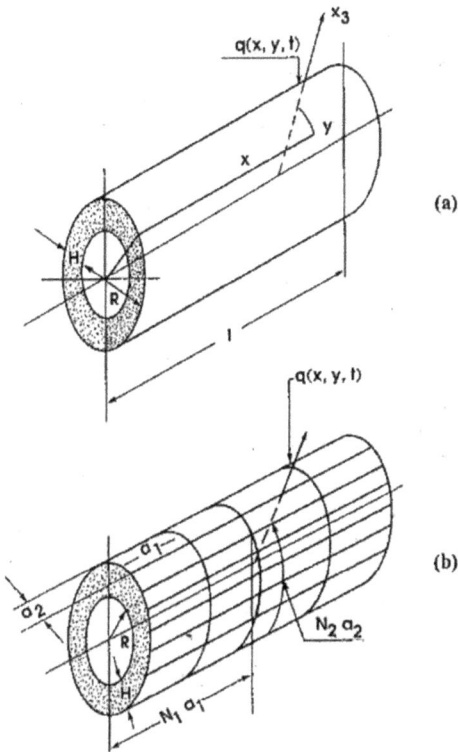

Fig. 2.1: Cylindrical shells containing acoustic medium under turbulent pressure fluctuations $q(x,y,t)$; hatched area denotes sound-insulation system: (a) unstiffened shell, (b) stiffened shell.

different question must be posed; let us try, 'What is the average dynamic response in a root-mean-square.'"

Fahy (1993) maintains that SEA currently is the only method for the prediction of high-frequency vibration of complex systems that are supported by commercial software (*e.g.*, SEAM, AutoSEA, VAPERS, GSSEACAL). Still, Fahy is critical of SEA: "It is superficially attractive because it is so much cheaper and quicker to apply than computationally intensive methods such as the finite element method (FEM) or the boundary element method (BEM). However, behind its simple façade lurk some little publicized traps. Attempts to apply SEA in ignorance of these traps can lead to dangerously uncertain conditions. SEA in its current stage of development can, in the hands of the unexperienced, truly be a 'Wolf in Sheep's Clothing.'"

Indeed, we cannot fully agree with these authors' viewpoint that the discretization techniques are inapplicable to large complex structures. In the old work by the present writer (Elishakoff, 1971), 6,500 coupled elastic-acoustic modes were taken into account to compute the noise levels in smooth (Fig. 2.1a) or stiffened (Fig. 2.1b) cylindrical shells with realistically modeled sound-insulation layers enclosing an acoustic medium and excited by the random fluctuations in the turbulent layer (see also paper by Bolotin and Elishakoff, 1971). In contrast to the overwhelming majority of studies, no assumptions were made on the auto-correlation function, but the existing experimental results (Fig. 2.2) were directly incorporated into the analysis.

Roozen (1992) was successful in utilizing a FEM model of approximately 550,000 degrees of freedom to study the vibration of a two-meter-long aircraft fuselage at frequencies up to 225 Hz. Straightforward applications of FEM often can be very cumbersome. In such circumstances, intelligent techniques, combining the power of FEM with some properties of the structure, must be utilized. An example of such a clever combination is described in the paper by Abdel-Rahman (1979), in which the FEM model and the wave propagation constants method by Mead (1971) were successfully superimposed.

Therefore, it appears to the present author that special FEM techniques must be developed to serve as sound alternatives to the SEA, asymptotic modal analysis (Kubota *et al.*, 1990), and Skudrzyk's mean-value method (1980, 1987). A combination of the FEM with the SEA can open an additional avenue to the modern random vibration analysis of complex structures (Steel *et al.*, 1994). An alternative technique by Soize (1986, 1993) (see also Soize *et al.*, 1992) for response of structures due to excitation in medium-range frequencies appears to be of promising potential (maybe instead of the term "internal fuzzy" one could use "appendages of uncertain properties," since no use is made of the theory of fuzzy sets).

2.3 Crandall's problem and cross-correlations

In the 1970s and 1980s, Stephen Harry Crandall (1920–2013), who is rightfully and unanimously recognized as the father of random vibrations, studied several problems of random vibrations of strings, beams, and plates excited by the wide-band excitation in time and applied at selected points of the structure (1974, 1977, 1979, 1980, 1988; see also Crandall and Wittig, 1972; Crandall and Kulvets, 1977, 1979; Itao and Crandall, 1978; Elishakoff *et al.*, 1979; Crandall and Zhu, 1983a, 1984). He has uncovered extremely interesting results of localization of the response. In the case of a uniform beam (Fig. 2.3a) with identical boundary conditions at its ends, the responses in the driven location $\xi = \frac{x}{L}$, $\alpha = \frac{a}{L}$ as well as in its symmetric counterpart $\xi = 1 - \alpha$ turned out to be identical (see Fig. 2.4, where $\alpha = 0.3$) and are about 50% higher than the response occurring in the regions far from the disturbance x is the axial coordinate, ξ is the nondimensional axial coordinate, $L =$ length. For plates of different configuration, he and his associates discovered lanes of intensified response.

Beams exhibit a special dynamic behavior: The natural frequencies are well apart; i.e.,the modal density is not high. In order to see if some additional effects may be important, Crandall's problem has been generalized (Elishakoff *et al.*, 1979) in the following manner: instead of discussing the beam, a thin cylindrical shell of uniform thickness (Fig. 2.3b) with identical shear diaphragms at its ends was considered.

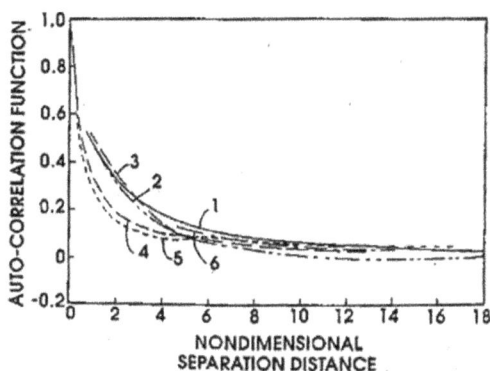

Fig. 2.2: Experimentally determined autocorrelation function of the turbulent pressure fluctuations in the supersonic flight (curves 1–6 denote various analytical approximations; for details see Elishakoff, 1971).

Instead of the point load, a uniform ring load, at the section $x = a$, was applied so that the axial symmetry was preserved. This was not done just for simplicity of analysis but rather from the considerations of the physics of the problem which would otherwise be obstructed by mathematical detail. (Following Einstein's dictum, we must make problems as simple as possible, but not simpler.) For our purposes the simplest problem will be the axisymmetrically vibrating shell.

I will concentrate on describing the final, numerical results for the mean-square velocity response.

The curve (Fig. 2.5) marked s_1 denotes the terms associated with direct correlations, whereas the curve marked d_v^2 denotes the terms associated with cross-correlations. The former sums up all the contributions derived as a result of the interplay of identical modes, whereas the cross-correlation terms sum up the contributions stemming from different modes. For the relatively thick shell the cross-correlations are relatively small and neglected for practical purposes. Yet, it is noteworthy that they do not possess any specific property of symmetry or anti-symmetry. We just note that in the driving location $\xi = a$ the cross-correlation terms make a positive contribution (Fig. 2.5), whereas for the mirror image cross-section ($\xi = 1 - a$) cross-correlations have a negative sign. This implies that even in the case of the innocent uniform beam, there is no perfect symmetry in the response distribution: the mean-square response in the driving location exceeds that in the mirror image location.

Thus, cross-correlation terms, although small in the case of the beam, still alter the picture qualitatively. With the cross-correlation terms taken into account, the space-wise distribution of the mean-square response is non-symmetric with respect to

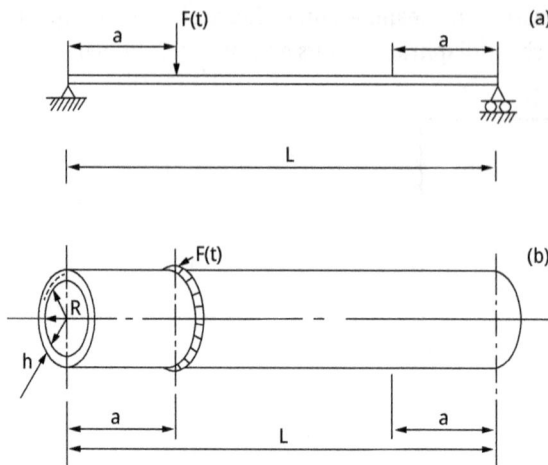

Fig. 2.3: Crandall's problem: uniform structures with identical boundary conditions subjected to wide-band excitation $F(t)$: (a) uniform beam simply supported at its ends under point load, (b) cylindrical shell of uniform thickness, supported by shear diaphragms, and subjected to a ring load $F(t)$.

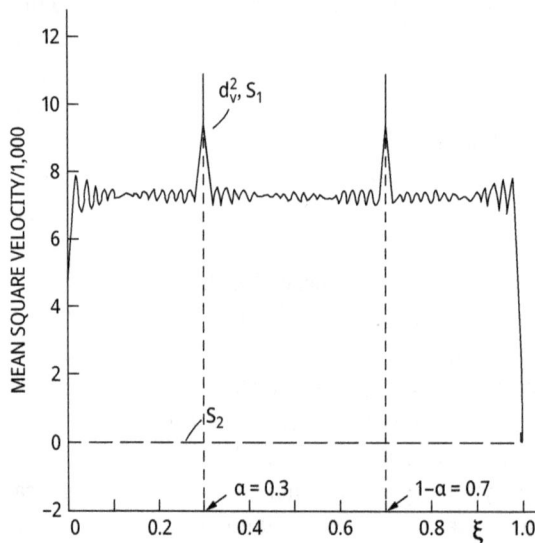

Fig. 2.4: Localization of the response in the beam under wide-band random excitation; $a = \dfrac{a}{L}$ driving location; $1 - a =$ mirror image of the driving location, with respect to the middle cross-section of the beam; $s_1 =$ direct correlation terms, $s_2 =$ cross-correlation terms, $d_v^2 =$ mean-square velocity response.

the middle cross-section of the structure. The cross-correlation terms I_{jk} are of the following magnitude, for the viscously damped structure (Bolotin, 1969; Der Kiureghian, 1980; Elishakoff, 1983):

$$I_{jk} = \frac{8\left(\varsigma_j \omega_j + \varsigma_k \omega_k\right)}{\left(\omega_j^2 - \omega_k^2\right)^{2+4}\left[\varsigma_j \varsigma_k \omega_j \omega_k \left(\omega_j^2 + \omega_k^2\right) + \left(\varsigma_j^2 + \varsigma_k^2\right)\omega_j^2 \omega_k^2\right]} \tag{2.1}$$

Here, ω_j are natural frequencies ς_j are damping coefficients, and i and j are serial numbers corresponding to ith and jth modes, respectively.

As early as in 1961, Bolotin (1926–2008) formulated the condition for negligibility of cross-correlations for the case of the viscously damped structure:

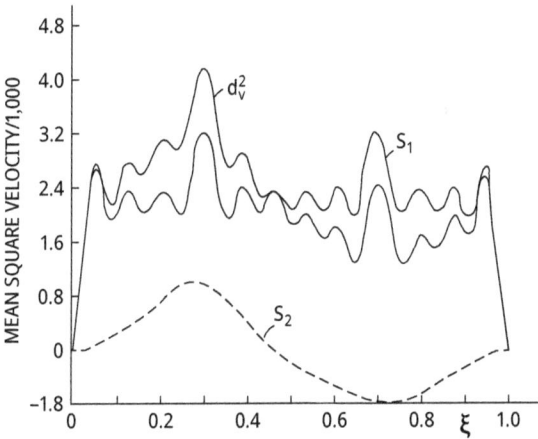

Fig. 2.5: Localization of the response in the shell under wide-band random excitations; s_1 = direct correlation terms, s_2 = cross-correlation terms; d_v^2 = mean-square velocity response, $\frac{B}{L} = 0.5$, $\frac{h}{L} = 0.01$, $\Omega_c = 2\pi$, $N_c = 14$ (after Elishakoff et al., 1979).

$$\left|\omega_j^2 - \omega_k^2\right| \gg \max\left(\omega_j \varsigma_j, \omega_k \varsigma_k\right) \tag{2.2}$$

For the structurally damped structures (Elishakoff, 1977),

$$I_{jk} = \frac{\pi}{\omega_j \omega_k} \frac{\sqrt{1+\mu}}{2} \frac{\left(\omega_j + \omega_k\right)\left[a\mu\left(\omega_j^2 + \omega_k^2\right) - \beta\left(\omega_j - \omega_k\right)^2\right]}{\left(\omega_j^2 + \omega_k^2\right)^2 + 2\mu^2\left(\omega_j^4 + \omega_k^4\right) + \mu^4\left(\omega_j^2 + \omega_k^2\right)^2}$$

$$a = \sqrt{\sqrt{1+\mu^2}+1}, \ \beta = \sqrt{\sqrt{1+\mu^2}-1} \tag{2.3}$$

Here, μ is the coefficient of structural damping. In 1977, the present author formulated a condition of negligibility of cross-correlation terms when structural damping is involved:

$$\left|\omega_j^2 - \omega_k^2\right| \gg \mu \max\left(\omega_j^2, \omega_k^2\right) \tag{2.4}$$

On the effects of cross-correlations, the reader may also consult Davies (1965), Garellik and Chayes (1978), and Hagedorn and Nascimento (1985).

In beams the natural frequencies are spaced well apart, whereas in the case of the shell there may be regions of their high concentration. For an axisymmetrically vibrating cylindrical shell, the frequencies ω_j are clustered around the first frequency:

$$\omega_1^2 = \frac{D}{\rho h} \left(\frac{\pi}{L}\right)^4 + \frac{E}{\rho R^2} \tag{2.5}$$

The thinner the shell, the more natural frequencies fall in the same frequency range, i.e. the greater the modal density. When the shell thickness decreases, there will be more natural frequencies that will violate the conditions of negligibility of cross-correlations. At some thickness the cross-correlation effect becomes more important than that of the direct correlation. As a result, the response becomes totally non-symmetric, although there is still a remnant of magnified response in the mirror image location $\xi = 1 - \alpha$.

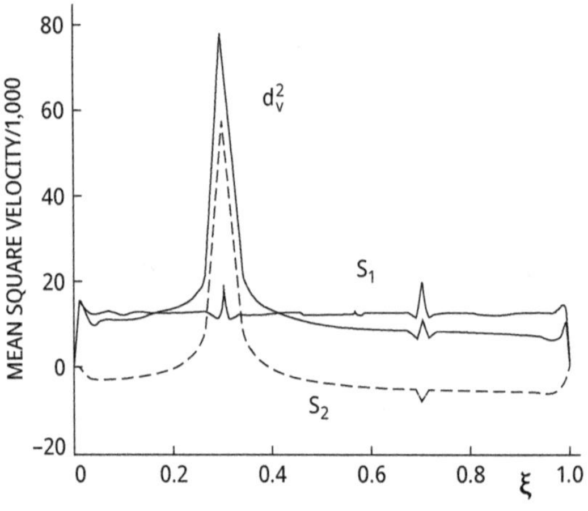

Fig. 2.6: Localization of the response in the shell under wide-band random excitation; $s_1 =$ direct correlation terms, $s_2 =$ cross-correlation terms; $d_v^2 =$ mean-square velocity response, $\frac{R}{L} = 0.5$, $\frac{h}{L} = 0.0003$, $\Omega_c = 2\pi$, $N_c = 81$ (after Elishakoff et al., 1979).

The solid curve is associated with the total response (i.e., sum of direct and cross-correlations). It indicates that the true response may exceed its counterpart obtained by the neglect of cross-correlations, by a factor four or more (Fig. 2.6).

However, the results for the shell are derived by utilizing extensive numerical analysis. This example is also included in the textbook by Preumont (1994). Could the results be the artifact of numeric analysis? To answer this question, the present writer devised a model of a two-degree-of-freedom system (Elishakoff, 1983) that may exhibit, as it were, the behavior of either of a beam or a shell. Consider a two-degree of freedom system, subjected to $F_1(t) =$ random excitation with zero mean and ideal white noise intensity S_0 (Fig. 2.7). The problem consists in finding the mean-square values of the displacements $X_1(t)$ and $X_2(t)$ velocities $\dot{X}_1(t)$ and $\dot{X}_2(t)$. The control parameter ε characterizes a degree of coupling of two masses. We immediately note that as $\varepsilon \to 0$, for the driven mass we should recover the result for a single-degree-of-freedom system, whereas for an un-driven mass the mean-square value will become zero as the response of an unexcited system. The analysis yields:

$$E\left(X_{1,2}^2\right) = \frac{\pi S_0}{4kc}\left[1 + \frac{1}{(1+2\varepsilon)^2} \pm \frac{\frac{2c^2}{km}}{\frac{\varepsilon^2}{(1+\varepsilon)} + (1+\varepsilon)/\left(\frac{c^2}{km}\right)}\right]$$

$$E\left(\dot{X}_{1,2}^2\right) = \frac{\pi S_0}{4mc}\left[1 + \frac{1}{1+2\varepsilon} \pm \frac{\frac{2c^2}{km}}{\frac{\varepsilon^2}{1+\varepsilon} + (1+\varepsilon)\left(\frac{c^2}{km}\right)}\right] \tag{2.6}$$

The plus sign is associated with the first mass, whereas the minus sign corresponds to the second mass. The first two terms in eqs. (2.5) and (2.6) are associated with direct correlations. For $\varepsilon \to 0, E\left(X_1^2\right) \to \frac{\pi S_0}{kc}, E\left(\dot{X}_1^2\right) \to \frac{\pi S_0}{mc}$, as they should be since the latter results pertain to a single DoF system. Also, $E\left(X_2^2\right) \to 0$, $E\left(\dot{X}_2^2\right) \to 0$, since the second mass becomes a separate unexcited system. When ε tends to zero, direct and cross-correlations contribute equally, and hence the neglect of cross-correlations leads to an error of 50% in estimating $E\left(X_1^2\right)$ (Fig. 2.8a). When $\varepsilon \to 0$, the mean-square response of an un-driven mass $E\left(X_2^2\right)$ tends to a value $\frac{\pi S_0}{2kc}$ instead of yielding zero if we neglect the cross-correlations. Thus, the error associated with the neglect of cross-correlation becomes unbounded (Fig. 2.8b).

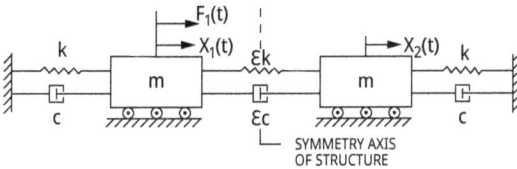

Fig. 2.7: Two-degree-of-freedom system subjected to a wide-band random force: for large values of ε system behaves on the beam; for small values of ε system behaves as the shell (after Elishakoff, 1983).

The natural frequencies of the system read

$$\omega_1^2 = \frac{k}{m}, \quad \omega_2^2 = \frac{k}{m}(1+2\varepsilon) \tag{2.7}$$

If ε is not small, the natural frequencies are far apart, and the system behaves like the beam. Indeed, even at moderate value of $\varepsilon = 1$ and $\frac{c^2}{km} = 0.01$ the contribution of direct correlations is 96.72%, whereas the contribution of cross-correlations is 3.28%. Then approximately, Crandall's interesting result

$$E(X_1^2) \approx E(X_2^2) \tag{2.8}$$

holds. However, for extremely small ε, natural frequencies shift toward each other, and the system behaves like a shell.

What do the vibrations textbooks say on the subject? Meirovich (1986) writes:

On the other hand, the integral

$$\int_{-\infty}^{\infty} (H_1^* H_2 + H_1 H_2^*)d\omega \tag{2.9}$$

requires once again the use of the residue theorem. Because no new knowledge is gained from the evaluation of the integral, we shall not pursue the subject any farther.

As we saw however, when natural frequencies are crowding together, significantly new knowledge can be derived from evaluating the cross-correlation integrals.

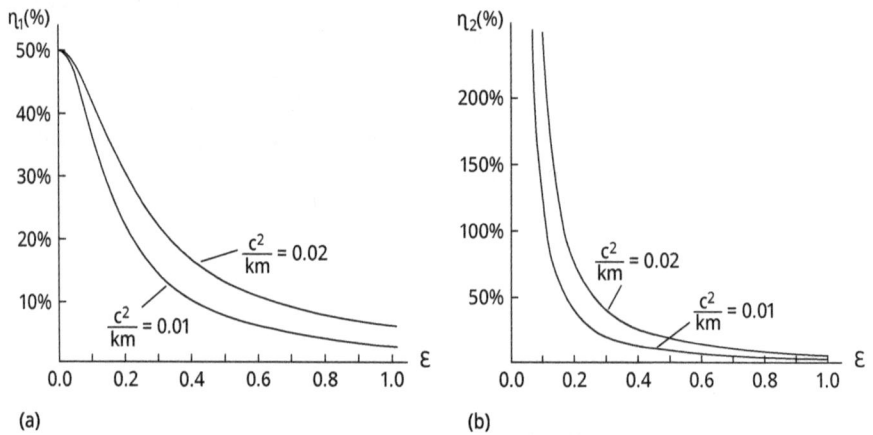

(a)   (b)

Fig. 2.8: Percentagewise error in estimation of mean-square responses of (1) first mass, (2) second mass.

Interestingly, when Joseph-Louis Lagrange (1736–1813) considered deterministic vibrations of conservative bodies, he also discussed the problem of equal frequencies. He arrived at the unexpected conclusion that when characteristic roots r_1 and r_2 are equal, the solutions must read $exp(irt)$ and $t \cdot exp(irt)$, which may mean, as Lagrange maintained, an unstable motion. Stunned with the possible instability of an innocent system, just due to having coinciding frequencies, Lagrange mentioned (see Panovko and Gubanova, 1973):

As these cases are not strictly relevant to the title problem, we shall not elaborate them here.

Another textbook, by C.Y. Yang (1986), when describing random vibration of plates, mentions:

We also assume that the average modal overlap ratio

$$R = \frac{\text{modal bandwidth}}{\text{average modal spacing}} \qquad (2.10)$$

is sufficiently small, to permit approximations analogous to Eq. (1.9) that is we only only include in the summation in Eq (1.10) those terms of I_{jpkq} for which $j = k$, $p = q$, and the associated natural frequency ω_{jp} is less than the out-off frequency. For all such terms

$$I_{jpjp} = \frac{\pi S_0}{m^2 \beta_{jp}}. \qquad (2.11)$$

As is seen, the author neglects the cross-correlations in plates. However, plates may have many coinciding frequencies. For example, the square plate possesses coalescent frequencies of type $\omega_{jp} = \omega_{pj}$; it also has three coinciding frequencies $\omega_{1,7} = \omega_{7,1} = \omega_{5,5}$, four coinciding frequencies $\omega_{5,9} = \omega_{9,5} = \omega_{11,1} = \omega_{1,11}$, etc., and the neglect of cross-correlation terms will be illegitimate. The effect of such frequencies in curved shell vibrations was studied by Elishakoff (1977), and for plates, by Crandal and Zhu (1983a) and Elishakoff (2020a).

Nigam (1983) mentions:

It follows from the orthogonality property of the normal modes that . . . cross terms do not contribute to the [spacewise-IE] integrated variance. Hence cross terms account only for a local variation of the response about its [spacewise-IE] average value.

As we saw, the cross-correlations may alter both qualitatively and quantitatively the variation of the response. As it was demonstrated the estimate of the noise level inside a stiffened shell with sound insulation layers may be underestimated by as much as 10 db. Der Kiureghian (1981), Wilson *et al.* (1981), and Singh (1973) have introduced the effect of cross-correlations into the response spectrum method for random vibration analysis of seismically excited structures. As Der Kiureghian showed, the so-called SRSS rule (square root of sum of the squares) may lead to serious errors for closely spaced modes. Cross-correlation terms turn out to be important when studying critical configurations of systems (Igusa, 1991). To sum up, the cross-correlations effects are most important when we are looking for the local maxima of the probabilistic responses, rather than their spatial averages. As Dowell (1971) mentions that these maxima govern the safe performance and integrity of structures. Crandall's problem has been extensively studied by Langley and Taylor (1979), Riemer and Wedig (1981), Elishakoff (1983), Wedig (1982, 1992), Nascimento (1983, 1989), Wallaschek (1986, 1987),

Nascimento and Wallaschek (1986), Zhu and Lei (1986), and Elishakoff (2020a). Interested readers can also consult with definitive work by Chou *et al.* (1982).

We had an opportunity to review another recent textbook for the publisher when the manuscript was submitted. Along with warm recommendations to publish the book, the authors were advised to include the topic of cross-correlations (as well as other topics). Nevertheless, the authors decided not to include the recommended topics in their book. The most recent textbook does not cover this topic. These books neither direct their readers to studies where the subject is discussed in detail. This non-inclusion of topics pertinent to random vibration of continuous structures is not surprising: One (recognized) stochastic dynamics' researcher has mentioned that classical engineering mechanics was a completed subject and that this was reflected, in his view, by federal agencies that were issuing many grants to stochastic studies. (Are some of the agencies attracted by mathematical fancy rather than relevance?) We feel that a fusion is needed between classical and neo-classical methods of analyses to address problems of practical interest. It appears that the recent developments in the random vibration of *continuous* structures should not be overlooked in modern textbooks and research monographs. Otherwise, the graduate students, especially the engineers, will be left in the middle of nowhere. It is essential to instruct them to the effect that the real action takes place in complex continuous structures, not in the single degree-of-freedom approximations! To the best of the author's knowledge, the random vibration books that deal with continuous structures are those by Zhu (1992) and Elishakoff (2020a).

2.4 Which deterministic theories to utilize?

The "supermarket" of deterministic theories contains numerous items; theories contains numerous items: theories for Bernoulli-Euler beams, Bresse-Rayleigh beams, Timoshenko-Ehrenfest beams, Kirchhoff-Love plates, Uflyand- Mindlin plates, Reissner-Stavsky composite plates, Kirchhoff-Love shells, Flügge shells, Donnell-Vlasov-Mushtari shells, Sanders-Koiter (first best approximate) theory of shells, Ambartsumian's composite shells, Reddy's layer-wise shells, Vekua's high-order shells, Bolotin-Herrmann's layered media; the list can go on. (Recently, Soldatos and Elishakoff (1990) put forward a theory for plates that considers both transverse shear and normal deformations and includes, as particular cases, many other theories.) An immediate question arises: Which structural theory is the most appropriate one so as to constitute a balance with the stochastic problem at hand?

Stochastic dynamicists only "flirted" with refined theories for just a short while. Samuels and Eringen (1958) studied the random vibrations of beams based on Timoshenko-Ehrenfest beam theory (Elishakoff, 2020b). The beam was subject to the white noise excitation in time. The result of the calculation was as follows: The difference between the computations based on the Bernoulli-Euler theory and those based on

Timoshenko beam theory was below 5%. This appears to be the main reason why stochastic dynamicists have discontinued their interest in integrating refined dynamic theories and stochasticity. Let us revisit Crandall's problem when the excitation is not a
band-limited white noise with a cut-off frequency ω_c. We are interested in the excitation whose spectral density possesses two cut-off frequencies $\omega_{c,1}$ and $\omega_{c,2}$, with the
spectral density around zero frequency vanishing (Fig. 2.9a). In other words, the spec-
tral density is constant and equals S_0, in the range $\omega_{c,1} \leq \omega \leq \omega_{c,2}$, and is zero otherwise.
  The random vibration response depends upon the quantity $\bar{\Phi}_{jk}$:

$$\bar{\Phi}_{jk} = \left[(AE)^2 / L^3 \left(\frac{E}{\rho} \right)^{1/2} \right] \left\{ \int_{-\omega_{c,2}}^{-\omega_{c,1}} H_j^*(\omega) H_k(\omega) d\omega + \int_{\omega_{c,1}}^{\omega_{c,2}} H_j^*(\omega) H_k(\omega) d\omega \right\} \qquad (2.12)$$

Here, $H_j(\omega)$ is the frequency response function corresponding to jth mode, A = cross-
sectional area, E = modulus of elasticity, L = length, ρ = material density.
  It can be shown (see Elishakoff and Lubliner, 1985) that $\bar{\Phi}_{jj}$ is close to unity for

$$\frac{\omega_{c,1}}{\omega_j} \leq 1, \ 1 \leq \omega_{c,2}/\omega_j \qquad (2.13)$$

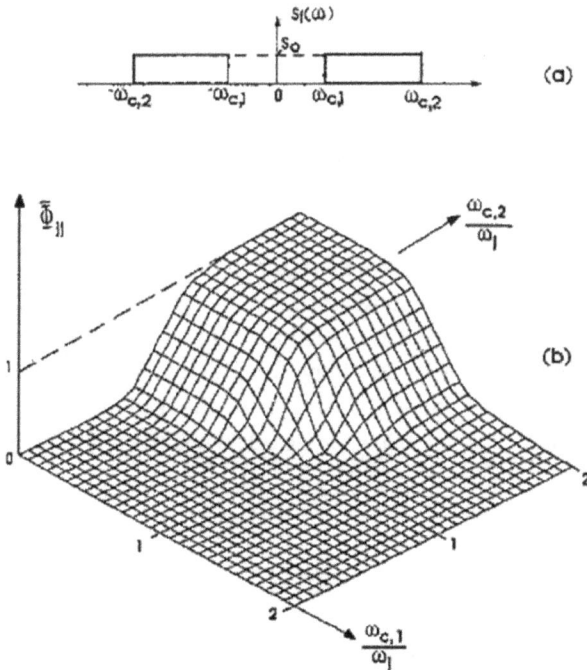

Fig. 2.9: (a) Band limited white noise with two cut-off frequencies (b) function Φ_{jj} is close to unity if
natural frequency belongs to the excitation band (after Elishakoff and Lubliner, 1985).

In other words, $\bar{\Phi}_{jj}$ is close to unity when or when $\omega_{c,1} \leq \omega_j < \omega_{c,2}$, the natural frequencies, fall within the excitation band. Otherwise, a good approximation is $\bar{\Phi}_{jj} \approx 0$ (Fig. 2.9b). This implies that instead of using infinite modal summations, one should take only terms between N_1 to N_2, where N_1 is the smallest number such that $\omega_{N_1} < \omega_{c,1}$, whereas N_2 is the largest number such that $\omega_{N_2} < \omega_{c,2}$. For ideal white noise excitation $\omega_{c,1} = 0, \omega_{c,2} \rightarrow \infty$; then, the classical and refined theories yield coincident results, which agrees with numerical findings of Samuels and Eringen (1958). However, when the excitation band tends to contain only high frequencies, the predictions of the two theories tend to diverge considerably. Elishakoff and Lubliner (1985) report a difference of about 50%. This is due to the fact that, on the one hand, the natural frequencies decrease as a result of the effects of shear deformation and rotary inertia and, on the other hand, the modal contributions are inversely proportional to ω_j^2. Decrease of natural frequencies leads therefore to increase of the estimate of the response mean-square values. Figure 2.10 clearly illustrates that the use of classical theories instead of refined theories is on the unsafe side.

In the study on the space shuttle weather protection systems (Elishakoff et al., 1995), the engineering approximation leads to replacement of the corrugated structure by the equivalent I-beam; random vibration bending moment response turns out to be approximately

50% more than that associated with the Bernoulli-Euler beam theory. The calculations are based on the experimentally measured statistical characteristics of the cross-spectral density of the excitation.

The necessity of use of refined theories in composite structures is even more pronounced. The response of a moderately thick cylindrical shell shows that refined theories must be used for wide-band excitation; moreover, neglect of cross-correlations may underestimate the true response by a factor greater than seven. These considerations must not be overlooked when studying random vibration of continuous structures. Actually, it is the *excitation* that decides on which theory to use. This seemingly surprising conclusion is fully understandable if we recall that the *level* of excitation often decides whether or not to use a refined theory in preference to the classical theory. For applications of refined theories in composite structures, the reader may consult the studies by Witt and Sobczyk (1980) and Singh et al. (1988) and the monograph by Cederbaum et al. (1992).

2.5 Is stochastic linearization a "dead" subject?

Some years ago, in his capacity as an associate editor of one of the journals, the present writer had to deal with reviewing a manuscript devoted to the state of the art of stochastic linearization technique. We had requested critical views of three recognized stochastic mechanicians on the manuscript. The first reviewer recommended to go ahead with the publication of the manuscript without any change; the second one

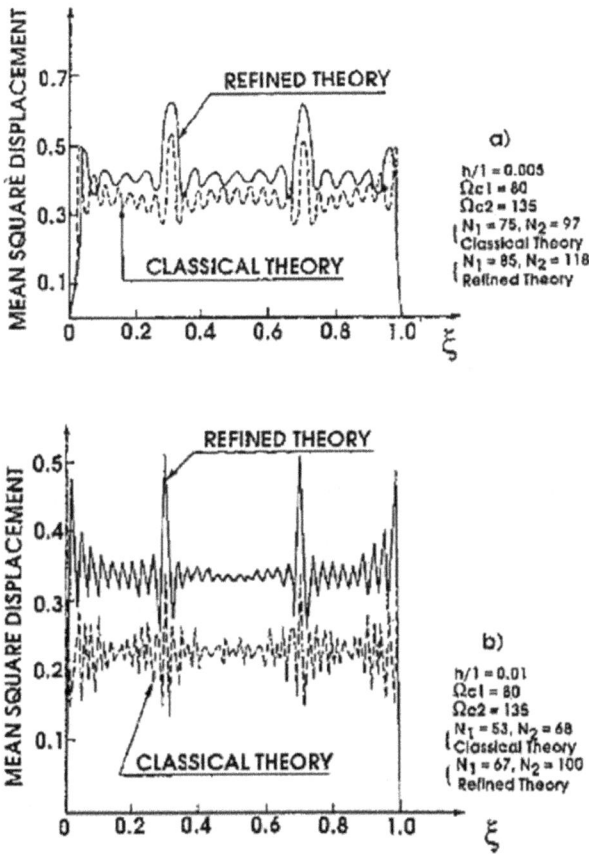

Fig. 2.10: Use of classical theory may significantly underestimate the stochastic response of a structure (after Elishakoff and Lubliner, 1985).

provided us with a list of several publications and recommended to accept the manuscript provided that these publications, mainly by the reviewer, be referenced in the final version of the review. We did not hear from the third reviewer for about six months. I have called him or her and asked for input (I certainly do not want to reveal the identity of this or other reviewers). The reply was as follows (almost verbatim): *"The review should not be published. The subject is dead."*

Can we indeed say that any subject is a "dead subject," even if it is an extremely "old" subject?

Indeed, the stochastic linearization technique is an old technique. It was suggested by Booton (1953) and Kazakov (1954, 1956). In the following, we will illustrate that it may be premature to declare this subject as a "dead" one.

According to the classical stochastic linearization technique, the nonlinear restoring force $f(X)$ is replaced by the linear one $k_{eq}^{(1)}$, where $k_{eq}^{(1)}$ is the spring coefficient chosen in such a way that the difference between the original nonlinear restoring force and the corresponding linear force attains a minimum in the mean-square sense, that is,

$$E\left\{\left[f(X) - k_{eq}^{(1)}X\right]^2\right\} = \min \tag{2.14}$$

The subscript *"eq"* indicates the equivalence in the sense of eq. (2.14) between the original nonlinear system and its linear counterpart. The monograph by Roberts and Spanos (1991) and the reviews by Sinitsyn (1974), Spanos (1981), and Socha and Soong (1991) list about 400 references on stochastic linearization. A classical version of the stochastic linearization to continuous structures was utilized by Seide (1975), Wentz *et al.* (1982), and Mei and Wolfe (1986).

A natural question arises: Can anything new be done and said about this old and seemingly "dead" topic?

In their papers, Zhang (1989), Zhang, Elishakoff, Zhang (1991) and Elishakoff and Zhang X.T. (1992) put forward a new stochastic linearization technique. Following this technique, the original nonlinear system is replaced by a linear one with a new stiffness coefficient $k_{eq}^{(2)}$ chosen so as to minimize the mean-square difference between the potential energies $U(X)$ and $k_{eq}^{(2)}X^2/2$ possessed by the original nonlinear system and its linear counterpart, respectively:

$$E\left\{\left[U(X) - \frac{1}{2}k_{eq}^{(2)}X^2\right]^2\right\} = \min \tag{2.15}$$

This new technique has been applied to a nonlinear oscillator subjected to broadband excitation. A detailed study showed that for extremely small nonlinearities the classical stochastic linearization performs better than the new technique. However, for the high nonlinearity the energy-based technique is superior to the classical scheme. Analogous findings have been recorded for the Duffing oscillator subjected to colored noise by Falsone and Elishakoff (1994).

For the nonlinearity damped structures, the criterion of minimization of the mean-square error between the Rayleigh's dissipation function $\Phi\left(\dot{X}\right)$ of the original nonlinear system and the one in the linearized system $(1/2)\, c_{eq}^{(1)}\dot{X}^2$ was suggested:

$$E\left\{\left[\Phi(\dot{X}) - \frac{1}{2}c_{eq}^{(1)}\dot{X}^2\right]^2\right\} = \min \tag{2.16}$$

Kazakov (1954) in his first paper also suggested the criterion that the mean-square values of the nonlinear restoring force in the original nonlinear system and its linear counterpart should be equal, i.e.,

$$E\left\{[f(X)]^2\right\} = E\left\{\left[k_{eq}^{(3)}X\right]^2\right\}$$ (2.17)

This criterion is mentioned, to the best of our knowledge, only in two monographs on random vibration (Popov and Pal'tov, 1965; Bolotin, 1971). This suggests that Mark Twain's (1835–1910) definition of the classic: "A classic is something everybody wants to have read, but no one wants to read," still holds.

We also generalized criterion of Kazakov (1954) to establish the following requirements:

$$E\left\{[U(X)]^2\right\} = E\left\{\left[\frac{1}{2}k_{eq}^{(4)}X^2\right]^2\right\}$$

$$E\left\{[\Phi(\dot{X})]^2\right\} = E\left\{\left[\frac{1}{2}c_{eq}^{(2)}\dot{X}^2\right]^2\right\}$$ (2.18)

Another approach, developed by Casciati *et al.* (1993), represents an additional alternative to the classical stochastic linearization method. It should be borne in mind that in principle one may have *numerous* equivalence criteria. The latter face is reminiscent of the notion of classical equivalent stress in deterministic applied elasticity, where several equivalent stress criteria coexist for different uses.

Fang and Elishakoff (1995) utilized a new energy-based technique on random vibrations of nonlinear beams. In Fig. 2.11, the energy-based stochastic linearization turns out to be much more accurate than the classical linearization technique for a specified set of system parameters. More recently, the method has been extended to elastic beams so that it can now deal with continuous structures rather than only with a single-degree-of-freedom system. It appears that more work should be performed for combining of the FEM and new stochastic linearization criteria to treat plate and shell structures.

To sum up, we cannot say in defense of the stochastic linearization technique that it offers mathematical sophistication or fancy; we must say, however, that it appears to be the only universal, analytical technique, thus supplementing the numerical Monte Carlo method, (For a combination of the two methods, the reader may consult the study by Elishakoff and Colombi, 1993c.) At present we do not have a mathematical proof on why the stochastic linearization technique performs reasonably well in some circumstances in the first place; but can anyone refuse to take a universal medicine, even if one does not know why it is often effective?

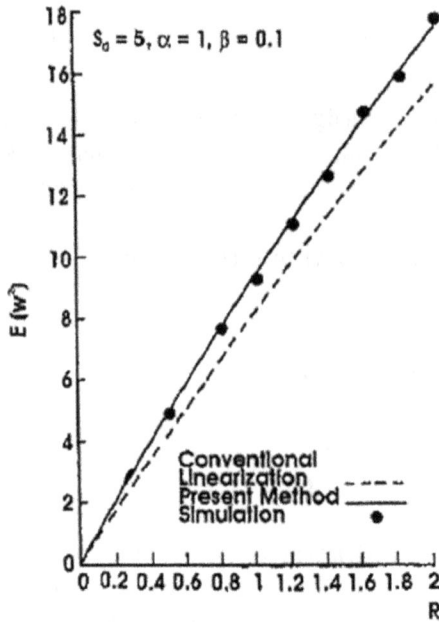

Fig. 2.11: Energy-based stochastic linearization may yield stochastic responses which are much closer with the results of exact solution or simulation, than those obtained through application of conventional stochastic linearization technique.

2.6 Arbitrary randomization and physical sense do not always go together

Spending at least one afternoon weekly in the library and browsing through journals will not be a bad idea to familiarize ourselves with recent developments and new thoughts in deterministic mechanics (even if one is a chairperson, director, or other administrator). This may prevent us from, at least, falling on the arbitrary "randomization" path. Attending just "stochastic" conferences may not be necessarily helpful in this respect. Indeed, as Michael Foot (1913–2010) remarks, ". . . *men of power have no time to read; yet the men who do not read are unfit for power.*"

One may maintain that all problems are uncertain in their nature from the very start, and the deterministic analysis simplifies them to neglect uncertainty. Therefore, they may argue that any "randomization" (even an arbitrary one) is a "return to roots" rather than giving "new clues" to the question under discussion. However, it appears that the researchers in stochasticity themselves do not view the matters in this way; otherwise, instead of "randomization" other terms such as "probabilistic setting" could have been utilized. We want to refer to the paper entitled "Hoff's problem in probabilistic setting" (Elishakoff, 1980). Originally, Hoff considered the problem at hand in a deterministic setting and explained the physical phenomenon; that is, he investigated the maximum load supported by an elastic column in a compression test (Hoff, 1951, 1954; Hoff *et al.*, 1951, 1955); this problem if of interest due to the following observation: The theoretical buckling load of a column is usually obtained by the Euler method, which reduces the job of stability analysis of a form of equilibrium to the simpler one of funding the minimum eigenvalue of a certain linear boundary-value problem. As for the experimental buckling load, it is usually determined as the maximum load recorded in a standard buckling testing machine. Hoff addressed an important question of the interrelation between the two loads. In order to explain the essence of the phenomenon, he has assumed that the initial imperfections formed a deterministic function. The effect of the unavoidable uncertainty in initial imperfections was addressed, but the modification of the deterministic analysis of Hoff (1951) forms a cornerstone of the realistic analysis. Thus, a probabilistic analysis of Hoff's problem does not constitute its arbitrary *randomization* but rather takes into account an additional important aspect of the problem, with ample experimental evidence that this uncertainty is important, whereas other uncertainties could be discarded. Also, one studies first a problem in a simplest, deterministic setting; only then some quantities may be considered in the probabilistic setting if justified by experimental observation on their scatter. Other extremes will be considering all variables (some use the term *primitive variables*) as random; for universality then π can also be considered a random variable with mean value 3.14159 . . . and zero variance in the formula for beam's frequency $\omega = \left(\frac{\pi}{L}\right)^2 \sqrt{\frac{EI}{\rho A}}$, where some authors consider all variables, namely the length L, modulus of elasticity E, moment of inertia I, material density ρ, and the cross-sectional area A to be randomly distributed. Computer codes have then a list of candidate distributions, and reliability, *i.e.*, the probability that the random frequency will be below a specified value, is quickly calculated. Such a combination of computers and "randomization" (i.e., without any substantiating experimental data) appears to be at least unattractive.

One may argue that stochastic dynamicists ought to develop theoretical techniques; when the experimental data becomes available, it would be incorporated into the theoretical derivations and numerical codes. However, as Blekhman *et al.* (1983) mention:

> [I]n the theory of probability, the probabilistic characteristics are actually primary-like lengths and angles in geometry-so that it would seem logical, in resorting to a stochastic approach, to regard them as given. The aim of applied mathematics, however, is not merely to determine some quantities from other (or, in particular, logically secondary quantities from logically primary ones) – but to find from quantities which may realistically be regarded as given (i.e., which are capable of direct measurement or calculation), those incapables of direct determination. Thus, in constructing a mathematical model, the question of obtainability of initial data, and of the effort involved, is a paramount-possibly decisive-importance.

As one can observe, applied mathematicians are well advised to pay an extreme importance to the obtainability of experimental data. How much more so this statement is applicable to the engineers?

Therefore, arbitrary analytical or computerized randomization, without resorting to the initial data, may turn out to be somewhat misleading and even disserving of the profession.

Indeed, the top engineering management may not grasp all the "nuances" of such an analytical or computerized randomization. We hope that the engineering management will still have enough common sense to appreciate the importance of the input data and to follow the advice of Sir Arthur Conan Doyle (1859–1930), who tells us, through Sherlock Holmes: *"We must never theorize without access to data."*

One of the anonymous reviewers of this work notes in his/her remarks to the effect that

> the work presents critical review of several fields of current research. It recommends greater emphasis on some and lesser emphasis on others. In my (rather long) experience, such recommendations are rarely followed. Academics usually persist in the directions of their research, regardless of whether or not they are meaningful.

2.7 Can we reliably predict small probabilities of failure?

In the reliability analysis of systems with many degrees of freedom the response surface method and the Monte Carlo Method appear to be universal approaches. Recently, the so-called Double and Clump method was suggested by Pradlwarter *et al.* (1994) to increase for each time step the relatively low number of response samples associated with high mechanical energy.

As the authors maintain:

> [T]he selective Monte Carlo Simulation procedure is tested by analyzing nonlinear SDoF oscillators, for which exact analytical solutions exist. A good agreement between the distributions F(x) in the tails is obtained covering a wide range, i.e., $10^{-7} < F(x) < 1.0 - 10^{-7}$.

In these circumstances, one wonders if indeed we are able to reliably predict the probability of failure of order 10^{-7}. The errors of input distributions, along with analytical modeling errors, may change the estimates considerably. One would want to see some sensitivity studies of the error in information and in analytical transfer functions in order to assess the accuracy of predicted probabilities of failure. Furthermore, we must stress that stochastic dynamics should not be considered complete until the final product of the analysis – the reliability or its complement probability of failure – is determined with sufficient accuracy.

Indeed, some specialists in stochastic mechanics paradoxically maintain: *"We do not work in reliability. We work in the field of random vibrations."* However, if as a result of the stochastic analysis, either regions of stochastic stability or the reliability will not be found, then natural question would be posed: What is the value of such an analysis? In respect, this is a step forward: It addresses the ever-burning, practical issue of small probabilities of failure.

Indeed, in order for the society to adopt probabilistic methods, the resulting reliability must be extremely high, or the resulting probabilities of failure must be extremely low. In this respect the following incident is of relevance: Recently, a program manager of some research agency maintained, at some conference, that the manager's superiors rightfully demanded the reliability of their systems to be doubled. I have mentioned to the manager, after the lecture, that hopefully his superiors want the probability of failure to be halved, which is a possible task, whereas the former one may turn out to be an impossible task, unless their systems are quite unreliable. The present writer was communicated (Joseph Kogan, 1980) that when the pioneer of stochastic mechanics in Russia, Nikolai Stanislavovich Streletski (1885–1967) – author of one of the first books (Streletski, 1947) on probabilistic analysis of safety factors – had suggested to calculate probabilities of failure of buildings, he was told: *"Our buildings must stand forever, without any probability of failure."* He has lost his job for several years . . . Fortunately, he was not imprisoned, since he also served as a General in the Red Army . . . Zero failures may turn out to be unachievable in all circumstances: If at all, extremely small probability of failure may be allowable. The question whether or not such a probability of failure can be safely estimated should be addressed and re-addressed.

2.8 Some relevant comments of A. Freudenthal

Professor Alfred Freudenthal (1906–1977) had a tremendous impact on the development of stochastic mechanics in the United States, in particular, and worldwide, in general. The American Society of Civil Engineers has established a Freudenthal Medal; the University of Innsbruck has established a Freudenthal Visiting Distinguished Professorship (Freudenthal was born in the then- Austro-Hungarian Empire). In our personal view, an important reason of the greatness of Freudenthal is the fact that, although a pioneer and developer of stochastic mechanics, he did not make "fetish" of his subject: He worked hard in stochastics, but he did not "idolize" it, as is often done by the present-day ever-energetic and ever-enthusiastic followers of his theory. Here I would like to give some quotations from Freudenthal, with the request for Freudenthal Medalists, Lecturers and Professors (as well as all stochastic dynamicists) to re-read the quotations from time to time:

Quotation 1

It seems absurd to strive for more and more refinement of methods of stress-analysis, if, in order to determine the dimensions of the structural elements, its results are subsequently compared with so-called working stress, derived in a rather crude manner by dividing the values of somewhat dubious material parameters obtained in conventional materials tests by still more dubious empirical numbers called safety factors (Freudenthal, 1968).

Quotation 2

When dealing with probabilities a clear distinction should be made between conditions arising on design of inexpensive mass products in which the probability figures are derived by statistical interpretation of actual observations or measurements (since a sufficiently large number of observations are actually attainable), and conditions arising in design of structures of complex systems. In the latter, probability figures are used simply as a scale or measure of reliability that permits the comparison of alternative designs. The figures can never be checked by observations or measurements since they are obtained by extrapolations so far beyond any possible range of observation that such extrapolation can no longer be based on statistical arguments but could only be justified by relevant physical reasoning. Under these conditions the absolute probability figures have no real significance

Quotation 3

An approach based on the direct specification of a very low failure probability alone suffers from a major shortcoming: there is no intrinsic significance to a particular failure probability since no a priori rationalization can be given for the adaptation of a specific quantitative probability level in preference to any other, so that the selection of this level remains an arbitrary decision (Freudenthal, 1972).

Quotation 4

Ignorance of the cause of variation does not make such variation random (Freudenthal, 1972).

This quotation gives us a hint that mechanicians should be tolerant and never confine themselves solely in the realm of *random* vibration; they may broaden their perspective by adopting the theories of fuzzy sets (*e.g.*, Yao, 1979; Blockley, 1980; Shiraishi and Furuta, 1985; Bernardini, 1992) and convex modeling (Ben-Haim and Elishakoff, 1990) since these newly developed theories represent alternatives and complements to the stochastic modeling. Recently, probabilistic and convex analyses have been combined for space shuttle applications (Elishakoff *et al.*, 1994). One should make a special mention of the non-stochastic set-theoretic methods pioneered by Leitmann (1979, 1993) in the context of optimal control.

We could complete our selection of Freudenthal's quotations with a closely related one due to Grandori (1991):

The probabilistic approach to structural safety is today a well-established paradigm. As to the current state of this paradigm, however, one can notice an asymmetry similar to that observed by Freudental, in the traditional approach. An overwhelming part of the research effort, in fact, has been and still is devoted to estimating failure probabilities. By contrast, only sporadic research deals with the problem of choosing an acceptable risk of failure. It is true that the adaptation of a probabilistic approach is in any case a progress, even in the case when acceptable risk levels are

conventionally defined, because it allows us to treat different structures with homogeneous crite-
ria. However, the concept of structural safety will not leave the 'realm of metaphysics' unless we
devise a method for justifying the choice of risk acceptability levels.

A word of caution is in order: Not all stochasticians overestimate the tools developed
by the probabilistic analysts. It appears instructive to quote from the 1989 paper of
Masanobu Shinozuka (1930–2018) Shinozuka:

[E]ngineers and researchers active in the field of structural reliability tend to expend a dispropor-
tional number of resources to deal with mathematical questions involving β [reliability index,
point of minimum distance between the limit state surface and origin of normalized basic varia-
bles-IE], although their motivation is quite understandable. Indeed, finding the β value becomes the
sole purpose of reliability analysis in many instances without carefully examining not only the as-
sumptions for structural behavior and loading models, but also the background against which the
analysis is to be used.

Professor Zdeněk P. Bažant (2018) of the Northwestern University made the following
comments, among others, when receiving the ASCE Alfred Freudenthal Medal on May
31, 2018: "Freudenthal's work epitomizes my favored dictum: that probabilistic mechanics of strength cannot ignore mechanics of the material failure process. Indeed, about
half of Freudenthal's numerous seminal papers deal with mechanics – residual stresses
in fatigue, effect of material flaws on strength, strength of airframes, work-hardening
laws for metals, viscoplasticity, plastic shells, shrinkage stresses, consolidating media,
shear dilatancy in rock, seismic waves, orthotropic sandwich plates and shells, relaxation spectra, etc. Besides, he wrote great books on solid mechanics and on viscoelasticity. At the same time, he became the father of structural safety. His works dealt with
the statistics of microscopic flaws, statistics of cumulative fatigue damage, integration
of joint material and load randomness (now called the Freudenthal integral), random
lifetime, random failure of structures with multiple load paths, reliability of nuclear reactor components, reliability of aircraft and of offshore platforms in seismic regions,
extreme-value risk assessment, safety of prestressed concrete, structural optimization,
risk control, etc. Freudenthal obviously perceived the fields of (1) structural safety and
(2) mechanics and physics of materials and structures, as inseparable. He mastered
both and tackled both at the front of research of his time. After him, unfortunately – a
schism. On one side, there have been outstanding probabilists who developed and successfully marketed sophisticated computer programs to assess the safety, reliability and
lifetime of concrete structures without noting that their failure probability cannot be
predicted with simplistic material models that eschew fracture mechanics and size effect. Or there have been experimenters who conducted extensive histogram testing of
strength of tough ceramics and fiber composites but did not realize that testing the size
effect would have invalidated their conclusions. On the other side, there have been mechanicians who constructed sophisticated constitutive and computational models for
the failure of concrete, geomaterials, composites, etc., without recognizing that, aside
from load randomness, big prediction errors stem from their assumed probability dis-

tribution conflicting with the material failure process. We have stochastic finite element codes, but they deliver hardly more than the standard deviation. They tell us nothing about extrapolation to the probability tail. Yet the tail where the devil is. It is agreed that bridges, aircraft, microelectronics, etc., must be designed for the failure probability of less than one in a million. But the tail of such a small probability is a major challenge, which cannot be overcome without modeling the mechanics of material failure. The distances from the mean to the tail of the Gaussian and Weibull distributions differ enormously, by about 2:1. For quasibrittle materials such as concrete, composites, toughened ceramics, the distance to the tail can be anywhere in between. This is especially important for developing new architected materials. A new material with a superior mean strength can be way inferior at the tail. Material scientists and NSF – please take note."

Likewise, Bažant and Le (2017) maintain: "Although some would vehemently deny it, many specialists would agree that, since the 1977 death of Freudenthal, the research field of structural safety and reliability has been in a schism. Alfred Freudenthal, the founder of this field of research in the 1960s, perceived the fields of (1) structural safety and (2) mechanics and physics of materials and structures as inseparable, He mastered both, and treated both to the depth of knowledge in his time. Since that time, unfortunately, most researchers have immersed themselves in one of these two fields in great detail and sophistication, while treating the other aspect simplistically and superficially. The connection has been week." The authors justifiably criticize those who "develop and successfully market complex computer programs to assess safety, reliability, and lifetime of concrete structures without recognizing that failure probability of concrete structures cannot be predicted with simplistic or obsolete material models that eschew fracture mechanics and energetic size effect." Moreover, Bažant and Le (2017) boldly state that "without realistic failure mechanics, probabilistic analysis of structural safety is a fiction." It appears that these statements ought to be seriously studied by probabilistic mechanicians, as well as considered at national and international conferences and journals with a view of strengthening the subject matter by intimately connecting mechanics of failure, on one hand, and uncertainty quantification and propagation techniques.

2.9 Summary and conclusion

The following conclusions appear to be relevant:
(1) One of the researchers who had read the preliminary version of this work told us to the effect: ". . . we, the analysts via stochasticity, live within models we built. Only some of us are looking for the connection with the outside real world. Therefore, those who demand relevance and close connections with the real world must leave science and do something which is close to life." Whereas I do not dislike the sophistry of this phrase, it reminds me of the other relatives (a) "If the experiments produce results which are far from my theory, it is too bad for ex-

periments," or (b) "Within the next decade, the only use aero-dynamicists will have for wind tunnels is a place to store computer equipment" (phrase attributed to one of the computational mechanics experts; for details the reader may consult with Oden and Bathe, 1978). Indeed, referring to the state in stochastic structural dynamics, Soong (1988) rightfully remarks:

> *The gap between research and practice, it appears, is widening, not only from the viewpoint of knowledge base, but also from that of thinking processes involved.*

I cannot refrain from stating that the engineering stochasticians should not live in ivory towers, but as the legendary Greek hero Hercules, should maintain a touch with the ground. We should not live in the world of "mental constructs" alone but produce useful works, which make a maximal use of the experimental data. The use of extensive data to validate the probabilistic approach, although extremely cumbersome, is not an impossible task, as shown by Simiu (1979).

Numerous experimental data is dealt with in the monograph by Nigam and Narayanan (1994), which represents apparently the first book integrating many applications of the random vibration theory. More books of this kind, going into the "nitty-gritty" of practical applications appear to be needed to justify to our professional colleagues (and to ourselves) that we have not lost touch with reality.

(2) Interrelation between the exact solutions and the Monte Carlo method, on one hand, and the finite element method for problems involving stochastic properties, needs further study.

(3) Recognition of several nagging problems in probabilistic analysis of structures has paramount importance for its further development. Indeed, if you know that you do not know, it is already quite a good start. However, if you don't know, that you don't know, cannot be a good indication, since you think that you know all aspects of the problem, and hence you cannot be ready to embark on other avenues of thinking. The so-called Bliss principle, *"What You Don't Know Won't Hurt You"* may well be inapplicable in structural reliability. Therefore, the effect of human errors should be extensively addressed in future studies.

(4) Additional comments on the granting agencies and the appropriate policies of selection appear in order, but these will be left for future occasions. One thing must be said however: It appears that the granting agencies must establish some cap (say 10 years), after which "cooling off" period of several years may follow, in order to allow the agencies to identify new and risky topics, rather than to almost grant "tenure" to several specific researchers.

It appears instructive to quote one of the anonymous reviewers of this study:

> *[W]hile peer review is the best method both for publishing papers and for funding research, serious flaws have developed during the past several decades. In particular, in the small random vibration*

community the research was set by very few 'experts' and by a larger group of 'followers'. . . thus we find funded efforts languishing about 'techniques in search of a problem to solve.' While this is fine for mathematicians, if one is an engineering faculty, one should nominally do 'engineering research.' This is rarely the case in random vibration community.

Reviewers of the proposals are well advised to distinguish between the engineering terminology (bridges, helicopters, aerospace structures, ocean structures, etc.) and their single or two-DoF analogs. Concentration solely on mathematical analysis of these simplified (and often, over-simplified) models appears, to say strongly (maybe even too strongly), not to constitute neither 'pure 'engineering nor 'pure' mathematics. The departments of engineering mathematics appear in need to be established, as it happens in Europe. It appears advisable to include researchers and engineers from industry to reviewers in order to get their responses.

Some mechanism should be developed so that proposals will be directed to unbiased and neutral reviewers in order to avoid, as it's felt to be in some cases, a "mutual admiration" situation.

It is a strong belief of this writer that the manner the university research is financially supported must be critically reevaluated.

The author will be appreciative if the readers of this book will communicate to him their comments in writing so that we could have a continuous and meaningful dialogue on the state-of-the-art and the future of stochastic structural mechanics. Indeed, as Benjamin Disraeli remarks ". . . *it is much easier to be critical than to be correct.*" I hope that incorporation of readers' remarks to the sequel of this book will make it "more correct." Random vibrations are expected to continue to flourish in coming decades, for, as David Mumford, a mathematician at Brown University says, "we live in the age of stochasticity."

We ought to also mention definitive books on random vibration by Ibrahim (2008), Wijker (2009), Li and Chen (2009), To (2012), Liang and Lee (2015), Cai and Zhu (2016).

Chapter 3
Essay on reliability index, probabilistic interpretation of safety factor, and worst-case design

> If you don't know, the thing to do is not to get scared, but to learn.
> Ayn Rand, *Atlas Shrugged*, 1986

This chapter represents a brief review of the concepts of the reliability of structures. Some closed form solutions, as well as approximate methods are elucidated. Different attempts to describe probabilistically the so-called safety factor are described, and some instructive counterexamples are given. High sensitivity exhibited by the reliability of the structure is documented in a model structure. The same structure is also analyzed on the basis of non-probabilistic, convex modelling. The latter is complementary to the probabilistic methods when only limited information is available on uncertain variables and functions.

3.1 Introduction

The first harbinger of the new discipline of "probabilistic mechanics" appeared in Germany, in 1926, as the book by Mayer (1926). This method then was developed in Russia by Khozialov (1929), Streletsky (1947), Rzhanitsyn (1959,1978); in England by Tye (1944); in France by Levi (1949); and in Sweden by Johnson (1953). It fully flourished due to the efforts by Freudenthal, first in Israel and then in the United States (1947,1956), and many researchers around the world who followed these investigators. Now this subject has passed the age of adolescence (Cornell, 1981) and became a widely accepted discipline with many monographs written and a number of periodical journals appearing. In this chapter, we will give a brief overview of the basics of probabilistic approach to structures. The pertinent but not exhaustive references are those by Ekimov (1966), Cornell (1969), Benjamin and Cornell (1970), Murzewski (1970), Ferry Borges and Casthanheta (1971), Tichy and Vorlicek (1972), Bolotin (1974), Hasofer and Lind (1974), Ghiocel and Lungu (1975), Kapur and Lamberson (1977), Haugen (1980), Bolotin (1981), Schuëller (1981), Thoft-Christensen and Baker (1982), Timashev (1982), Hart (1982), Elishakoff (1983, 2017), Shinozuka (1983), Ang and Tang (1984, 2007), Augusti, Baratta and Casciati (1984), Madsen, Krenk and Lind (1986), Melchers (1987, 1999), Madsen, Krenk and Lind (2006), Lemaire Chateauneuf and Mitteau (2009), Verma, Ajit and Karanki (2010), Nowak and Collins (2013), Leira (2013), Murzewski (1989), Ditlevsen (1981, 1990), Nowak and Collins (2013), Melchers and Beck (2018), Der Kiureghian (2021), Zao and Lu (2021), and Raizer and Elishakoff (2022).

https://doi.org/10.1515/9783111354231-003

3.2 Basic concepts

Consider the situation where the state of a structure in use can be described by a finite number of probabilistically dependent or independent parameters X_1, X_2, ..., X_n, part of which characterizes the loadings acting on the structure, and the other part is associated with the strength of the materials. For some combinations of its parameters, the system is "acceptable" for use (in which case it is said to be in the safe state), whereas for other combinations it is "unacceptable" (in the failed state). The function $f(x_1, x_2, ..., x_n)$, x_j being the possible value of the random variable X_j may take on, which vanishes at the transition surface between the two states, is so defined that its positive values

$$f(x_1, x_2, ..., x_n) > 0 \tag{3.1}$$

represent the safe state, while its negative values

$$f(x_1, x_2, ..., x_n) < 0 \tag{3.2}$$

represent the failed state.

For example, if parameters $X_1, X_2, ..., X_m$ represent the strength of the materials, and X_{m+1}, X_{m+2}, ..., X_n – the actual stresses, the failure surface could be put in the form

$$f = R(x_1, x_2, ..., x_m) - S(x_{m+1}, x_{m+2}, ..., x_n) \equiv M(x_1, x_2, ..., x_n), \tag{3.3}$$

with $M(x_1, x_2, ..., x_n)$ representing the "safety margin."

The reliability of the structure – the probability of its being in the safe state is obtained as

$$R = \int_0^\infty f_M(m)dm, \tag{3.4}$$

and the probability of failure, or the unreliability of the structure, as

$$P_f = \int_{-\infty}^0 f_M(m)dm, \tag{3.5}$$

where $f_M(m)$ is the probability density of the safety margin that can be found through the familiar expression for the probability density of the difference of the random variables. That is,

$$f_M(m) = \int_{-\infty}^\infty f_R(s+r)f_S(s)ds, \tag{3.6}$$

where $f_R(r)$ is the probability density of the strength and $f_S(s)$ – that of the stress. Irrespective of the specific densities of R and S, we have for the safety margin the mean

$$E(M) = E(R) - E(S).\qquad(3.7)$$

For the uncorrelated R and S we get the following expression of the variance

$$Var(M) = Var(R) + Var(S).\qquad(3.8)$$

The number of standard deviations of the safety margin in the interval $S = 0$ to $s = E(S)$ is called reliability index (Fig. 3.1):

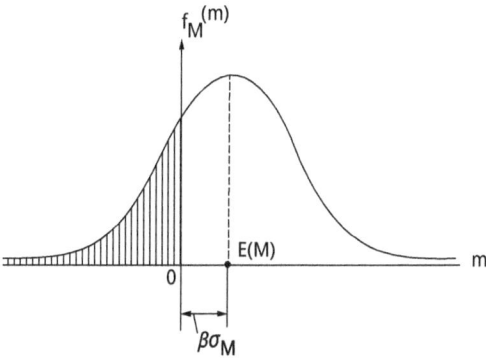

Fig. 3.1: Probability density of the safety margin.

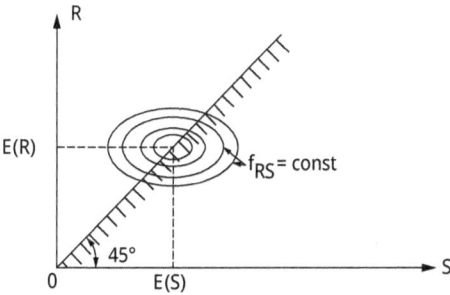

Fig. 3.2: Curves of equal joint probability density of stress and strength.

$$\beta = \frac{E(M)}{\sigma_M},\qquad(3.9)$$

where $\sigma_M = \sqrt{Var(M)}$ is the mean square deviation of the safety margin.

For the case where R and S are correlated, we have instead of eq. (3.9),

$$\beta = \frac{E(R) - E(S)}{[Var(R) - 2Cov(R, S) + Var(S)]^{1/2}}.$$ (3.10)

If R and S are normally distributed, then the reliability and probability of failure equal, respectively,

$$R = \Phi(\beta),$$ (3.11)

$$P_f = \Phi(-\beta),$$ (3.12)

where $\Phi(x)$ is the normal cumulative distribution function

$$\Phi(x) = \frac{1}{\sqrt{2\pi}} \int_{-\infty}^{x} e^{-t^2/2} dt.$$ (3.13)

Figure 3.2 shows the iso-probability curves, ellipses for the general case of $\sigma_\beta \neq \sigma_S$. The reliability index β has an interesting geometrical interpretation using the standard independent normal.

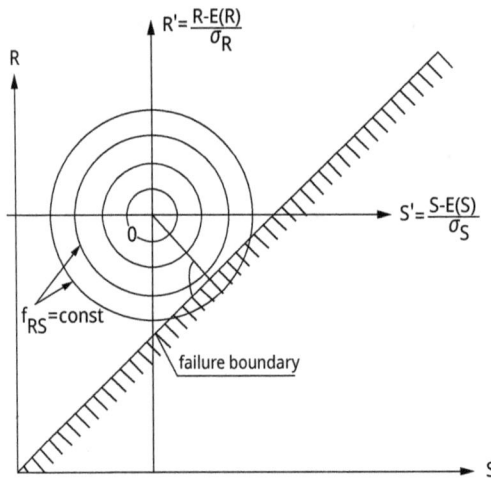

Fig. 3.3: Curves of equal probability density in the standard space and the failure boundary.

$$R' = \frac{R - E(R)}{\sigma_R}, \quad S' = \frac{S - E(S)}{\sigma_S}.$$ (3.14)

The failure boundary (Fig. 3.3) is rewritten as

$$\sigma_R R' - \sigma_S S' + [E(R) - E(S)] = 0.$$ (3.15)

According to the analytical-geometry formula, the distance from the origin to the failure surface is

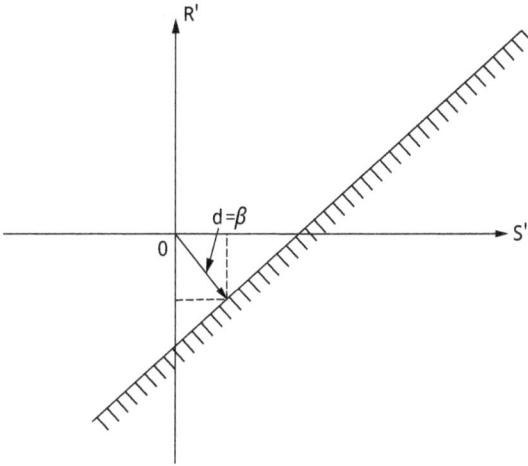

Fig. 3.4: Interpretation of minimum distance for the linear failure boundary.

$$d = \frac{E(R) - E(S)}{\sqrt{\sigma_R^2 + \sigma_S^2}}, \tag{3.16}$$

which is formally identical with the expression for the reliability index β (eq. 3.10) (Fig. 3.4). The point where d meets the failure surface is called the design point. Thus, we arrive at

$$R = \Phi(\beta) = \Phi(d), \tag{3.17}$$

$$P_f = \Phi(-\beta) - \Phi(-d). \tag{3.18}$$

There are a few other situations in which exact expressions can be obtained for the reliability. Let, for example, the strength and stresses be independent and the marginal densities exponential:

$$f_R(r) = \frac{1}{E(R)} exp\left[-\frac{r}{E(R)}\right], \tag{3.19}$$

$$f_S(s) = \frac{1}{E(S)} exp\left[-\frac{s}{E(S)}\right], \tag{3.20}$$

where $E(R)$ and $E(S)$ are the mean strength and stress, respectively. We obtain the following expression for the reliability:

$$R = \frac{E(R)}{E(R) + E(S)}. \tag{3.21}$$

Analogously, if R and S are independent random variables, having Rayleigh distribution

$$f_R(r) = \frac{\pi r}{2[E(R)]^2} \, exp\left\{ -\frac{\pi r^2}{4[E(R)]^2} \right\},$$ (3.22)

$$f_S(s) = \frac{\pi s}{2[E(S)]^2} \, exp\left\{ -\frac{\pi s^2}{4[E(S)]^2} \right\},$$ (3.23)

the reliability becomes

$$R = \frac{[E(R)]^2}{[E(R)]^2 + [E(S)]^2}.$$ (3.24)

An additional important case is when both the stress and the strength have a log-normal distribution:

$$f_S(s) = \frac{1}{s\sigma_1\sqrt{2\pi}} \, exp\left[-\frac{(\ell ns - a)^2}{2\sigma_1^2} \right], \quad (s \geq 0)$$ (3.25)

$$f_R(r) = \frac{1}{r\sigma_2\sqrt{2\pi}} \, exp\left[-\frac{(\ell ns - b)^2}{2\sigma_2^2} \right], \quad (r \geq 0)$$ (3.26)

where a, b, σ_1 and σ_2 are the density parameters, so that

$$E(S) = exp\left(a + \frac{1}{2}\sigma_1^2 \right),$$

$$E(R) = exp\left(b + \frac{1}{2}\sigma_2^2 \right),$$ (3.27)

$$Var(S) = exp\left(2a + \sigma_1^2 \right)\left[exp\left(\sigma_1^2 \right) - 1 \right],$$

$$Var(R) = exp\left(2b + \sigma_2^2 \right)\left[exp\left(\sigma_2^2 \right) - 1 \right].$$

The reliability is then

$$R = Prob\left(V = \frac{S}{R} \leq 1 \right) = F_V(1),$$ (3.28)

where $F_V(v)$ is the probability distribution of the random variable V. Equation (3.28) may be rewritten as

$$R = Prob(\ell nV \leq 0) = F_{\ell nV}(0).$$ (3.29)

Note that

$$\ell nV = \ell nS - \ell nR,$$ (3.30)

and since ℓnR, both have a normal distribution, specifically ℓnS is $N(a_1\sigma_1)$ and ℓnY is $N(b_1\sigma_2^2)$, ℓnV is also normal, as a difference of normal variables $N(a - b, \; \sigma_1^2 + \sigma_2^2)$, implying that V is log-normal. Reliability becomes

$$R = \Phi\left(-\frac{a - b}{\sqrt{\sigma_1^2 + \sigma_2^2}}\right). \tag{3.31}$$

For other cases where exact solutions are obtainable, one should consult the monographs by Ferry Borges and Castanheta (1971), Rzhanitsyn (1959), Bolotin (1981), Ang and Tang (1984), Augusti, Baratta and Casciati (1984), and Elishakoff (1983, 1992, 2017).

How does one calculate the reliability of a structure where exact solutions are unavailable? Such is unusually the case if the relationship (3.1) is nonlinear. Under these circumstances, if the basic variables are still normally distributed, the formulas (3.17) and (3.18) are used, but now as approximations to the exact reliability and probability of failure, respectively (Fig. 3.5). Equations (3.11) and (3.12), when the failure boundary is nonlinear, are commonly referred to as "the Hasofer and Lind index" (1974) on account of their systematic developments, which lead to the widespread characterization of the reliability index as the minimal distance from the origin to the nonlinear failure surface. It appears instructive to give a quote from the paper by Shinozuka (1983): ". . . it is worth noting that the checking format for a modified design, recommended on the basis of these recent developments was in essence suggested by A. M. Freudenthal in this 1956 paper (1956). In this chapter, referring to what is now known as the checking point, he wrote "because the critical condition $(x°, y°)$ has the highest probability of occurrence along the line $r = 0$, it represents the combination to be used in design." The critical failure condition $(x°, y°)$ indicates the point on the limit state (or failure) surface $r = 0$ and located at the shortest distance, p, from the origin on the two-dimensional rectangular Cartesian coordinate space of the standardized Gaussian variables x and y. The critical point is also the point of maximum likelihood due to the Gaussian property assumed in the design variables. L. S. Lawrence (1956) and J. M. Corso (1956) in their discussion

Freudenthal paper pointed out, and Freudenthal concurred, that the limit state probability (probability of failure) is a function of the shortest distance, p. This is now known as the safety index and can be obtained as $\Phi(-p)$, where $\Phi(.) = $ the standardized Gaussian distribution function . . . his paper did suggest that the checking point used for design and that the checking point of shortest distance from the origin in a standardized Gaussian or transformed Gaussian variable space and at the same time is the point of maximum likelihood. Interestingly enough, the description of the minimum distance method appears in the monograph by Olszak et al. (1961). Murzewski (1989) is attributing this method to Levi (1949). In light of these comments, it appears more appropriate to designate β in eqs. (2.11) and (2.12) as a minimum distance reliability index.

3.3 How accurate is minimum distance reliability index?

To best answer this question, we will study the circular shaft-case amenable to exact solution (Elishakoff and Hasofer, 1987). Given a circular shaft of radius a subjected simultaneously to a bending moment M and torque T, characterized as random variables with joint probability density function $f_{MT}(m, t)$; (Fig. 3.6) the yield stress σ_y is constant with probability unity. According to the maximum shear stress theory of failure, the strength requirement reads

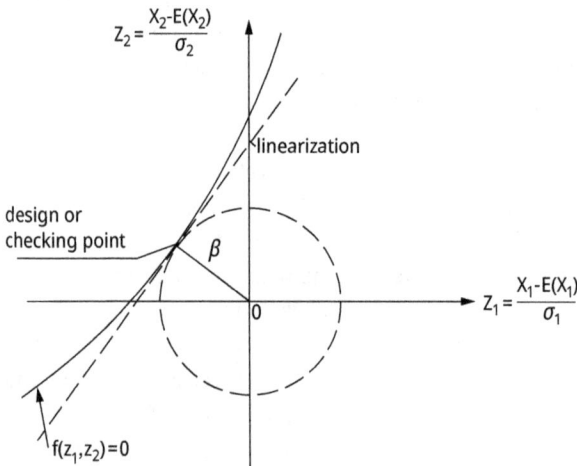

Fig. 3.5: Minimum distance as the reliability index.

Index terms: reliability index; maximum shear stress theory;

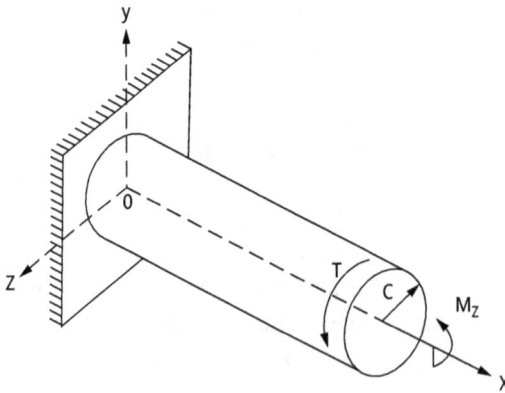

Fig. 3.6: Circular shaft subjected to random bending moment and torque.

$$\frac{M_{eq}c}{I} \le \sigma_y, \tag{3.32}$$

where M_{eq} is the "equivalent" moment:

$$M_{eq} = \sqrt{M^2 + T^2}, \tag{3.33}$$

$$R = Prob\left(\sqrt{M^2 + T^2} \le \frac{\pi}{4}\sigma_y c^3\right). \tag{3.34}$$

Consider first the simplest case treated by Bolotin (1974): M and T are independent normal variables with zero means $(a = b = 0)$ and equal variance $\sigma_M = \sigma_T = \sigma$. Then, the equivalent moment has a Rayleigh distribution

$$f_{M_{eq}}(m_{eq}) = \frac{m_{eq}}{\sigma^2} \exp\left(-\frac{m^2 eq}{2\sigma^2}\right), \tag{3.35}$$

with the attendant reliability

$$R = 1 - \exp\left(-\frac{\pi^2 \sigma_y^2 c^6}{32\sigma^2}\right). \tag{3.36}$$

To compare this exact expression with the minimum distance method, we introduce the basic variables

$$Z_1 = \frac{M}{\sigma}, \quad Z_2 = T/\sigma. \tag{3.37}$$

The failure boundary becomes

$$Z_1^2 + Z_2^2 \le p^2, \tag{3.38}$$

where

$$p = \frac{\pi}{4}\frac{\sigma_y}{\sigma}c^3. \tag{3.39}$$

Equation (3.38) represents a circle with radius p. The minimum distance to the circle equals the radius itself, so that $\beta = p$. Hence, under the minimum distance approximation we have

$$R = \Phi(p), \quad P_f = \Phi(-p), \tag{3.40}$$

whereas the exact solution eq. (3.36) in terms of p is

$$R = 1 - \exp\left(-\frac{1}{2p^2}\right), \quad P_f = \exp\left(-\frac{1}{2p^2}\right). \tag{3.41}$$

For highly reliable structures, which is where our interest lies, the approximate solution is remarkably close to the exact one.

Consider now the more realistic case $a^2 + b^2 \neq 0$ with the former restriction $\sigma_M = \sigma_T = \sigma$ still retained. In terms of the basic variables, the probabilistic counterpart of eq. (3.31) reads:

$$R = Prob\left[A \equiv \left(Z_1 + \frac{a}{\sigma}\right)^2 + \left(Z_2 + \frac{b}{\sigma}\right)^2 \leq p^2\right]. \tag{3.42}$$

The random variable A has a non-central chi-square distribution with two degrees of freedom, and the noncentrality parameter

$$r = \left(a^2 + b^2\right)/\sigma^2. \tag{3.43}$$

Thus, the reliability can be found via the extensive tables available. In the case $\frac{a}{\sigma} \gg 1, b/\sigma \gg 1$, an asymptotic expression is available, valid for $p > 5$:

$$R = Prob\left(A \leq p^2\right) \cong \Phi\left(\sqrt{p^2 - 1} - r\right). \tag{3.44}$$

Let us compare this result to that obtained by the minimum distance method. The failure boundary is again a circle with radius p but now centered at $\left(-\frac{a}{\sigma}, -\frac{b}{\sigma}\right)$. The minimum distance from the coordinate origin to the circle is (Fig. 3.7):

$$d = p - r. \tag{3.45}$$

Hence,

$$R \cong \Phi(p - r), \quad P_f \approx \Phi[-(p - r)]. \tag{3.46}$$

Comparison of eqs. (3.44) and (3.45) suggests that the minimum distance approximation is an excellent one, as for $p \gg 1$ the asymptotically exact eq. (3.44) and approximate expression.

3.4 Remarks on safety factor

Numerous attempts at probabilistic interpretation of the safety factor have been made in the literature despite the fact that the "spirits" of these two approaches are different. Before reviewing them, it is instructive to quote the following excerpts from popular textbooks concerning its definition:

a) "To allow for accidental overloading of the structure, as well as for possible inaccuracies in the construction and possible unknown variables in the analysis of the structure, a factor of safety is normally provided by choosing an allowable stress (or working stress) below the proportional limit."

b) "Although not commonly used, perhaps a better term for this ratio is factor of ignorance."
c) "The need for the safety margin is apparent for many reasons: stress itself is seldom uniform; materials lack the homogeneous properties theoretically assigned to them; abnormal loads might occur; manufacturing processes often impart dangerous stresses within the component. These and other factors make it necessary to select working stresses substantially below those known to cause failure."
d) "A factor of safety is used in the design of structures to allow for (1) uncertainties of loading, (2) the statistical variation of material strengths, (3) inaccuracies in geometry and theory, and (4) the grave consequences of failure of some structures."

Freudenthal (1968) remarks to repeat, ". . . it seems absurd to strive for more and more refinement of methods of stress-analysis if in order to determine the dimensions of the structural elements, its results are subsequently compared with so called working stress, derived in a rather crude manner by dividing the values of somewhat dubious material parameters obtained in conventional materials tests by still more dubious empirical numbers called safety factors." Indeed, it appears to the present author that in addition to its role as a "safety" parameter for the structure, it is intended as "personal safety" factor of sorts for the design companies.

A probabilistic interpretation of the safety factor is not unique. We will discuss two possible approaches toward such an interpretation. The "straightforward" safety factor itself, as the ratio R/S, is a random variable. The question is how to define it in probabilistic terms. A possible answer is the so-called central safety factor:

$$c = \frac{E(R)}{E(S)}, \tag{3.47}$$

which in certain situations is in direct correspondence with the reliability level. Indeed, for exponentially distributed strength and stress, the reliability, as per eq. (3.21), reads

$$R = \frac{c}{1+c}. \tag{3.48}$$

However, to achieve the reliability level of say, 0.999, the required central safety factor should be 999!

The situation is somewhat "better" for the case when R and S are independent Rayleigh distributed variables; in terms of the central safety factor c, eq. (3.24) rewrites as

$$R = \frac{c^2}{1+c^2} \tag{3.49}$$

Under new circumstances, in order to achieve reliability of 0.999, the central safety factor should be $\sqrt{999} = 31.61$.

For normally distributed strength and stress, the safety index could be written as

$$\beta = d = \frac{c-1}{\sqrt{\gamma_S^2 + c^2 \gamma_R^2}} \qquad (3.50)$$

where $\gamma_S = \frac{\sigma_S}{E(S)}$, $\gamma_R = \sigma_R/E(R)$ are coefficients of variation of the stress and strength, respectively. The central safety corresponding to the reliability level r satisfies the quadratic equation

$$\omega_1 c^2 + \omega_2\, c + \omega_3 = 0, \qquad (3.51)$$

where

$$\omega_1 = 1 - \gamma_R^2 \left[\Phi^{-1}(r)\right]^2,$$

$$\omega_2 = -2,$$

$$\omega_3 = 1 - \gamma_S^2 \left[\Phi^{-1}(r)\right]^2, \qquad (3.52)$$

where $\Phi^{-1}(.)$ is the inverse of $\Phi(.)$. Under these circumstances, c depends not only on the reliability level r but also on the coefficients of variation of the strength and stress.

Analogously, for the stress and strength, which have a log-normal distribution, the central safety factor is given by Elishakoff (1983, 2017):

$$c = \frac{\exp(a + \sigma_1^2/2)}{\exp\left(b + \frac{\sigma_2^2}{2}\right)}. \qquad (3.53)$$

For $\sigma_1/a \ll 1$ and $\sigma_2 \ll 1$, Leporati (1977) derived the following approximation

$$c = \exp\left\{\beta\left[\left(\frac{\sigma_1}{a}\right)^2 + \left(\frac{\sigma_2}{b}\right)^2\right]^{\frac{1}{2}}\right\}. \qquad (3.54)$$

The alternative safety factor is introduced as follows

$$t = E\left(\frac{R}{S}\right) \qquad (3.55)$$

instead of eq. (3.43). The following counterexample is constructed by Elishakoff (1983, pp. 243–246), in which both R and S have an identical, uniform distribution over interval $[0, a]$. Then, on the one hand, reliability is just one half, but the factor of safety t turns out to tend to infinity (!).

One can conclude that for the reliability calculations one should have an information on the required reliability of the structure, with additional parameters $E(R)$, $E(S)$,

σ_R, and σ_S specified, not necessarily in their direct connection with the "safety factor." In this connection it is instructive to quote Freudenthal (1972): "The predictive use, in structural design and analysis, of the theory of probability implies that the designer, on the basis of his professional competence, is able to draw valid conclusions from the probability figures obtained, as to justify design decisions which in most cases, hinge on considerations of economy and utility. It is not implied that this use is in itself sufficient to make a design more reliable or more economical, any more than that the avoidance of the probabilistic approach makes it safer."

In fact, an approach based on the direct specification of a very low failure probability alone suffers from a major shortcoming: "There is no intrinsic significance to a particular failure probability since no a priori rationalization can be given for the adaptation of a specific quantitative probability level in preference to any other so that the selection of this level remains an arbitrary decision."

In the following section, we will illustrate the high sensitivity exhibited by the failure probability.

3.5 Sensitivity of failure probability

Consider an elastic bar compressed by an axial force. The uncertainty parameter is described by the nonvanishing eccentricities e_1 and e_2 of the force (Fig. 3.8).

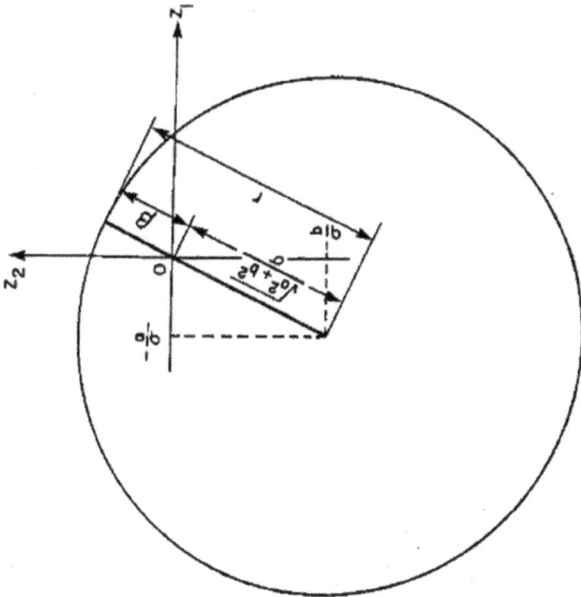

Fig. 3.7: Exact calculation of the minimum distance.

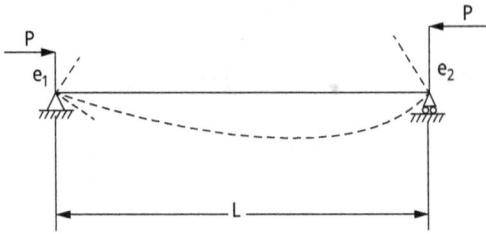

Fig. 3.8: Beam-column subjected to eccentric forces.

The differential equation describing the deflection of the bar reads:

$$EI\frac{d^4w}{dx^4} + P\frac{d^2w}{dx^2} = 0, \ (0 \le x \le L) \tag{3.56}$$

where EI is the flexural stiffness, P the axial force, w the displacement, and L the length of the bar. Denoting

$$k^2 = \frac{P}{EI}. \tag{3.57}$$

The boundary conditions in terms of the bending moment are

$$M_z(0) = Pe_1, \ M_z(L) = Pe_2. \tag{3.58}$$

Compliance with the boundary conditions yields the final expression for $M_z(x)$:

$$M_z(x) = -\frac{P}{\sin kL} \ (e_2 - e_1 coskL) \ sinkx + Pe_1 coskx. \tag{3.59}$$

The maximal bending moment M_z^* is

$$M_z^* (e_1, e_2) = \frac{P}{sinkL} \sqrt{e_1^2 + e_2^2 - 2e_1 e_2 coskL}. \tag{3.60}$$

This expression coincides with eq. (3.44) in Pikovsky (1961). The problem of a bar in compression with two eccentricities was also studied by Young (1932).

One can show that for the maximal bending moment to take place inside the bar, $0 < x^* < L$, the following conditions must be satisfied:

$$\cos kl < \frac{e_2}{e_1} < \frac{1}{\cos kL}, \tag{3.61}$$

and

$$0 < P < \frac{\pi^2 EI}{4L^2}, \ 0 < kL < \frac{\pi}{2} \tag{3.62}$$

It can be shown that the maximal bending moment occurs inside the bar, and condition eq. (2.59) is dispensed with, in the following range of load variation,

$$\frac{\pi^2 EI}{4L^2} < P < \frac{\pi^2 EI}{L^2} \tag{3.63}$$

We assume now that the eccentricities constitute a random vector with a jointly exponential distribution and the following distribution function by Gumbel (1960):

$$F_{E_1 E_2}(e_1, e_2) = \left[1 - exp\left(-\frac{e_1}{\beta}\right)\right]\left[1 - exp\left(-\frac{e_2}{\gamma}\right)\right] * * \left[1 + \alpha\ exp\left(-\frac{e_1}{\beta} - \frac{e_2}{\gamma}\right)\right], \tag{3.64}$$

where e_1 and e_2 take on only positive values and $\beta = E(E_1)$, $\gamma = E(E_2)$, $E()$ denoting the mathematical expectation.

For the sake of simplicity, we will concentrate on the case represented by eq. (3.61). We are interested in the reliability of the bar, defined as the probability of non-exceedance of a limiting value m^* by the random variable M^*.

$$R = Prob\left(M^* = \frac{P}{sin\ kL}\sqrt{E_1^2 + E_2^2 - 2E_1 E_2\ cos\ kL}\ \le m^*\right) \tag{3.65}$$

The integration results in

$$F_M^*(m^*) = 1 - \frac{2\beta^3 + \alpha\beta^2\gamma - 5\beta^2\gamma - 2\alpha\beta\gamma^2 + 2\beta\gamma^2}{2\beta^3 - 7\beta^2\gamma + 7\beta\gamma^2 - 2\gamma^3}exp\left(-\frac{m^*\ sin\ kL}{\beta P}\right)$$

$$- \frac{2\alpha\beta^2\gamma - 2\beta^2\gamma - \alpha\beta\gamma^2 + 5\beta\gamma^2 - 2\gamma^3}{2\beta^3 - 7\beta^2\gamma + 7\beta\gamma^2 - 2\gamma^3}exp\left(-\frac{m^*\ sin\ kL}{\gamma P}\right)$$

$$- \frac{\alpha\beta\gamma(2\gamma - \beta)}{2\beta^3 - 7\beta^2 + 7\beta\gamma^2 - 2\gamma^3}exp\left(-\frac{2m^*\ sin\ kL}{\gamma P}\right)$$

$$- \frac{\alpha\beta\gamma(\gamma - 2\beta)}{2\beta^3 - 7\beta^2\gamma + 7\beta\gamma^2 - 2\gamma^3}exp\left(-\frac{2m^*\ sin\ kL}{\beta P}\right), \tag{3.66}$$

when the following restriction holds

$$2\beta^3 - 7\beta^2\gamma + 7\beta\gamma^2 - 2\gamma^3 \neq 0.$$

In the particular cases where instead of the above inequality we have an equality, the expressions for the reliability read:

$$F_M^*(m^*) = 1 - \left[\left(1 + \frac{m^* \sin kL}{\beta P} \right) exp \left(-\frac{m^* \sin kL}{\beta P} \right) \right.$$

$$- \alpha \left[\left(\frac{m^* \sin kL}{\beta P} - 3 \right) exp \left(-\frac{m^* \sin kL}{\beta P} \right) \right.$$

$$\left. \left. + \frac{2m^* \sin kL}{\beta P} + 3 \right) exp \left(-\frac{2m^* \sin kL}{\beta P} \right) \right], \text{ for } \beta = \gamma. \tag{3.67}$$

and

$$F_M^*(m^*) = 1 - 2 \left(1 + \frac{\alpha}{3} \right) exp \left(-\frac{m^* \sin kL}{\beta P} \right) + \left(1 + \frac{2\alpha m^* \sin kL}{\beta P} \right) exp \left(-\frac{2m^* \sin kL}{\beta P} \right)$$

$$+ \frac{2\alpha}{3} exp \left(-\frac{4m^* \sin kL}{\beta P} \right), \text{ for } \beta = 2\gamma.$$

$$\tag{3.68}$$

Finally,

$$F_M^*(m^*) = 1 - 2 \left(1 + \frac{\alpha}{3} \right) exp \left(-\frac{m^* \sin kL}{2\beta P} \right) + \left(1 + \frac{\alpha m^* \sin kL}{\beta P} \right) exp \left(-\frac{m^* \sin kL}{2\beta P} \right)$$

$$+ \frac{2\alpha}{3} exp \left(-\frac{2m^* \sin kL}{\beta P} \right), \text{ for } \beta = \gamma/2.$$

$$\tag{3.69}$$

It has been shown by Gumbel (1960) that the following restrictions should be met $F_M^*(m^*)$ to serve as the distribution function:

$$-1 \leq \alpha \leq 1,$$

$$\alpha = 4p,$$

where p is the coefficient of correlation between E_1 and E_2.
Also, k on eq. (3.2) could be expressed as

$$k = \frac{\pi}{L} \sqrt{\frac{P}{P_{cl}}}, \tag{3.70}$$

where P_{cl} is the classical buckling load of the simply supported bar $P_{cl} = \pi^2 EI/L^2$. The reliability equals

$$R = Prob \left(\sum \leq \sigma_Y \right) = Prob \left[\frac{M^* c}{I} \leq \sigma_Y \right] = F_M^* \left[\frac{\sigma_Y I}{c} \right], \tag{3.71}$$

where $\sum$ is the maximum stress, σ_Y is the yield stress assumed to be constant, I is the moment of inertia and c is the distance between the centroidal line and the extreme fiber where the maximum stress occurs. Reliability of the structure is obtainable from eqs. (3.66)–(3.69) by replacing m^* by $\sigma_Y I/c$.

Figures 3.9–3.11 depict the reliability of the structure versus the yield stress of the bar's material σ_Y. Figure 3.9 is associated with $P = 15$ kN, $\frac{P}{P_{cl}} = 0.569$. The following data is used in Fig. 3.10: $P = 20$ kN, $\frac{P}{P_{cl}} = 0.759$, whereas in Fig. 3.11 the data is fixed at $P = 23$ kN, $\frac{P}{P_{cl}} = 0.873$. In all three figures $\beta = 2mm$, $\gamma = 1.5mm$, and $\frac{I}{c} = 1.333.3$ mm^3; $L = 1,000$ mm, $E = 200,000$ MPa. As we see, the increase in the applied loading results in reduced reliability of the structure. The coefficient a is varied in Figs. 3.9–3.11, in the range $-0.99 \le a \le 0.99$. Whereas data on β and γ may be reliable, information on the coefficient a could be insufficient; hence, Figs. 3.9–3.11 demonstrate the possible scatter in the reliability of the structure due to imprecise knowledge of parameter a. This is illustrated in Fig. 3.12, which addresses the design problem: Find the radius of the cross-section c so that the required codified reliability r, or codified probability of failure $P_f^* = 1 - r$ will be achieved.

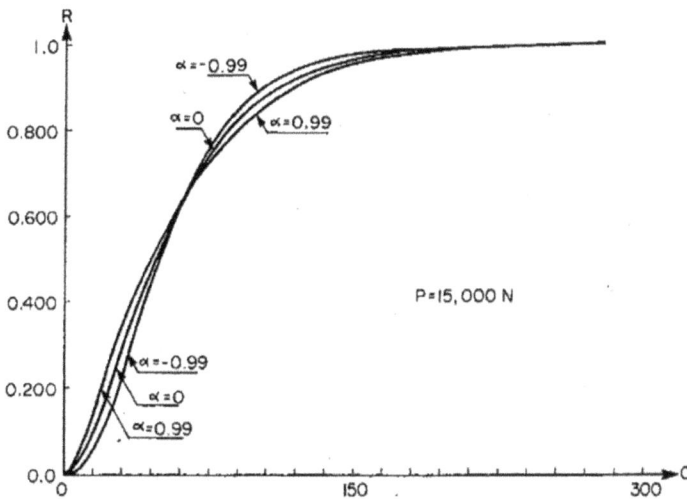

Fig. 3.9: Structural reliability versus radius.

If we fix value of P_f^* at 0.01, then, if the calculations are based on $a = 0.99$, the design value of the radius is $c = 12.257$; now, if the true value of a is -0.99, then the actual probability of failure of the system at this radius is $3.74 \cdot 10^{-3}$, i.e. is lower than the codified probability of failure. This implies that we had a case of the "favorable" imprecision. However, if our calculations are based on value $a = -0.99$, then the minimum required radius of the cross section should be $c = 12.029$. If, however, the true value of a is 0.99, then the chosen value of the radius corresponds to actual probability of failure 0.0234 instead of $P_f^* = 0.01$. This corresponds to the underestimation of the probability of failure by more than twice.

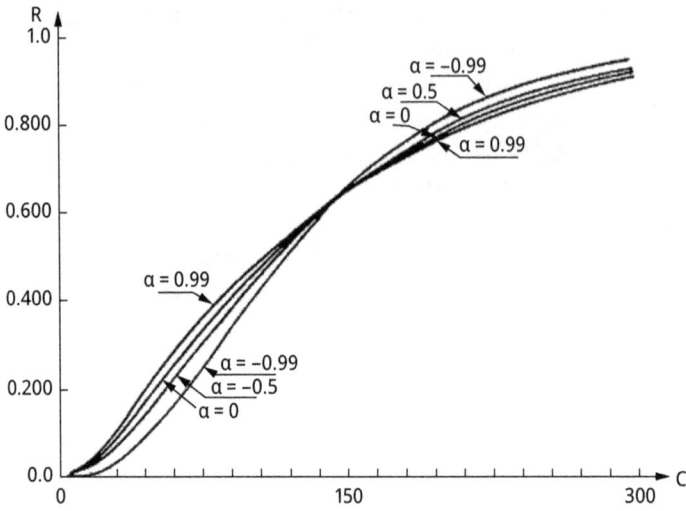

Fig. 3.10: Dependence of reliability upon parameters.

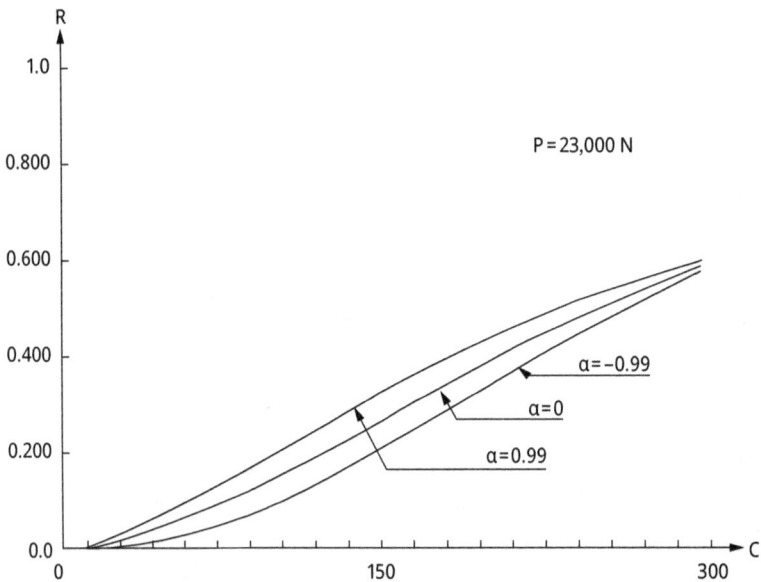

Fig. 3.11: Influence of the correlation coefficient.

The situation is more severe for highly reliable structures. To get more insight we define the underestimation factor as the ratio of the actual-to-codified probabilities of failure

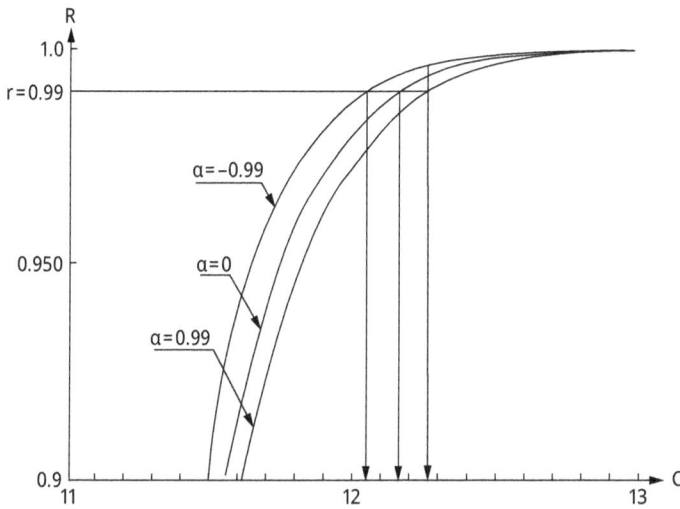

Fig. 3.12: Solution of the design problem.

$$\eta = \frac{P_f}{P_f^*}.$$

For $P_f^* = 10^{-3}$, the underestimation factor is over three; for $P_f^* = 10^{-4}$, $\eta = 3.47$ for $P_f^* = 10^{-5}$, $\eta = 3.705$ and finally for $P_f^* = 10^{-6}$ the underestimation factor reaches 3.82.

Thus, one would conclude that the system is acceptable for use, whereas the actual probability of failure is exceeding the codified one, and the system in fact is in a failed state, since the actual reliability is lower than the codified one, and the system in fact is in a failed state.

Under these circumstances of the high sensitivity of probability of failure, the natural question arises on how the probabilistic analysis could be used for design purposes. To attempt to answer this question, we will visualize that the cost C of production of the column is expressible as

$$C = \frac{q_1}{E(E_1) + E(E_2) + q_2}$$

where q_1 and q_2 are constants. Such a postulation maintains that more cost is associated with finer manufacturing, i.e., the one with less $E(E_j)$. Figure 3.12 depicts the reliabilities of the columns associated with different mean imperfections $E(E_j)$, but their sum $E(E_1) + E(E_2)$ is kept constant. The Figure demonstrates that the maximum reliability is achieved for the equal mean imperfection parameters $E(E_1) = E(E_2)$. Thus, reliability studies could be utilized for comparative purposes; under other conditions being equal, one prefers the manufacturing process, with higher reliability.

In the following section, we will devise an alternative non-probabilistic method to deal with uncertainty in the same problem.

3.6 Remarks on convex modelling of uncertainty

A number of linear problems have been considered under set-theoretical, convex modelling of uncertainty in structures, put forward in monograph by Ben-Haim and Elishakoff (1990). Particularly, impact failure of bars (Ben-Haim and Elishakoff, 1989) and shells (Elishakoff and Ben-Haim, 1990) was studied in detail, as was the response of a vehicle in uneven terrain (Ben-Haim and Elishakoff, 1990). By contrast, the only non-linear problem considered in applied mechanics literature within the set-theoretical, convex modelling is buckling of shells with uncertain initial imperfections (Ben-Haim and Elishakoff, 1989). The latter paper studied the first- and second-order approximation for the nonlinear function, since the exact solution was unavailable. The present chapter contrasts the first- and second-order approximations discussed in detail by Ben-Haim and Elishakoff (1990), with the exact analysis presented here for the first time, for the model structure of the bar with two eccentricities.

Consider again an elastic bar under an axial compressive force, applied with eccentricities e_1 and e_2. The maximum bending moment M_z^* is given by eq. (3.60).

With e_1 and e_2 specified, the maximum value of the moment can be directly evaluated from eq. (3.60). Assume now that the initial eccentricities are uncertain. In contrast to the previous section, we do not propose to model this uncertainty as randomness, under a probabilistic approach, but use an alternative, set-theoretical description, called by Ben-Haim (1985), "convex modelling." The nominal values of the eccentricities are $e_1°$ and $e_2°$, respectively, and the deviations from these nominal values are denoted ζ_1 and ζ_2. We assume that these deviations vary within the ellipsoidal set:

$$Z(a,\omega_1,\omega_2) = \left\{ (\zeta_1,\zeta_2) : \left(\frac{\zeta_1}{\omega_1}\right)^2 + \left(\frac{\zeta_2}{\omega_2}\right)^2 \le a^2 \right\}, \tag{3.72}$$

where ω_1 and ω_2 are semi-axes of the ellipsoid, and a is its size parameter. We are interested in finding the maximum $\mu(a,\omega_1,\omega_2)$, with respect to the uncertainty in the eccentricity, of the spacewise maximum bending moment

$$\mu(a,\omega_1,\omega_2) = maxM_z^* (e_1° + \zeta_1, e_2° + \zeta_2); \{\zeta_1,\zeta_2 \varepsilon Z(a,\omega_1,\omega_2)\} \tag{3.73}$$

where $\mu(a,\omega_1,\omega_2)$ is the bending moment of the weakest bar in the ensemble Z. The maximum bending moment for uncertain eccentricities ζ_1 and ζ_2 to the first order in ζ_1 and ζ_2 is

$$M_z^* (e_1^0 + \zeta_1, \ e_2^0 + \zeta_2) = M_z^* (e_1^0, \ e_2^0) + \frac{\partial M_z^*}{\partial e_1}\bigg|_{e_i=e_i^0}\zeta_1 + \frac{\partial M_z^*}{\partial e_2}\bigg|_{e_i=e_i^0}\zeta_2. \tag{3.74}$$

We will evaluate the maximum bending moment as ζ_1 and ζ_2 vary in an ellipsoidal set $Z(a, \omega_1, \omega_2)$. For convenience, we define the vector ϕ as follows:

$$\phi^T = \left\{ \frac{\partial M_Z^*}{\partial e_1} \bigg|_{e_i = e_i^0}, \ \frac{\partial M_Z^*}{\partial e_2} \bigg|_{e_i = e_i^0} \right\}, \tag{3.75}$$

where the superscript T denotes matrix transposition. Equation (3.73), in view of eqs. (3.74 and 3.75), becomes

$$\mu(a, \omega_1, \omega_2) = max_{\zeta_1, \zeta_2 \varepsilon Z(a, \omega_1, \omega_2)} \left[M_Z^* \left(e_1^0, e_2^0 \right) + \phi^T \zeta \right], \tag{3.76}$$

where

$$\zeta^T = (\zeta_1, \zeta_2). \tag{3.77}$$

Define Ω as 2×2 diagonal matrix

$$\Omega = \begin{bmatrix} \frac{1}{\omega_1^2} & 0 \\ 0 & \frac{1}{\omega_2^2} \end{bmatrix}. \tag{3.78}$$

Then, eq. (3.72) can be rewritten as

$$A(a, \Omega) = \left\{ \zeta : \zeta^T \Omega \zeta \leq a^2 \right\}. \tag{3.79}$$

Equation (3.76) calls for finding the maximum of the linear functional $\phi^T \zeta$ on the convex set $Z(a, \omega_1, \omega_2)$. According to the well-known theorem (see, e.g., Leunberger (1984) and Arora (1989), a linear functional, considered on the convex set Z, assumes the maximum on the set of extreme points of Z. The latter is the collection of vectors $\sigma = (\zeta_1, \zeta_2)$ in the following set:

$$C(a, \Omega) = \left\{ \sigma : \sigma^T \Omega \sigma = a^2 \right\}. \tag{3.80}$$

Thus, the maximum bending moment becomes

$$\mu(a, \Omega) = max_{\sigma \varepsilon C(a, \Omega)} \left[M_Z^* \left(e_1^0, e_2^0 \right) + \phi^T \sigma \right] \tag{3.81}$$

To solve the problem, we use the method of Lagrange multipliers. For details of derivation, the reader should consult the paper by Elishakoff (1991a). The probabilistic analysis of the identical problem is given in the latter paper.

For the maximum bending moment, we arrive at the following expression

$$\mu(a, \omega_1, \omega_2) = M_Z^* \left(e_1^0, e_2^0 \right) + \phi^T \sigma_1 = M_Z^* \left(e_1^0, e_2^0 \right) + a \sqrt{\phi^T \Omega^{-1} \phi}. \tag{3.82}$$

In an analogous manner we arrive at the following expression for the minimum bending moment

$$\mu_{min}(a, \omega_1, \omega_2) = M_Z^*\left(e_1^0, \ e_2^0\right) + \phi^T \sigma_2 = M_Z^*\left(e_1^0, e_2^0\right) - a\sqrt{\phi^T \Omega^{-1}\phi}. \qquad (3.83)$$

For the problem under consideration, the elements of vector ϕ can be found analytically:

$$\frac{\partial M_Z^*}{\partial e_1} = \frac{P\beta_1}{\sqrt{\beta_3}sinkL}, \qquad (3.84)$$

$$\frac{\partial M_Z^*}{\partial e_2} = \frac{P\beta_2}{\sqrt{\beta_3}sinkL}, \qquad (3.85)$$

where

$$\begin{aligned}
\beta_1 &= e_1 - e_2 coskL, \\
\beta_2 &= e_2 - e_1 coskL, \\
\beta_3 &= e_1^2 + e_2^2 - 2e_1 e_2 \ coskL.
\end{aligned} \qquad (3.86)$$

Hence, the maximum bending moment reads

$$\mu_{max}(a, \omega_1, \omega_2) = M_Z^*\left(e_1^0, e_2^0\right) + a\sqrt{\left[\omega_1 \frac{\partial M_Z(e_1, e_2)}{\partial e_1}\bigg|_{e_i=e_i^0}\right]^2 + \left[\omega_2 \frac{\partial M_Z(e_1, e_2)}{\partial e_2}\bigg|_{e_i=e_i^0}\right]}, \qquad (3.87)$$

whereas the minimum bending moment is

$$\mu_{max}(a, \omega_1, \omega_2) = M_Z^*\left(e_1^0, e_2^0\right) - a\sqrt{\left(\omega_1 \frac{\partial M_Z(e_1, e_2)}{\partial e_1}\bigg|_{e_j=e_j^0}\right)^2 + \left(\omega_2 \frac{\partial M_Z(e_1, e_2)}{\partial e_2}\bigg|_{e_j=e_j^0}\right)^2}. \qquad (3.88)$$

The detailed second-order analysis can be found in the paper by Elishakoff, Gan-Shvili, and Givoli (1991).

We will show below that the maximum bending moment is a convex function of its argument ζ_1 and ζ_2. Indeed, according to a theorem stated in Leunberger (1984) and Arora (1989), a function of n variables defined on a convex set S is convex of and only if its Hessian matrix is positive and semi-definite at all points in S. In our case the elements of the Hessian matrix Y_{ij} are

$$Y_{11} = \left(e_2^1 + e_2^2 - 2e_1 e_2 coskL\right)^{-1/2}\left[1 - \frac{(e_1 - e_2 coskL)^2}{e_2^1 + e_2^2 - 2e_1 e_2 coskL}\right], \qquad (3.89)$$

$$Y_{12} = -\left(e_1^2 + e_2^2 - 2e_1 e_2 coskL\right)^{-\frac{1}{2}}\left[coskL + \frac{(e_1 - e_2 coskL)(e_2 - e_1 coskL)}{e_1^2 + e_2^2 - 2e_1 e_2 coskL}\right], \qquad (3.90)$$

$$\gamma_{22} = \left(e_1^2 + e_2^2 - 2e_1 e_2 coskL\right)^{-1/2} \left[1 - \frac{(e_2 - e_1 coskL)^2}{e_1^2 + e_2^2 - 2e_1 e_2 coskL}\right].$$ (3.91)

Direct calculation yields

$$\det[\Gamma] = \gamma_{11}\gamma_{22} - \gamma_{12}^2 \equiv 0.$$ (3.92)

Also,

$$\gamma_{11} > 0, \quad \gamma_{22} > 0.$$ (3.93)

Equations (3.92) and (3.93) imply that the function $M_{\hat{z}}^*$ is convex. Therefore, we can apply the following theorem stated by Leunberger (1984): "Let f be a convex function defined on a bounded closed convex set S. If f has a maximum over S, this maximum is achieved at an extreme point of S."

Such being the case, we deduct that the maximum moment is achieved on the ellipse

$$\left(\frac{\zeta_1}{\omega_1}\right)^2 + \left(\frac{\zeta_2}{\omega_2}\right)^2 = a^2$$ (3.94)

We express ζ_2 from the latter equation as

$$\zeta_2 = \pm \omega_2 \sqrt{a^2 - \frac{\zeta_1^2}{\omega_1^2}}$$ (3.95)

and substitute in eq. (3.73), to yield

$$M_{\hat{z}}^* \left(e_1^0 + \zeta_1, \ e_2^0 + \zeta_2\right) = M_{\hat{z}}^* \left(e_1^0 + \zeta_1, \ e_2^0 \pm \omega_2 \sqrt{a^2 - \frac{\zeta_1^2}{\omega_1^2}}\right)$$ (3.96)

Now we seek the maximum of $M_{\hat{z}}^*$ with respect to ζ_1 alone. For the maximum moment we demand that

$$\frac{\partial M_{\hat{z}}^*}{\partial \zeta_1} = 0.$$ (3.97)

This equation defines the ζ_1^* at which $M_{\hat{z}}^*$ assumes maximum; then eq. (3.95) determines the value of ζ_2. Substituting ζ_1^* and ζ_2^* into eq. (2.73), we obtain the maximum value of the bending moment. Figure 3.14 shows the variation of the bending moment over the region

$$\left(\frac{\zeta_1}{\omega_1}\right)^2 + \left(\frac{\zeta_2}{\omega_2}\right)^2 \leq a^2$$ (3.98)

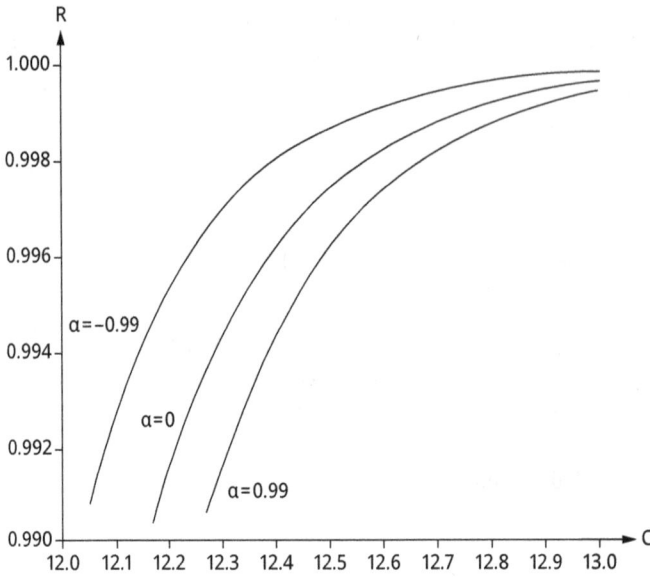

Fig. 3.13: How does the high required reliability influence the decision-making.

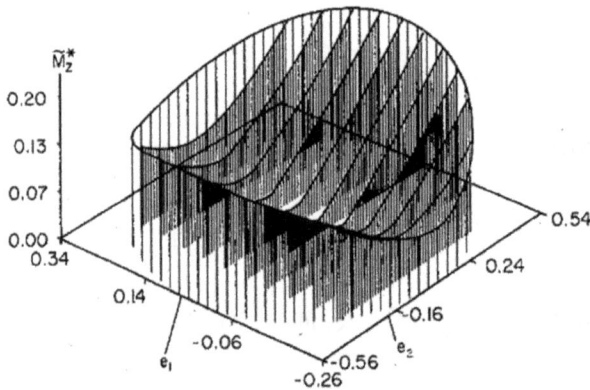

Fig. 3.14: Variation of the bending moment over the uncertainty ellipse.

for $\omega_1 = 0.3$, $\omega_2 = 0.5$, at the nondimensional load level

$$\upsilon = \frac{P}{P_E}, \quad P_E = \frac{\pi^2 EI}{L^2} \tag{3.99}$$

equal $\upsilon = 0.3$. Figure 3.15 is associated with $\omega_1 = 0.3$, $\omega_2 = 0.6$ and $\upsilon = 0.3$, whereas Fig. 3.16 illustrates the variation of M_Z^* for the values $\omega_1 = 0.3$, $\omega_2 = 0.6$ and $\upsilon = 0.5$. As we see in all these three instances the maximum value is achieved at the exterior point of the ellipse.

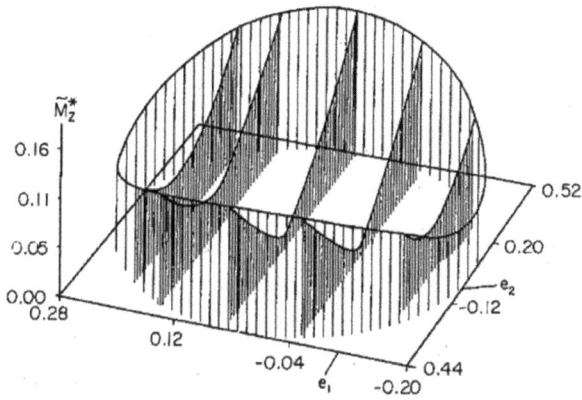

Fig. 3.15: Moment uncertainty as a function of uncertainty in eccentricities.

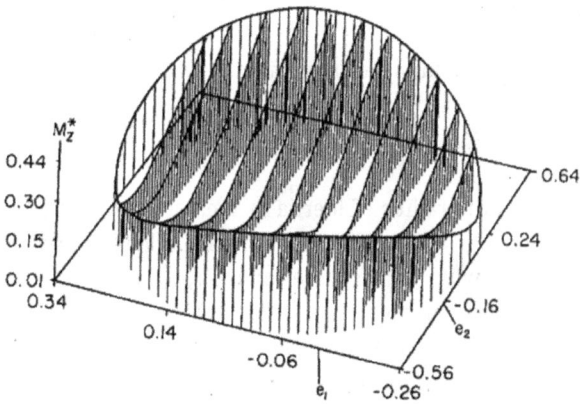

Fig. 3.16: Moment uncertainty as a function of uncertainty in eccentricities.

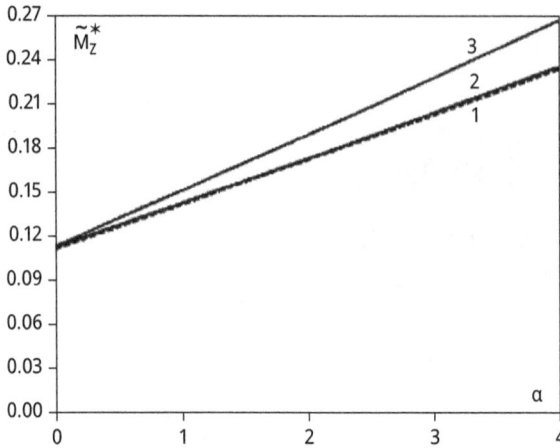

Fig. 3.17: Comparison of first-order and second-order solutions with exact solution.

Figure 3.17 portrays the variation of the nondimensional maximal moment

$$\bar{M}_Z^* = \frac{M_Z^*}{P_e c} \tag{3.100}$$

versus the ellipse's size a, where c is the radius of inertia of the bar's cross section. Moreover, $\omega_1 = 0.01$, $\omega_2 = 0.02$, $e^0 = 0.04$, $e^0 = 0.04$. The nondimensional load is fixed at $v = 0.3$. Broken curves 1 are associated with the first-order analysis, curves marked 2- with the second-order analysis and curves marked 3- with the exact results. The maximum moment increases with the size a ; the agreement between the first order, second order, and the exact analysis is excellent. It turns out that with the increase of the non-dimensional applied load, the percentage-wise disagreement between the first-order and the second-order analyses decreases considerably. On the other hand, for the larger nondimensional applied loads, the difference between the low-order approximations and the exact analysis becomes wider.

We conclude, therefore, that the lower-order approximation yields acceptable results for small uncertainties and smaller loads. This modifies the conclusion, based on the comparison of the first-order and the second-order analyses, drawn by Gumbel (1960) that the first-order approximation was acceptable for small uncertainties and greater loads. It is remarkable that the similarity of the first-order and second-order approximations that occurs for the elastic bar does not generally suggest that these approximations are in good agreement with the exact solution.

Chojaczyk et al. (2015) and Farid (2022) provide a review and application of artificial neural networks models in reliability analysis. Afshari *et al.* (2022) review machine learning-based methods in structural reliability analysis. Deep learning methodologies are discussed by Zhu Y. *et al.* (2019) and Sepahvand (2021). Depina *et al.* (2016) treat reliability evaluation with meta model line sampling. Song *et al.* (2021) deal with active learning line sampling for rare event treatment. Echard *et al.* (2013) and Liu X-X and Elishakoff (2020) combined importance sampling and active learning kriging model for reliability analysis. Gayton *et al.* (2003) presented a new response surface methodology for reliability estimation. Gayton *et al.* (2003, 2004) dealt with calibration methods for reliability-based design codes. Bourinet *et al.* (2009) discuss details of the software SERUM. Yuan, Liu *et al.* (2021) discuss estimation of the failure probability in augmented space. Morio and Balesdent (2015) provide practical approach for rare event probability in complex structures. Bonstrom *et al.* (2011) discussed the role of uncertainty in the political process for decision-making associated with risk investment. Corotis (2019) dealt with adopting life-cycle concepts within the political system.

Maes and Breitung (1994) treated reliability-based tail estimation. Asymptotic considerations on reliability are given by Breitung (2006, 2015). The interested reader can consult also the study by Hohenblicher *et al.* (1987). Issues associated with the small probability of failure were addressed in the comprehensive review by Lee (2022). Rackwitz (2001) provided both a review and some perspectives. He, specifically, noted: "In 1981 Cornell called structural reliability a healthy adolescent. In 1990 structural reliability then certainly was an efficient executive in all his virulence. But has this man in the meantime just got more years but no more wisdom? The meantime saw two ICOSSAR-and two ICASP-conferences, several other national and international conferences on more specialized topics and quite a number of papers on the subject in different journals."

He concluded: "FORM/SORM methodology is well established. Apart from some technical aspects little can be added and very little can be removed. If FORM/SORM does not work alternatives still avoiding numerical integration have been proposed. The errors we make when using FORM/SORM are acceptable in view of the large uncertainty in selecting the appropriate stochastic model and its parameters. It is still much better than crude Monte Carlo methods that have been the yardstick to compare with or in some cases the only way to obtain results. It has been argued that hardware development will soon make the intellectually more demanding methods somewhat obsolete. This is not true and will remain so for still some time. There are several promising fields of potential application, which need to be developed. Especially, the connection to stochastic mechanics needs to be strengthened. Design for reliability and some other objective is still a challenging task. In the end optimization with respect to reliability and/or underweight or cost constraints will concentrate on the question: how safe is safe enough? This question, no doubt, is crucial when applying any probabilistic method."

3.7 Summary and conclusion

This chapter reviews some pertinent questions associated with uncertainty modelling in analysis of structures. It gives a critical appraisal of the probabilistic method and describes a new, non-probabilistic philosophy. The former is valid when plentiful information (probability densities) is available on the uncertain quantities involved. The latter, convex modelling, is appropriate when the existing data is scarce, and no valid probabilistic models can be constructed. Both of these approaches deal with different facets of uncertainty treatment. The theory of probability and random processes is not the only way to deal with uncertainty. Indeed, as Freudenthal (1956) notes, "Ignorance of the cause of variation does not make such variation random." These two methods appear to successfully complement each other to make useful judgments based on reliably available experimental information.

There are a multitude of books on structural reliability. These include monographs and textbooks by Bolotin (1961, 1969, 1981), Tichy and Vorlicek (1972), Kapur and Lamberson (1977), Elishakoff (1983, 1999e, 2017), Ang and Tang (1984, 2007), Rao (1992), Nowak and Collins (2000, 2019), Haldar A. and Mahadevan (2000a), Melchers (1987, 1999), Raizer (2009), Melchers and Beck (2018), Thoft-Christensen and Baker (1982), Tichy (1993), Der Kiureghian (2021), and Raizer and Elishakoff (2022).

Chapter 4
Finite element method based on either stochasticity, fuzziness, or anti-optimization

The finite element method has been proven to be well suited for a large class of engineering problems. Finite element algorithms have a sound and well developed theoretical basis and their efficiency has long been proven and tested on a variety of problems. The formulation presented herein [of the stochastic finite element method] is, indeed, a natural extension of the basic ideas of the deterministic finite element method to accommodate random functions. Roger G. Ghanem and Pol D. Spanos (1991, p. 3)

4.1 Introduction

The first paper to utilize the finite element method in conjunction with uncertainty was apparently by Cambou (1975). This chapter will deal with various modes of finite element analysis, but we will start with digression. In 1990, the present writer introduced the notion of the *uncertainty triangle* to classify methods utilized for the description of uncertain external excitations and/or inner parameters of structures (Fig. 1.1). The first corner of this triangle constitutes a probabilistic method that treats uncertain parameters, processes, or fields as random variables, random processes, or random fields, respectively. Whereas the basic ideas of probability go to Pascal and Fermat, in structural applications these methods were apparently first developed by Mayer (1926), who tried to introduce a safety factor based on probabilistic analysis.

The probabilistic method became a popular tool of engineering analysis due to the efforts of numerous researchers, whereas Freudenthal (1947, 1956) probably can be considered as a main architect of modern structural reliability.

Probabilistic methods now have reached considerable maturity (Cornell, 1981), with numerous technical articles and books written. Indeed, as is mentioned in Ecclesiastics (12:12), *Of the making of many books there is no end.* This remark is not necessarily intended in the negative sense; most of the books have unique contributions showing the subject from different perspectives.

Probabilistic methods have been widely applied to numerous problems, irrespective of the presence or absence of experimental data, which is needed to substantiate the method. Indeed, in order to apply the probabilistic method, one needs to have as inputs probability density functions of uncertain variables, functions, or fields. When such information is available, the probabilistic method may be a viable procedure to predict structural reliability or, its complement, probability of failure. Since one looks for extremely small values of the probability of failure, it is easy to understand that input information will have paramount effect on the values of the probability of failure. When input information is missing, investigators are making various assump-

https://doi.org/10.1515/9783111354231-004

tions on the nature of the uncertain quantities. Extreme sensitivity of the output results to the assumptions on the inputs have been vividly demonstrated in the literature (see, for example, Chapter 1 in the monograph by Ben-Haim and Elishakoff,
1990). The necessity to make numerous assumptions on the nature of stochastic quantities (stationary or homogeneity; ergodicity; normal distribution, and the like) appears to be in some contradiction with the following statement of Albert Einstein:

> *The ground aim of all sciences is to cover the greatest possible number of empirical facts by logical
> deduction from the smallest possible number of hypothesis or axioms.*

These and other considerations led some scientists to look elsewhere. They developed
a new approach based on often available, vague, rather than crisp, knowledge on the
variables. In particular, Zadeh (1965, 1973) introduced his principle of incompatibility:

> *As the complexity of a system increases, our ability to make precise and yet significant statements
> about its behavior diminishes until a threshold is reached beyond which precision and significance
> (or relevance) become almost mutually exclusive characteristics.*

In particular, Zadeh (1973) remarks: A corollary principle may be stated succinctly as
"The closer one looks at a real-world problem, the fuzzier becomes its solution." In the
structural mechanics context, fuzzy-sets-based approaches were developed by several
investigators, the earliest ones probably were Brown (1979), Yao (1980), Blockley (1980),
and Shiraishi and Furuta (1985).

The theory of fuzzy sets was not met with excessive friendliness by the probabilistic analysts. Dennis Lindley remarked in this respect:

> *Anything that can be done with fuzzy logic . . . can better be done with probability.*

Yet fuzzy sets are finding wider and wider applications (see, for example, papers by
Klir, 1991; Zadeh, 1994). Note that although the fuzzy set theory was founded by Zadeh
in 1965, hints of some key ideas of the theory were envisioned about two decades
prior to that by Black (1937). Recent industrial developments of fuzzy sets in Japan are
given by Hirota (1991).

Third, the alternative treatment of uncertainty is anti-optimization, or Wald's
max-min approach, or Wald's max-min approach. For a detailed review and appropriate bibliography, the reader may consult Elishakoff (1995a).

In order for these methods to be practicable, it is necessary to combine them with
the most powerful analysis technique in structural mechanics, namely the finite element method. This chapter deals with three different versions based on stochasticity,
fuzziness, or anti-optimization.

4.2 Large-deviation FEM based on stochastic variational principles

Structures involving spatially random material and/or geometric parameters are re-ferred to as *stochastic structures*. The analysis of stochastic structures has attracted significant interest in recent decades. However, difficulties arise in obtaining exact solutions since the governing equations constitute random differential equations, pos-sibly with random boundary conditions. Therefore, several approximate analytical and numerical methods have been developed to address this problem. Among these methods, the FEM developed recently for stochastic problems is an example of pertur-bation-based numerical methods. Apparently, the first works in finite element method for stochastic structures were those by Su et al. (1969) and Cambou (1975). Presently, there are seven monographs devoted to this subject (Nakagiri and Hisada, 1985; Gha-nem and Spanos, 1991, 2003; Haldar and Mahadevan, 2000b; Kleiber and Hien, 1993; Elishakoff and Ren, 2003; To, 2013; Papadopolulos and Giovanis, 2017). For critical journal reviews, the readers may consult the studies of Yamazaki et al. (1986); Shino-zuka and Yamazaki (1988); Benaroya and Rehak (1988); Elishakoff *et al.* (1996); Gha-nem and Spanos (1997); Deodatis and Graham (1997); Sudret and Kiureghian (2000, 2002); Keese (2003); Berveiller *et al.* (2006); Stefanou (2009); Kandler *et al.* (2015); Arre-gui-Mena *et al.* (2016); Aldosary *et al.* (2018); Matthies (2004, 2008); Matthies *et al.* (1997) to provide an inexhaustive list.

Consider, for example, the beam bending problem with spatially stochastic stiff-ness $D(x) = E(x)I(x)$ subjected to loads. The governing differential equation reads:

$$\frac{d^2}{dx^2}\left[D(x)\frac{d^2w}{dx^2}\right] = q(x) \tag{4.1}$$

where $w(x)$ = displacement, $q(x)$ = transverse distributed force, $D(x)$ = the bending stiff-ness, which is a spatially random field, $E(x)$ = Young's modulus, and $I(x)$ = moment of inertia. Equation (4.1) can be rewritten as

$$\frac{d^2}{dx^2} = \frac{m(x)}{D(x)} \tag{4.2}$$

where

$$m(x) = -\int_0^x\int_0^v q(u)dudv + Q_0 x + M_0 \tag{4.3}$$

is the bending moment in the beam, M_0 and Q_0 are constants of integration represent-ing bending moment and shear force at the end $x = 0$, respectively. Assume that mo-ments and shear forces in the beam are statically determinate, namely that M_0 and Q_0

are independent from stochastic stiffness. By taking expectation of eq. (4.2), we get the governing equation for the mean displacement $\bar{w}(x)$

$$\frac{d^2\bar{w}}{dx^2} = \frac{m(x)}{D_0(x)} = f_0(x)m(x) \tag{4.4}$$

Here, $\bar{w}(x) = E[w(x)], E[\cdot]$ stands for the mathematical expectation and

$$f_0(x) = \frac{1}{D_0(x)} = E\left[\frac{1}{D(x)}\right] \tag{4.5}$$

Pre-multiplying equation (4.6) by $D_0(x)$ and then differentiating it twice, we get another form of governing equation for the mean displacement $\bar{w}(x)$ as follows

$$\frac{d^2}{dx^2}D_0(x)\frac{d^2\bar{w}}{dx^2} = q(x) \tag{4.6}$$

The governing equation eqs. (4.4) or (4.6) is identical in its form with that of a bending beam with equivalent deterministic stiffness $D_0(x)$ or equivalent deterministic flexibility $f_0(x)$. It is remarkable that D_0 is defined as the reciprocal of the mean flexibility of the beam, rather than simply the mean stiffness as in the first-order perturbation finite element method for stochastic structures (.g., see Kleiber and Hien, 1993). Subtracting eq. (4.4) from eq. (4.2) and multiplying the resulting equation by itself, we get

$$\frac{d^2[w(x) - \bar{w}(x)]}{dx^2}\frac{d^2[w(y) - \bar{w}(y)]}{dy^2} =$$

$$m(x)m(y)\left[\frac{1}{D(x)} - \frac{1}{D_0(x)}\right]\left[\frac{1}{D(y)} - \frac{1}{D_0(y)}\right] \tag{4.7}$$

Taking expectation of the above yields

$$\frac{\partial^4 C(x,y)}{\partial x^2\partial y^2} = \frac{m(x)m(y)}{D_1(x,y)} = f_1(x,y)m(x)m(y) \tag{4.8}$$

where $C(x,y)$ is the covariance function of displacements $w(x)$ at position x and $w(y)$ at position y; $f_1(x,y)$ and $D_1(x,y)$ are defined as

$$f_1(x,y) = \frac{1}{D_1(x,y)} = Cov\left[\frac{1}{EI(x)}, \frac{1}{E(y)}\right] \tag{4.9}$$

By partially differentiating eq. (4.8) twice with respect to x and twice with respect to y, we obtain an alternative form of the governing equation for covariance function $C(x,y)$

$$\frac{\partial^4}{\partial x^2\partial y^2}D_1(x,y)\frac{\partial^4 C(x,y)}{\partial x^2\partial y^2} = q(x)q(y) \tag{4.10}$$

The boundary conditions for the mean displacement $\bar{w}(x)$ at $x = 0$ and $x = L$ read

$$\bar{w} = 0 \quad \text{or} \quad \frac{d\bar{w}}{dx} = 0$$

$$\frac{d}{dx}\left[D_0(x)\frac{d^2\bar{w}}{dx^2}\right] = \bar{Q} \quad \text{or} \quad D_0\frac{d^2\bar{w}}{dx^2} = \bar{M} \tag{4.11}$$

where $\bar{M}$ and $\bar{Q}$ are prescribed moment and shear force at ends, respectively. The boundary conditions for the covariance function $C(x,y)$ are as follows: at $x = 0$ and $x = L$

$$\frac{\partial C}{\partial x} = 0 \quad \text{or} \quad C = 0$$

$$D_1\frac{\partial^4 C}{\partial x^2 \partial y^2} = \bar{M}m(y) \quad \text{or} \quad \frac{\partial}{\partial x}\left[D_1\frac{\partial^4 C}{\partial x^2 \partial y^2}\right] = \bar{Q}m(y) \tag{4.12}$$

At $y = 0$ and $y = L$

$$\frac{\partial C}{\partial y} = 0 \quad \text{or} \quad C = 0$$

$$D_1\frac{\partial^4 C}{\partial x^2 \partial y^2} = \bar{M}m(x) \quad \text{or} \quad \frac{\partial}{\partial y}D_1\frac{\partial^4 C}{\partial x^2 \partial y^2} = \bar{Q}m(x) \tag{4.13}$$

It is worth pointing out that only geometric boundary conditions need to be satisfied when eqs. (4.4) and (4.8) are used. It is seen from eq. (4.4) [or eq. (4.6)] and eq. (4.11), that the mean displacement of the beam with stochastic stiffness $D(x)$ is exactly the same as the displacement of a beam with equivalent deterministic stiffness $D_0(x)$. It is remarkable that although the differential eqs. (4.6) and (4.10) are extremely simple, they were not derived until 1995 (Elishakoff *et al.*).

Since the deterministic finite element method is mostly presented through variational formulation, it was thought (Elishakoff *et al.*, 1996) that analogously, variational formulation must be pursued for stochastic problems. As Besseling (1985) writes:

> *A variatonal formulation of a theory for physical processes has several advantages. As it has been stated by Washizu (1968), first, the functional which is subject to variation usually has a definite physical meaning and is invariant under coordinate transformation. Second, by the Lagrange multipliers method a given problem may be solved more easily than the original or that may give more insight in the nature of the process. Third, the variational method provides a sound basis for an approximate formulation of the problem. Theories of beams, plates and shells, but also finite element models are typical examples of such approximate formulations. If the functional which is subject to variation processes a minimum or maximum it is, moreover, possible to construct upper and lower bounds of the exact solution of the problem under consideration.*

The variational principle for the mean displacement $\bar{w}(x)$ corresponding to eq. (4.4) and boundary conditions eq. (4.11) is given by minimizing the following functional

$$\pi_1 = \int_0^L \left[\frac{1}{2} D_0(x) \left(\frac{d^2 \bar{w}}{dx^2} \right)^2 - q(x)\bar{w} \right] dx - \left[M \frac{d\bar{w}}{dx} - Q\bar{w} \right]_0^L \tag{4.14}$$

The above functional is similar to that of a deterministic beam, which has an equivalent deterministic stiffness $D_0(x)$, as it should be. The variational principle for the covariance function $C(x,y)$ corresponding to eq. (4.10) and boundary conditions eqs. (4.12) and (4.13) is given by minimizing the following functional

$$\pi_2 = \int_0^L \int_0^L \left[\frac{1}{2} D_1(x,y) \left(\frac{\partial^4 C}{\partial x^2 \partial y^2} \right)^2 - q(x)q(y)C \right] dxdy$$

$$- \left[\int_0^L (\bar{M} \frac{\partial C}{\partial x} - \bar{Q}C)q(y)dy \right]_{x=0}^{x=L}$$

$$- \left[\int_0^L (\bar{M} \frac{\partial C}{\partial y} - \bar{Q}C)q(x)dx \right]_{y=0}^{y=L} \tag{4.15}$$

$$- \left[\bar{M}\bar{M} \frac{\partial^2 C}{\partial x \partial y} - \bar{M}\bar{Q} \frac{\partial C}{\partial x} - \bar{M}\bar{Q} \frac{\partial C}{\partial y} + \bar{Q}\bar{Q}C \right]_{y=0 \, x=0}^{y=L \, x=L}$$

The finite element formulation is obtainable through established variational principles. For the mean displacement, one can utilize the two-node cubic Hermitian beam element. As far as the covariance function is concerned, even for the one-dimensional problem, one must utilize a four-node rectangular element. The covariance function $C(x,y)$ is the element $x_1 \leq x \leq x_2$ and is interpolated as follows:

$$C = \sum_{i=1}^4 N_i \delta_1 = N\delta \tag{4.16}$$

where δ is the vector of nodal degrees of freedom

$$\delta_i = \left[C_i, \left(\frac{\partial C}{\partial x} \right)_i, \left(\frac{\partial C}{\partial y_i} \right) \right]^T \tag{4.17}$$

and N is the vector of shape functions

$$[N] = [N_{i1}, N_{i2}, N_{i3}], N_{i1} = \xi_i \eta_i \left(2 + \xi_i \xi + \eta_i \eta - \xi^2 - \eta^2 \right) \times (\xi + \xi_i)(\eta + \eta_i)/8 \tag{4.18}$$

$$N_{i2} = \eta_i a \left(\xi^2 - 1 \right)(\xi + \xi_i)(\eta + \eta_i)/8$$

$$N_{i3} = \xi_i b (\eta^2 - 1)(\xi + \xi_i)(\eta + \eta_i)/8 \tag{4.19}$$

$$\xi = \frac{x - x_1}{x_2 - x_1} - \frac{x_2 - x}{x_2 - x_1}$$

$$\eta = \frac{y - y_1}{y_2 - y_1} - \frac{y_2 - y}{y_2 - y}$$

where ξ and η are local coordinates, $a = (x_2 - x_1)/2$ and $b = (y_2 - y_1)/2$ are side length of the rectangular element.

Calculations have been performed for the variances of the mid-displacement, obtained by the present finite element method and the first-order perturbative finite element method, for different coefficients of variation of the stochastic bending stiffness (Ren et al., 1997). It turns out that the results obtained by the first-order perturbative method were for small variation. For larger variation, the classical perturbation solution underestimates the exact result respectively by 3%, 6%, or 28%, for the corresponding coefficients of variation 0.1, 0.2, and 0.3. This preliminary study clearly demonstrates the superiority of the proposed variational principle-based finite element method. For other most recent developments, the reader may consult the review by Elishakoff et al. (1996).

Stochastic variational principles for shear beams and attendant finite element formulation were developed by Impollonia and Elishakoff (1998).

The generalization of the above variational principle for general problems and to the higher dimensions (plates and shells) appears to be imperative. The main developments will include (a) development of variational-principles-based stochastic FEM for Timoshenko-Ehrenfest beams, (b) development of variational principles based stochastic FEM for thin plates, (c) development of the variational-principles-based stochastic FEM of thin shells; comparison with available first-order perturbation solutions; development of a stochastic post processor to deal with stochastic FEM, as an addition to deterministic codes.

Definitive works on stochastic finite element method appear to be those by Nakagiri and Hisada (1985), Ghanem and Spanos (1991, 2003), Haldar and Mahadevan (2000b), Deodatis and Graham (1997), and Elishakoff and Ren (1997, 1999, 2003).

4.3 Interval finite element method based on anti-optimization idea

Archimedes (circa 287–212 BC) was apparently the first scientist who engaged himself with interval mathematics. He denoted an irrational number π through an interval $3\frac{10}{71} < \pi < 3\frac{1}{7}$ (see Heath, 1953). This was done by approximating the circle with the inscribed and circumscribed 96-side angular polygons. Now, numerous fields of engineering use techniques that provide upper and lower bounds of solutions. Such methods for natural frequencies and buckling loads are well known. Yet, a revolutionary step from such considerations to introducing interval numbers was accomplished by Young (1908) and Young (1931). Sunaga (1958) introduced the term *interval algebra*. The methods of interval algebra are discussed in the books by Moore (1966), Alefeld und Herzberger (1983), Neumaier (1990), and many others. These investigators mainly concerned themselves with roundoff errors, and guaranteed calculations. Yet this beautiful branch of mathematics appears to be particularly important for uncertainty modeling from the following perspective. It was demonstrated in numerous studies (for books, one may consult Schweppe, 1973; Eliasberg, 1976; Bakhasan *et al.*, 1980; Ben-Haim and Elishakoff, 1990; Chernousko, 1994; Elishakoff, 1995; Ben-Haim, 1996; Kurzhanskii and Valyi, 1997) that the classical probabilistic analysis of structures has a major drawback (see also Elishakoff and Hasofer, 1996). Indeed, the main product of such an analysis, the probability of failure, is extremely sensitive to a) deterministic modeling of the underlying physical phenomenon, b) probabilistic densities of input parameters, details of which may be unknown, and c) distribution tails whose prediction is especially difficult, if at all possible, to predict. These observations led to the development of the alternative approach, namely *convex modeling* in applied mechanics (Ben-Haim and Elishakoff, 1990; Elishakoff *et al.*, 1994; Ben-Haim, 1996). This approach postulates that at least the input parameters or functions possess (or can be approximated as having) the convexity property and is intimately associated with classical convex optimization. It should be noted that the input parameters or functions, and in most cases the output parameters too, may lack the convexity property, and then numerical nonlinear programming should be applied to determine the extremal behavior of the structure. Therefore, the present writer dubbed a new term *anti-optimization* which stresses the physical aspect rather than the mathematical one. Since we have only the sets, be they convex or not, of input parameters or functions, we are looking for the most favorable, and the least favorable responses are of prime interest. In contrast to the genuine optimization when the structure is designed in some best way, here we are looking for the worst responses. Thus, the idea of *anti-optimization* has appeared (Elishakoff, 1990). The ideas of optimization and anti-optimization can be combined, following the folk wisdom *to make the best out of the worst*. A hybrid technique was developed by Elishakoff *et al.* (1994). Such analyses, however, will not become practical if they are not cast in the context of the finite element method. We will consider here, as an example, the free vibrations of an uncertain string.

We determine the frequencies of the string when its mass per unit length $m(x)$ and the extension force $T(x)$ are interval variables

$$\underline{m} \leq m(x) \leq \bar{m}, \ \leq T(x) \leq \bar{T} \tag{4.20}$$

The equation of motion reads

$$\frac{\partial}{\partial x}\left[T(x)\frac{\partial u}{\partial x}\right] = m(x)\frac{\partial^2 u}{\partial t^2} \tag{4.21}$$

We use the finite element method. The stiffness and mass matrices read:

$$k_{ij} = \int_0^L T(x)N_i'(x)N_j'(x)dx$$

$$m_{ij} = \int_0^L m(x)N_i(x)N_j(x)dx \tag{4.22}$$

where $N_i(x)$ are shape functions

$$N_1(x) = 1 - \xi, \quad N_2(x) = \xi, \quad \xi = x/L \tag{4.23}$$

The lower and upper elemental stiffness matrix elements can be written

$$\underline{K_e} = \frac{1}{L}\begin{bmatrix} \underline{T} & -\bar{T} \\ -\bar{T} & \underline{T} \end{bmatrix}$$

$$\overline{K_e} = \frac{1}{L}\begin{bmatrix} \bar{T} & -\underline{T} \\ -\underline{T} & \bar{T} \end{bmatrix} \tag{4.24}$$

The appropriate mass matrices read

$$\underline{M_e} = \frac{\underline{m}L}{6}\begin{bmatrix} 2 & 1 \\ 1 & 2 \end{bmatrix}, \quad \overline{M_e} = \frac{\bar{m}L}{6}\begin{bmatrix} 2 & 1 \\ 1 & 2 \end{bmatrix} \tag{4.25}$$

Consider the string of length b represented by N elements with span length $L = \ell N, n \geq 2$. Considering boundary conditions $u(u = 0, t) = u(x = l, t) = 0$, we get following global matrices of dimension $(N - 1) \times (N - 1)$. The global stiffness matrix reads

$$\underline{K} = \frac{1}{L}\begin{bmatrix} 2\underline{T} & -\bar{T} & 0 & 0\cdots & 0 & 0 \\ -\bar{T} & 2\underline{T} & -\bar{T} & & & \\ 0 & -\bar{T} & 2\underline{T} & -\bar{T}\cdots & 0 & 0 \end{bmatrix}$$

$$\begin{bmatrix} \vdots \end{bmatrix} \tag{4.26}$$

$$\begin{bmatrix} 0 & 0 & 0 & 0 & \cdots & 2\underline{T} & -\bar{T} \\ 0 & 0 & 0 & 0 & \cdots & -\bar{T} & 2\underline{T} \end{bmatrix}$$

In order to obtain matrix $\bar{K}$, we must replace $\underline{T}$ by $\bar{T}$ and $\bar{T}$ by $\underline{T}$. The lower bound mass matrix reads

$$M = \frac{mL}{6} \begin{bmatrix} 4 & 1 & 0 & 0\cdots & 0 & 0 \\ 1 & 4 & 1 & 0\cdots & 0 & 0 \\ 0 & 1 & 4 & 1\cdots & 0 & 0 \\ & \begin{bmatrix} \vdots \end{bmatrix} & & & \begin{bmatrix} \vdots \end{bmatrix} & \\ 0 & 0 & 0 & 0\cdots & 4 & 1 \\ 0 & 0 & 0 & 0\cdots & 1 & 4 \end{bmatrix} \tag{4.27}$$

To obtain $\bar{M}$, we must formally change $\underline{m}$ by $\bar{m}$.

In order to obtain natural frequencies, we must use two Rayleigh-quotient type equations,

$$\underline{K}\underline{u}_i = \underline{\lambda}_i \underline{M}\underline{u}_i, \ (i = 1, 2, \ldots, n) \tag{4.28}$$

$$\bar{K}\bar{u}_i = \bar{\lambda}_i \underline{M}\bar{u}_i, \ (i = 1, 2, \ldots, n) \tag{4.29}$$

following the theorem established by Qiu *et al.* (1995).

For example, for $n = 3$, we get the following equation for λ:

$$\det \left[\frac{3}{\ell} \begin{bmatrix} 2\underline{T} & -\bar{T} \\ -\bar{T} & 2\underline{T} \end{bmatrix} - \frac{\bar{m}\ell}{18} \begin{bmatrix} 4 & 1 \\ 1 & 4 \end{bmatrix} \lambda \right] = 0 \tag{4.30}$$

With notation

$$\underline{a} = \frac{54\underline{T}}{\bar{m}\ell^2}, \quad \bar{a} = \frac{54\bar{T}}{\bar{m}\ell^2}, \quad \underline{\beta} = \frac{T}{\bar{T}}, \quad \bar{\beta} = \frac{\bar{T}}{\underline{T}} \tag{4.31}$$

we get

$$\det \left[\underline{a} \begin{bmatrix} 2 & -\bar{\beta} \\ -\bar{\beta} & 2 \end{bmatrix} - \begin{bmatrix} 4 & 1 \\ 1 & 4 \end{bmatrix} \lambda \right] = 0 \tag{4.32}$$

or

$$\underline{\lambda}_1 = \frac{(\underline{a}2 - \bar{\beta})}{5} = \frac{54}{5} \frac{\underline{T}}{\bar{m}\ell^2} (2 - \bar{\beta}) \tag{4.33}$$

Analogously,

$$\bar{\lambda}_1 = \frac{54}{5} \frac{\bar{T}}{\bar{m}\ell^2} (2 - \underline{\beta}) \tag{4.34}$$

For the interval FEM for beam problems, the reader should consult the study by Köylüoğlu *et al.* (1995) (see also the discussion of this article by Modaressi, 1997). The latter work maintains that *". . . the lower and upper limits of these intervals should not appear simultaneously in the definition of an elementary stiffness matrix."*

In the enclosure, Köylüoğlu, Çakmak, and Nielsen stressed *". . . the need for verification studies for the goodness and validity of the bounds obtained from interval finite element discretization, and for the goodness of the interpolation by comparison with closed-form solutions for the interval deflection field for simple structural systems."*

The development of special methods yielding sharp bounds appears to be necessary. For most recent works on interval mathematics in applied mechanics problems, readers may consult the works of Nakagiri and Yoshikawa (1996), Dimarogonas (1994), Nakagiri and Suzuki (1997), and Rao and Berke (1996). Interval finite elements are exposed by Nakagiri and Suzuki (1996, 1997), Nakagiri and Yoshikawa (1996), Köylüoglu Elishakoff (1998), Muhanna and Mullen (2001, 2011), Dessombz *et al.* (2001), Moens and Vandepitte (2005), Sofi *et al.* (2019), van Mierlo *et al.* (2022). Muhanna *et al.* (2013) and Qiu *et al.* (2017), offer a comprehensive review of the interval finite element method. The book by Nayak and Chakraverty (2018) is exclusively devoted to this topic.

4.4 Fuzzy finite element method

Fuzzy-sets-based analysis is now considered by many not to be a *poor relative* of probabilistic analysis. Numerous applications, including in engineering, strikingly demonstrate that *fuzzy sets* do work, especially in the AI and/or expert systems context. The fuzzy-sets-based finite element method is referred to hereinafter as *fuzzy finite elements*, following Rao and Sawyer (1995). The methodology applied by them is the first to treat the deterministic problem and to write the appropriate *crisp finite* element equations. Then in the equations either the loads of the stiffness or both are treated as fuzzy numbers with triangular membership functions. These equations are subsequently solved following the procedure by Zhao and Goving (1991). This procedure results in membership functions of fuzzy displacement and fuzzy stress. An alternative fuzzy finite element method is under development (Ren *et al.*, 1997) and will be reported elsewhere.

The reader also may consult the works by Sparrow *et al.* (1994), Yagawa *et al.* (1992), Valliappan and Pham (1993, 1995), Chao and Ayyub (1996), Dixit and Dixit (1996), Elishakoff (1997, 1998), Sawyear and Rao (1995), Weintraub (1997), Yagawa *et al.* (1995), Ichihashi *et al.* (1995), and Fang *et al.* (1998).

4.5 Hybrid techniques should be sought

We may face a situation when based on experimental evidence some of the uncertain parameters can be treated as random variables, functions, or fields, whereas others cannot be treated in the probabilistic setting. There is no sufficient information available for the latter parameter to be dealt probabilistically. In these circumstances, these parameters may be considered in either the anti-optimization context or fuzzy-set-based modeling.

The hybrid probabilistic and fuzzy-sets-based analyses have been developed by Shiraishi and Furuta (1983), Bernardini and Modena (1987), Ou and Wang (1989), Haldar and Reddy (1992), Ayyub and Lai (1992), Li and Zhao (1995), Shih and Wangsawidjaja (1996), Cai (1996), and others.

The hybrid probabilistic and convex modeling was developed by Elishakoff and Colombi (1993a, 1993c, 1994), Zhu and Elishakoff (1996b), and Elishakoff *et al.* (1994) in the context of the space shuttle weather protection system, which is subjected to a random excitation with unknown-but-bounded parameters. The hybrid fuzzy-sets-based analysis and convex modeling was presented recently by Fang *et al.* (1998).

4.6 Comparison of probabilistic modeling and anti-optimization

It appears important to compare the techniques available in the *uncertainty supermarket*. Indeed, when facing a necessity to solve a specific engineering problem, different designers may utilize different techniques, namely probabilistic analysis, fuzzy-sets-based analysis, or anti-optimization. The pertinent question arises: In which interrelation will the results of such analyses be?

Such an investigation has been undertaken by Elishakoff *et al.* (1994). The model problem of the beam on the nonlinear elastic foundation was studied. The problem was solved both in the probabilistic setting and via the convex modeling. For the probabilistic treatment, the initial imperfection's Fourier coefficients were assumed to have a truncated normal distribution. The reliability of the structure was derived through the numerical analysis.

The reliability for the column with a random imperfection to a prescribed axial load α is defined as the probability that the structure does not fail prior to α. In other words, reliability equals the probability that the limit load α^*, which turns out to be a random variable due to stochasticity assumption for $\bar{X}_m$, exceeds α:

$$R(\alpha) = Prob[\alpha^* > \alpha] \tag{4.35}$$

If we design a column based on the stochastic approach, the value of the load corresponding to reliability R equal to a codified, required reliability r is the maximum admissible axial load. The latter load is referred to as a *design load* within the stochastic modeling. The design loads associated with $r = 0.9$, $r = 0.99$, $r = 0.999$, *or* $r = 0.9999$ have been calculated by Elishakoff *et al.* (1994).

The same problem has also been analyzed by the anti-optimization modeling, in order to compare the designs obtained by alternative processing of the same information. If we expand the initial deflection as a Fourier series, then a simple non-probabilistic model for the initial imperfection is that its Fourier coefficients vary in a hyper-cuboid set (a solid *box*):

$$\{Z(\bar{X}):|\bar{X}_m| \le a_m(a_m \ge 0, m = 1, 2 \ldots, M)\} \tag{4.36}$$

The objective in these new circumstances is to find the minimum, the least-favorable limit load for all possible initial imperfections $\bar{X} = \{\bar{X}_1, \bar{X}_2, \ldots, X_M\}$ that belong to set $Z(\bar{X})$. If we design the column based on the anti-optimization approach, the above minimum limit load is the maximum admissible value of the axial load. In this manner, we arrive at an alternative way of determining the admissible axial load, which can be applied to an ensemble of columns with bounded Fourier coefficients.

It is remarkable that in some circumstances both approaches, although being of fundamentally different nature, may yield close values for the design axial loads. If probabilistic information is unavailable, one should not propose a probabilistic model, based on an arbitrary assumption of the distribution on the Fourier coefficients. Rather, in such circumstances, one should use the non-stochastic approach to model uncertainty. On the other hand, when the full probabilistic information is available and the initial imperfection's Fourier coefficients have a relatively small deviation, use of the non-stochastic approach will be inadvisable, and purely stochastic analysis should be conducted.

Remarkably, even when probabilistic information is available to substantiate the probabilistic analysis, if the density of the initial imperfections is rather *flat* (i.e., if the distribution is close to the uniform one), one may prefer a simpler non-stochastic, convex analysis, since it yields admissible axial loads comparable with the results of stochastic approach. This result is in agreement with the recent study conducted by Barmish and Lagoa (1997), who justify the use of the uniform distribution in the robustness analysis.

Comparison of the fuzzy-sets-based design with the probabilistic design was performed by Maglaras *et al.* (1996).

4.7 In different circumstances, either probabilistic analysis or convex modeling can be more robust than the other

In the literature, there is a claim that the bounding techniques like the convex modeling (Ben-Haim and Elishakoff, 1990; Elishakoff *et al.*, 1994; Ben-Haim, 1996) or ellipsoidal modeling (Schweppe, 1973; Leitman, 1979; Chernousko, 1994; Kurzshanskii and Valyi, 1997) are more robust than the probabilistic analysis. Such a conclusion appears to be questionable. Let us bring an engineering example.

The Netherlands is a low-flying country bordering on the North Sea with nearly half of its land below the mean sea level and only partially shielded by natural sand dunes. In these circumstances, permanent habitation became possible only when the population learned how to set up protective dikes against the incursions of the sea and control the water level in the areas so protected. The protective system of dikes and dunes has to sustain extreme loads, especially during winter storms; and as perfect safety was unattainable, the country was repeatedly devastated by floods throughout the course of history.

As a result, Dutch engineers adopted an approach somewhat reminiscent to convex modeling of uncertainty. According to their rule of thumb, every new dike built had to be one meter higher than the ones that were demolished by previous disasters.

After the disaster which occurred on February 1, 1953, a governmental committee (known as the *Delta Committee*) was appointed with a view of re-answering the age-old question: How high must the dikes be? The probabilistic solution, devised by Van Danzig (1956), in conjunction with probabilistic experimental information on the annual exceedance frequencies during the high tide in the period 1888–1937 provided by Wemelsfelder (1939), constituted 1.57 meters of elevation. As is seen, the probabilistic analysis resulted in the necessary to elevate the dikes more than the historical rule of thumb would suggest. In this case, the convex-type analysis turned out to be less robust than the probabilistic one. Galambos *et al.* (1994) write:

> We know that records must be broken in the future, so if a flood design is based on the Lechner and Simiu (1992) worst case of the past then we are not really prepared against floods.

Indeed, if the probability densities assumed extend to infinity, then for the extremely small, allowed probability of failure, the design loads predicted by the probabilistic method may turn out to be less than those based on the model with bounded uncertainty.

Only when probabilistic analysis involves truncated densities, the anti-optimization modeling may turn out to be more robust than the probabilistic one. Results of the anti-optimization modeling can be obtained by probabilistic analysis if one applies the uniform distributions, known in the literature as *ignorance* distributions. Yet, performance of the anti-optimization analysis may turn out to be much simpler, both conceptually and numerically, than the probabilistic calculations. For the details of probabilistic calculations in the dike example, one can consult the text by Elishakoff (1983).

4.8 More discussion

One of the reviewers of this book, when it was at the proposal stage, remarked to the effect that, most probably, there was a lively discussion after the presentation of this chapter at the workshop on *Uncertainty Models and Measures*. This reviewer requested the above discussion to be reproduced here. Indeed, the discussion that took place after the presentation of this chapter at the conference appears to be not uninstructive.

a) One of the participants asked about the finite element method for stochastic structures: *Are the stochastic functions describing the variation of material properties large indeed, to justify the non-perturbative approach?* The answer to this question is affirmative. Whereas in many cases the variation may be small, in many reported instances the coefficients of variation are quite large. One may consult works by Virkler, Milberg, and Goel (1979), Beran *et al.* (1996), Ikeda *et al.* (1997). These investigators reported coefficients of variation ranging between 0.4 to unity and more. In such circumstances, application of the first-order perturbation technique appears to be questionable.

This point appears to need additional elucidation. One of the reviewers of our recent study devoted to the FEM of stochastic problems remarked with self-assurance, *Perturbation methods are universal.* Hopefully, this reviewer meant something more acceptable than the literal meaning of this statement. *If* the coefficients of variation of the involved quantities are *small, then* the perturbation method can directly be applied to the *majority* of problems, as over 300 (or more) published papers on the stochastic version of the FEM may attest. Some stochastic researchers appear to have a misunderstanding on this issue: The perturbation method can be applied to any structure if the coefficient of variation of the random quantity is small, yet it cannot be applied to the problems in which the coefficients of variation are moderate or large. Researchers are limiting themselves with perturbation analysis because this is what they can do. Yet, not dwelling on the large coefficients of variation limits the very spirit of the numerical analysis which tries to cover wide range of parameter variations. Obviously, large coefficients of variation can be dealt with the Monte Carlo method; yet the availability of non-simulation finite element methods for stochastic structures appears very attractive.

A contrast of the asymptotic analysis with other types of analyses, in a broad context, is performed in the papers by Kruskal (1963), Segel (1966), Barantsev (1989), to name just a few, and in the recent popular book by Andrianov and Manevich (1994). Kruskal notes:

> *Asymptotics is the science which deals with such questions as the asymptotic evaluation of integrals, of solutions of differential equations, etc, in various limiting cases.*

Indeed, what most stochastic mechanicians do is to develop FE techniques for the *limiting* case when the coefficient of variation tends to zero. One undoubtedly needs more general numerical techniques that are not based on this far-going assumption.

Let us *strengthen* the argument of our possible opponent. In their book, Andrianov and Manevich (1994) quote from Gell-Mann (1985), maintaining that in actuality theoreticians in their own works put some parameters to be small, yet they criticize others who do the same and attack them in making *unnatural* assumptions.

Obviously, when using theories of beams, plates, or shells, we use the assumption of smallness of the radius of gyration of the beam, or of the thickness of the plate or the shell, in comparison with other characteristic dimension(s); yet why should one *add* an *additional* assumption of smallness of the coefficient of variation of a stochastic quantity? Indeed, Blekhman *et al.* (1990, p. 38) quote from Courant, who maintained: *"The same mathematical problem can be solved by different means."* How much more so this applies to an engineering problem! (The reader may also consult the paper by Verhulst, 1984, and the book by Manevich *et al.* (1981), for elucidation of the role of asymptotic methods in applied mechanics.)

b) One of the organizers of the workshop remarked to the effect that since the probabilistic methods and the convex modeling (or, more generally, the anti-optimization method) are vastly different, these two methods cannot be compared either analytically or numerically. To reply to this question, the present writer suggested visualizing those two researchers who are employed in the same design company. One of them advocates the use of probabilistic methods, whereas the other one propagates the ideas of anti-optimization. Imagine that they were requested to design a structure and the same input data was provided to both of them; they came up with their designs. The client who ordered the project appears to have a right to compare two designs and make a choice.

Still, a skeptic may ask: "Probabilistic methods provide, as final result of the analysis, the reliability of the structure; by its very definition, reliability is a probability. The anti-optimization methods, however, do not provide information about the probability, but rather, yield the guaranteed designs. Thus, two approaches cannot be compared!"

In order to reply to such a possible argument, the present writer developed a method of comparison of two methods and solved several problems through two competing methods (Elishakoff, 1997). It turned out that when the required reliability tends to unity from below, the probability-based design tends to the anti-optimization results. Yet the method of anti-optimization is simpler both conceptually and computationally than the stochastic approach. Therefore, advice given by Henry David Toureau (1817–1826), 'Simplify, simplify may be understood in the sense that the researchers should engage in nonprobabilistic, anti-optimization analyses of uncertainty with much enthusiasm. Experimental validation of the anti-optimization approach was given by Senmeister *et al.* (1995) for detecting delamination damage.

Note that Kim *et al.* (1993) conducted a direct numerical comparison between the probabilistic and the unknown-but-bounded (guaranteed) model. They fixed the reliability at the level corresponding to the mean value plus three mean square deviations. In actuality, some calibration was established between these two seemingly contradictory approaches. One may argue that possibly one can set a level of reliability which would correspond to the anti-optimization results with suitably chosen bounds of variation, if the probabilistic version of the problem involves random functions or varieties with probability density that is not bounded.

Analogous homology between the probabilistic and fuzzy-sets-based designs was conducted by Maglaras *et al.* (1996). More research combining two or all three *musket-*

eers of uncertainty appear to be needed. To those researchers who think that the stochasticity is the only game in the town of uncertainty, one could recommend, as Charles A Dana did:

Fight for your opinions, but do not believe that they contain the whole truth, or the only truth.

This advice appears to be equally applicable to those who pursue other characterizations of uncertainty, or even hybrid approaches.

Yet, even if one agrees that there is a *supermarket* of uncertainty, still some intelligent choices must be made, most suitable for the circumstances at hand. Energetic research appears to be preceded to pinpoint when one should prefer a specific avenue of picking a problem that involves uncertainty. This essay does not attempt to establish *truth* but engages, to quote from Herman Hanover, *"in finding the best operational theories, based on the best possible generalization of the empirical data available at the time. A scientific theory is always in a state of transition. It is acceptable only until a better one is found."*

Chapter 5
Seven decades of progress in stochastic linearization technique

Stochastic linearization technique is a versatile method of solving nonlinear stochastic boundary value problems. It allows obtaining estimates of the response of the system when an exact solution is unavailable; in contrast to the perturbation technique, its realization does not demand smallness of the parameter; on the other hand, unlike the Monte Carlo simulation, it does not involve extensive computational cost. Although its accuracy may not be supreme, this is remedied by the fact that the stochastic excitation itself need not be known quite precisely. Although it was advanced about seven decades ago, during which several hundreds of papers were written, its foundations, as exposed in many monographs, appear to be still attracting investigators in stochastic dynamics. This chapter considers the methodological aspects of its exposition.

5.1 Introduction

This chapter follows two articles, namely by Villaggio (2011) and Maugin (2013). The former reviewed sixty years of solid mechanics whereas the latter dealt with the configurational forces. Villaggio (2011) writes: "The end of the Second World War marked a turning point of the history of Solid Mechanics. The reasons for this abrupt change are due to two causes: the opening of national frontiers, and a wave of enthusiasm for applied science, motivated by the technical achievements obtained in the production of new weapons." The method of stochastic linearization technique was proposed more or less simultaneously on both sides of the Atlantic: By Booton (1953, 1954) and Caughey (1963) in the United States and by Kazakov (1954) in the former Soviet Union.

Around the method's thirtieth anniversary, in his review, Spanos (1981) wrote: "It can be stated, with only minor reservations, that the method of stochastic or statistical or equivalent linearization, has proved, over the period of the last three decades, the most useful approximate method for probabilistic analysis of nonlinear structural dynamical systems." Around the method's half-century, Crandall (2006) noted: "The procedure has been very popular with investigators in the field of random vibration. In 1998 it was estimated (Elishakoff,) that there had been over 400 papers published on the subject of statistical linearization."

The method's essence can be demonstrated on the simple problem of a single-degree-of-freedom structure, governed by the following differential equation:

https://doi.org/10.1515/9783111354231-005

$$m\ddot{X} + f(X, \dot{X}) = P(t) \tag{5.1}$$

where m = mass, X = displacement, $\dot{X}$ = velocity, $\ddot{X}$ = acceleration, f = nonlinear function, $P(t)$ = stationary random process in time.

The autocorrelation function and hence the spectral density of $P(t)$ are given. The problem consists in finding the probabilistic characteristics of X and $\dot{X}$. The simplest characteristics would be mathematical expectations of response quantities, $E(.)$ indicating operation of mathematical expectation, namely $E(X)$, $E(\dot{X})$, $E(X^2)$ and $E(\dot{X}^2)$. Were $f(X, \dot{X})$ a linear function

$$f(X, \dot{X}) = k_0 X + c_0 \dot{X} \tag{5.2}$$

with k_0 the stiffness coefficient and c_0 the damping coefficient, the solution would be straightforward. For the case of the correlation function of $P(t)$

$$K_p(t, t') = E[P(t)P(t')] = 2\pi S_0 \delta(t_2 - t_1) \tag{5.3}$$

(S_0 being the intensity of the noise), one obtains

$$E(X) = E(\dot{X}) = 0 \tag{5.4}$$

$$E(X^2) = \pi S_0 / c_0 k_0 \tag{5.5}$$

$$E(\dot{X}^2) = \pi S_0 / c_0 m \tag{5.6}$$

We are not concerned with the linear case, however. The closed-form solution for arbitrary nonlinearity as well as arbitrary excitation of the nonlinear oscillator is an unsolved problem. The pioneers of the stochastic linearization technique posed a question on possible linearization of the nonlinear function in eq. (5.1), i.e., replacing the nonlinear function $f(X, \dot{X})$ by

$$f(X, \dot{X}) = k_{eq} X + c_{eq} \dot{X} \tag{5.7}$$

and finding equivalent values of the stiffness coefficient k_{eq} and the damping coefficient c_{eq}, so that the solution of the thus obtained linear system

$$m\ddot{X} + k_{eq} X + c_{eq} \dot{X} = P(t) \tag{5.8}$$

would produce sufficiently good approximations for the desired quantities. The question is: How to determine k_{eq} and c_{eq}? There is a gallery of answers. We dispense with a historical overview of these answers and direct the interested reader to various reviews, old and new, the most recent perhaps being that by Crandall (2006); the reader may also consult with the earlier reviews by Spanos (1981), Roberts (1981), Socha and Soong (1991), Socha (2005a, 2005b), Falsone and Ricciardi (2003), Elishakoff (2000b), and Proppe et al. (2003). There are two special monographs written on this subject, that by Roberts and Spanos (1991, 2004) and by Socha (2008). It must be noted that Crandall (2006) writes that some explanations provided in the literature since 1967 were "confusing." To deal with controversial topics is beyond this study, however. We concern ourselves with a suggested explanation of the technique which hopefully will be free of "confusion," on one hand, and will lead to rigorous pedagogical explanation of it for the novice. It is hoped that two alternative expositions proposed in this study will be adopted in future stochastic dynamics and random vibration textbooks.

5.2 A system possessing a nonlinear stiffness

Consider first the simplest form of nonlinearity that is exhibited by the system through its stiffness. In other words, the special form of eq. (5.1) is studied

$$m\ddot{X} + c\dot{X} + f(X) = P(t) \tag{5.9}$$

We replace eq. (5.9) by its "equivalent" given in eq. (5.8). Since the dumping is linear in both eq.(5.8) and (5.9), $c_{eq} = c$. We are looking for the equivalent linear stiffness k_{eq}. We evaluate the difference between the original nonlinear stiffness $f(X)$ and its linear equivalent $k_{eq}X$. Since $f(X)$ is in general a nonlinear function (no one would linearize a linear function!) the difference $f(X) - k_{eq}X$ does not vanish. At this stage one forms the mean-square difference

$$E(D^2) = E\left\{ \left[f(X) - k_{eq}X \right]^2 \right\} \tag{5.10}$$

and demands it to attain minimum with respect to k_{eq}; in eq.(5.10) the operator $E(.)$ is that of mathematical expectation. Thus,

$$E(D^2) = \int_{-\infty}^{\infty} \left[f(x) - k_{eq}x \right]^2 \varphi(x) dx \tag{5.11}$$

where $\varphi(x)$ is the probability density function of $X(t)$. It makes sense to recall that we do not know the probability density function of the solution; indeed, had we known it, we would not use the approximate technique of stochastic linearization; rather, the desired probabilistic characteristics $E(X)$ and $E(X^2)$ would be evaluated by straightforward integration

$$E(X) = \int_{-\infty}^{\infty} x\varphi(x) dx$$

$$E(X^2) = \int_{-\infty}^{\infty} x^2\varphi(x) dx \tag{5.12}$$

i.e., without resorting to linearization technique.

There are two possibilities to proceed at this juncture. One possibility is to recognize our lack of knowledge of the exact $\varphi(x)$ and to employ some approximate probability density function of the linearized system $\psi(x, k_{eq})$. To obtain a linearized system we do not simply drop the nonlinear term: we replace the entire expression of restoring force that may contain linear and nonlinear expressions by an equivalent linear force $k_{eq}x$.

Here it must be remarked that since the linearized system in eq. (5.8) inevitably depends on k_{eq}, so does the probability density $\psi(x, k_{eq})$. Then eq. (5.11) is replaced by the approximate mean-square deviation $E_a(D^2)$ defined as

$$E_a(D^2) = \int_{-\infty}^{\infty} [f(x) - k_{eq}x]^2 \psi(x, k_{eq}) dx \qquad (5.13)$$

The demand that this mean-square deviation to attain minimum with respect to k_{eq} leads to

$$\frac{dE_a(D^2)}{dk_{eq}} = -2 \int_{-\infty}^{\infty} [f(x) - k_{eq}x] x \, \psi(x, k_{eq}) dx + \int_{-\infty}^{\infty} [f(x) - k_{eq}x]^2 \frac{d\psi}{dk_{eq}} dx = 0 \qquad (5.14)$$

Another possibility is to assume that we know the exact probability density function in eq.(5.11); proceeding then with the minimization leads to

$$\frac{dE_a(D^2)}{dk_{eq}} = 2 \int_{-\infty}^{\infty} [f(x) - k_{eq}x] x \varphi(x) dx = 0 \qquad (5.15)$$

This demand reduces to the following expression for the equivalent stiffness coefficient k_{eq}:

$$k_{eq} = \frac{\int_{-\infty}^{\infty} f(x)x\varphi(x)dx}{\int_{-\infty}^{\infty} x^2\varphi(x)dx} \qquad (5.16)$$

Had we known the exact probability density $\varphi(x)$ eq. (5.16) would be replaced by

$$k_{eq} = \frac{E[Xf(X)]}{E(X^2)} \qquad (5.17)$$

At this juncture it's recommended to ask ourselves to comment on this equation. Some are realizing the seemingly paradoxical situation we find ourselves in: We are looking for $E[X^2]$, yet we know it the stochastic linearization leads us to determine the equivalent linear stiffness k_{eq} whose determination demands the knowledge of the above sought quantity for eq. (5.17) contains $E[X^2]$ in the denominator. Thus, the

eq. (5.17) must appear to the initial reader as totally useless, and the method of stochastic linearization as nonsensical, for it leads, as it were, to catch 22, not less!

Such a situation is not pertinent solely to the stochastic linearization technique. It
occurs even in deterministic problems. For example, analogous situations take place
while using the Rayleigh quotient method for the natural frequency evaluation.
Whereas the quotient is derived in view of knowledge of exact mode shape, the better
is approximate to obtain the estimate for the natural frequency.

Recalling that the exact density is not known, we instead of $\varphi(x)$ in eq. (5.16) thus
utilize its approximation $\psi(x, k_{eq})$:

$$k_{eq} = \frac{\int\limits_{-\infty}^{\infty} f(x)x\psi(x)dx}{\int\limits_{-\infty}^{\infty} x^2\psi(x)dx} \tag{5.18}$$

As can be observed by comparing eqs. (5.14) and (5.15), the former contains an additional term. A natural question arises: "Which version of the stochastic linearization
technique should be preferred?" As a popular proverb maintains, proof of the pud-
ding is in eating. Thus, the above question must be changed into the following:
"Which technique performs better?" In a series of studies, Socha and Pawleta (1994),
Elishakoff and Colajanni (1997, 1998), Colajanni and Elishakoff (1998a, 1998b) utilized
eq. (5.14) to derive k_{eq}. In the above papers it was shown that the mean-square values
of the responses of several oscillators, the approach based on eq. (5.14), led to results
that were farther from the exact solution than those obtained by employing eq. (5.15).
Only one oscillator, originally studied by Booton (1953, 1954), Elishakoff and Colajanni
(1998), demonstrated that both techniques led to coincident results. Thus, in balance
one has to prefer, due to pragmatic reasons, eq. (5.15) to eq. (5.14).

5.3 Discussion of eqs. (5.14) and (5.15)

It must be noted that the above derivation of the two possible approaches when one
minimizes the mean-square error is presented herein. The approach given by eq.
(5.14) was given in the paper by Elishakoff and Colajanni (1997, 1998) (see also Elishakoff and Crandall 2017). Crandall (2001) calls it a SPEC alternative, acronym being associated with the first letters in last names of the authors of papers by Socha and
Pawleta (1994); Elishakoff and Colajanni (1997, 1998). How can one explain, *post fac-
tum*, the success of the second approach? It appears that in the second approach we
carry as much as possible the attributes of exact analysis.

Crandall (2006) characterizes eq. (5.17) as "the recipe for selecting k_{eq}." Indeed,
according to Paul Valéry, a French poet, essayist, and philosopher, "Science is a collection of successful recipes." The approach based on eq. (5.14) was proposed due to the
absence in the literature of a specific statement that the recipe in eq. (5.17) is associ-

ated with the assumption that until its derivation it was assumed that the exact probability density as known. This led, according to Crandall (2001) to the fact that "there has been some confusion concerning the standard (i.e., second) procedure." He also noted:

> It must be admitted that the literature on this point has been confusing. Many descriptions of the standard procedure fail to explain why the expectation . . . are considered to be independent of k before the differentiation . . ., but immediately afterward are taken to be k dependent.

Likewise in the personal communication by late Professor Caughey (1998) to this author, he writes:

> Thank you for the papers that you sent me, I found them very interesting and a little disturbing. After reading both papers carefully, I have the following comments:
> a) It's surprising that both techniques lead to exactly the same first order corrections, it should be noted that perturbation theory also leads to the same first order correction term. As far as I know nobody has carried out the perturbation technique to obtain the higher corrections.
> b) It's also surprising that the improved minimization technique, i.e., eq. (5.14), leads to poorer results than the naïve technique. One thinks of asymptotic series where the best approximation given by the first couple of terms.
> c) If the naïve technique is applied to Duffing's equation with *Sinusoidal Excitation*, it predicts the same first order correction that is given by the *Harmonic Balance*. I have not repeated the problem using your minimization technique. Duffing's Equation with white noise excitation appears to be the simplest example to illustrate your technique; all other examples appear to be much more complex.

Likewise, Li and Chen (2009) in their book note:

> Although the above analysis (derivation of eq. (5.14)) is reasonable, the effect is not as good as expected. First, deduction is much more difficult and might be impossible for complex or multidimensional problems. Second, even for simple problems, it was shown that the accuracy of the 'error-free' linearization is sometimes lower than that of standard linearization (Elishakoff and Colajanni, 1997).

Crandall (2006) stressed that "The SPEC alternative has some interesting features (Crandall, 2001), but unfortunately it is more labor intensive and, almost always, less accurate than the standard procedure."

Another question arises on the role that the papers by Socha and Pawleta (1994) and Elishakoff and Colajanni (1997, 1998), Colajanni and Elishakoff (1998) had played in the elucidation of stochastic linearization technique. Crandall (2006) gives the following credit to the above studies:

> the inconsistency of applying recipes based on nonlinear response statistics independent of k to linear system statistics which were functions of k was recognized and corrected by Socha and Pawleta (1994, 1999) and by Elishakoff and Colajanni (1997, 1998a, 1998b).

These papers also inspired investigations by Crandall (2001, 2003, 2004a, 2004b, 2006), Socha (2005a, 2005b, 2008), Socha and Pawleta (1994,1999), Proppe *et al.* (2003), Elishakoff (2000a, 2000b) and possibly others.

Here the method of Gaussian closure (Iyengar and Dash, 1978; Wu, 1987) should be mentioned. It is widely known classical stochastic linearization technique coalesces with Gaussian closure technique. On the other hand, as we assume $\psi(x, k_{eq})$ to be Gaussian (and this is mandatory since the system is linearized and the input is Gaussian) then eq. (5.14) will return to Gaussian closure technique.

5.4 Stochastic linearization via Galerkin technique

We would like to start discussion on the title topic by a comment that appears to be instructive on derivation of eq. (5.14), or classical recipe for k_{eq}. One resorts to stochastic linearization as an approximate technique, knowing *a priori* that the exact probability density of the response is unknown. Yet, in order to derive eq. (5.17) one has to assume the knowledge of the exact probability density. Therefore, the derivation of eq. (5.17) may appear inconsistent. Consistency, naturally, is a desirable attribute to any derivation. According to William James's philosophy, truth is associated with the term 'leading' in the sense that true beliefs "lead to consistency, stability." However, the importance of consistency should not be overestimated. In words of Aldous Huxley, an English writer,

> Too much consistency is as bad for the mind as it is for the body. Consistency is contrary to nature, contrary to life. The only completely consistent people are dead.

It appears instructive to reproduce here the quote from Levinson (1987), commenting on his and Bickford's (1982) theories and the associated issues of consistency:

> It would seem that the Bickford's work has relegated the earlier work of the present writer to the status of an intellectual artifact in the history of applied mechanics whose importance is limited to providing the motivation for the work of Bickford; from a certain theoretical point of view this is clearly so. What is vexations, however, is that Bickford's theory, in the two elastostatic and one elastodynamic problems he consider, provides inferior results in two cases and essentially the same results in the remaining case when compared to the results of the present writer's theory; exact elasticity solutions being available for purposes of comparison in all three of the examples considered.

As is seen here too, the less consistent theory turned out to produce better results!

Let us turn now to recasting stochastic linearization technique via the Galerkin method. It is naturally not possible to replace the nonlinear force $f(X)$ by a linear counterpart $k_{eq}X$ in eq. (5.9). There is a difference between $f(X)$ and $k_{eq}X$. We refer to this difference as error $\varepsilon(X)$. Whereas we do not possess a magic wand to make it zero, we can try to make it as small as possible in some sense. We demand the first moment of $E(\varepsilon X)$ this error to vanish

$$E(\varepsilon X) = 0 \qquad\qquad (5.19)$$

where $E(\cdot)$ denotes mathematical expectation.

The fact that we do not know the probability density of the response to evaluate eq. (5.19), does not prevent us from realizing that our condition (5.19) in fact is orthogonality condition between the error ε and the system's displacement X. Yes, the error ε is not zero as we wish it to be, but at least we cannot see it, as it were, in the "direction" of X. We could metaphorically refer to condition (5.19) as the overlooking of one's "misbehavior," exhibiting itself in absence of being error-free, by parents, grandparents, and friends (as a proverb maintains, "Friend is one who tolerates our success and accepts us with our mistakes"); one usually has a better grasp of having condition (5.19) presented as a "friendship's" attribute. Thus, the condition (5.19) becomes:

$$E\big[\{f(X) - k_{eq}X\}X\big] = 0$$

For the coefficient k_{eq} we get

$$k_{eq} = \frac{E[f(X)X]}{E[X^2]} \qquad\qquad (5.20)$$

Thus, by the Galerkin method we arrive at the same expression as eq. (5.17).

Developments in stochastic linearization since its inception in 1953 are summarized in this chapter along with the new, statistical orthogonality-based derivation of the method. The developments that are described in this review mostly took place after extensive accounts on the classical version of the stochastic linearization technique, such as the monographs by Roberts and Spanos (1990, 2003), Donley and Spanos (1991), and by Socha (2008), and the review articles by Caughey (1963), Spanos (1981), Socha (2005a, 2005b), Elishakoff (1995, 2000b), Crandall (2006), and Elishakoff and Crandall (2017), have been published the recent decade.

Falsone (1992a, 1992b) demonstrated that when parametric excitations are present, it is preferable to measure the difference on the coefficients of the Itô differential rule. Remarkably, when only purely external excitations are present, Falsone's approach coincides with classical stochastic linearization. Recently, for the last-named case, new stochastic linearization techniques have been suggested in which the differences between the nonlinear original system and the linearized one are considered in terms of potential energy (Zhang, 1989). In particular, Elishakoff (1991c) investigated a Duffing oscillator subjected to white-noise excitation; it was shown that the best results are obtained when the linearized system parameters are obtained through minimizing the mean-square error between the potential energies of the two systems. The exact solution for the stationary probability density of the oscillator is readily available and approximate solutions are not needed for this problem; therefore, the validity of the modified stochastic linearization technique is easily checked for this case.

Moreover, the results obtained by the stochastic linearization technique are compared with those yielded by Monte Carlo simulation where exact solutions are unattainable.

Elishakof (1991) has demonstrated, that for some combinations of the parameters of the Duffing oscillator the proposed linearization criteria may yield results in perfect agreement with the exact solutions. For other sets of parameters, the proposed linearization yields result that is slightly higher than the exact probabilistic responses, whereas conventional linearization yields responses that are below the exact values. Since engineers generally resort to the notion of safety factors, structures realized through conventional stochastic linearization may turn out to be "over-designed." Where the Duffing oscillator is extremely weakly nonlinear, the conventional approach may result in less error than its energy-wise counterpart. However, for these almost linear systems the percentage-wise error is under two percent for either criterion. Therefore, for the Duffing oscillator, the energy-wise approach is almost always preferable.

The above coincidence of the stochastic linearization result with exact solution suggests that for a specific set of parameters (namely $e_0^2 \approx 0.54$) the new version of stochastic linearization constitutes *"true"* linearization, in terminology of Kozin (1987a, 1987b).

Some decades ago, Roberts (1981) noted: "Because linear systems are so much easier to analyze than nonlinear ones, a natural method of attacking nonlinear problems is to replace a given set of nonlinear equations by an equivalent set of linear ones; the difference between the sets of equations is minimized in some appropriate sense." Likewise, one of the three pioneers of the stochastic linearization technique, Kazakov (1954, p. 51) mentioned over half a century ago: "the sufficiently accurate approximation of essentially nonlinear characteristics is extremely difficult." Hence, it appears especially important to search for the *nonclassical* criteria that may lead to better approximations.

The effectiveness of the energy concepts in the nonlinear stochastic dynamics is not accidental. Indeed, the expression of the exact probability density contains the expression of the potential energy.

In this chapter, we investigate a system with nonlinear restoring force. Extension of energy concepts can be performed for the systems with nonlinear damping. Such approaches have been initiated in papers by Li *et al.* (2006) and Liang and Feeny (2006). It appears interesting to combine the present approach with that developed by Casciati *et al.* (1993).

5.5 Generalization of Anh and Di Paola version

In terminology of Kozin (1987), "the method of statistical linearization has remained a surprisingly popular tool over the many years since it was first formulated." The method is based on replacing the original nonlinear system by a linear one that is equivalent to the original one in some probabilistic sense. Several criteria have been suggested to arrive at the expressions of the equivalent stiffness and equivalent damping.

Anh and Di Paola (1995) suggested a new realization of the stochastic linearization, which appears to be extremely unusual at the first glance. Instead of simplifying a nonlinear expression appearing in the differential equation, they suggested to seemingly first complicate it by replacing it by higher order terms. These higher order terms then were replaced by the linear approximation, in several steps. This indirect linearization certainly prolongs, as it were, the linearization process. Yet it considers the higher order statistics and, as such, has more of a possibility of capturing the behavior of the system. It turned out that this long way towards linearization leads to results that are closer to those obtained via exact solution, when the latter is available, or Monte Carlo simulation, when the exact solution is not available. Commenting on this method as exemplified on a Duffing oscillator, Anh (2006a) stresses that "in the [usual] linearization we go from X^3 [term] to X. That will yield some error, and we should do something to balance. For regulated Gaussian equivalent linearization (RGEL) we should go back [to balance the error]. Since [the difference of the powers of the original cubic and replacing linear terms is] $3-1=2$, so we go back also 2 degrees, i.e. from X^3 to X^5 and come back to the first place X^3 but will regulated coefficient (7/9 in this case)."

For the details of implementation of RGEL for the Duffing oscillator the readers may consult with the study by Anh and Di Paola (1996). Anh (2006b) provides an additional justification of this method: "The natural explanation [of RGEL] is that when we want to go through a thing ahead we should move the hand back as how far ahead so far back. That is why we go first from X^3 to X^5." Since this method produces more accurate results than the classic linearization, one way metaphorically refers to as a "long shorter way," versus classical technique, that can be dubbed as a "short, longer way." This metaphor stems from an ancient story about a young boy who was asked by a stranger how to find the road to the big city. The boy asked: "Do you want a long, shorter way, or a short, longer way?" The stranger chose the latter, since the first adjective was a word "short." Yet, after several hours of wondering the man returned to the boy and told him: "The way is short, but there are unsurpassable rocks. Tell me the whereabouts of the long, shorter way." This time the stranger succeeded in getting to his destination.

Anh and Di Paola's (1996) derivation can be viewed as a "long shorter way" for it yields much more satisfactory results than the direct linearization technique ; the latter being a "short, longer way." Elishakoff (2000b, 2000c) demonstrated that the expressions for the equivalent stiffness and damping coefficients adopted in the literature can be obtained by alternative means, namely via modified orthogonality criterion.

These two ideas, those by Anh and Di Paola (1996) and by Elishakoff (2000) are combined in this study. We first apply the Anh and Di Paola procedure to the Atalik and Utku oscillator with attendant dramatic decrease in error in comparison with the classical stochastic linearization. Then we extend the Anh and Di Paola methodology to two-step regulation. The latter extension shows considerable improvement of results in comparison with both classical scheme as well as the single step regulation, in the Lutes and Sarkani (2004) oscillator.

In this section we will derive the results of Anh and Di Paola (1996) by the orthogonality requirement. Anh and Di Paola (1996) studied the following nonlinear random vibration problem

$$\ddot{X} + 2h\dot{X} + \omega_0^2 X + \varepsilon g\left(X, \dot{X}\right) = f(t) \tag{5.21}$$

where $X(t)$ is the displacement, $\dot{X}(t)$ is the velocity, $\ddot{X}(t)$ = acceleration of a single degree of freedom system, h = damping coefficient, ω_0= natural frequency of the system obtained when $h \equiv 0$, $\varepsilon \equiv 0$, $f(t) \equiv 0$; $g\left(X, \dot{X}\right)$ is a nonlinear function, ε = amplitude of nonlinearity, $f(t)$= random excitation. A polynomial expression of X and $\dot{X}$. The nonlinear function $g\left(X, \dot{X}\right)$ is taken as

$$g\left(X, \dot{X}\right) = \sum_{k=0}^{N} \sum_{j=0}^{N} \left(\alpha_{kj} \dot{X}^{2k} X^{2j+1} + \beta_{kj} X^{2k} \dot{X}^{2j+1}\right) \tag{5.22}$$

Classical linearization would perform following replacements of the nonlinear terms by the linear ones:

$$\alpha_{kj} \dot{X}^{2k} X^{2j+1} \longrightarrow \lambda_{kj} X \tag{5.23}$$

$$\beta_{kj} X^{2k} \dot{X}^{2j+1} \longrightarrow \mu_{kj} \dot{X} \tag{5.24}$$

Instead, most unusually, at least at the first glance, Anh and Di Paola (1996) suggested to replace non-linear terms by *higher-order nonlinear* ones

$$\alpha_{kj} \dot{X}^{2k} X^{2j+1} \longrightarrow c_{kj} \left(\dot{X}^{2k} X^{2j+1}\right) \left(\dot{X}^{2k} X^{2j}\right)$$
$$= c_{kj} \dot{X}^{4k} X^{4j+1} \tag{5.25}$$

$$\beta_{kj} X^{2k} \dot{X}^{2j+1} \longrightarrow d_{kj} \left(X^{2k} \dot{X}^{2j+1}\right) \left(X^{2k} \dot{X}^{2j}\right)$$
$$= d_{kj} X^{4k} \dot{X}^{4j+1} \tag{5.26}$$

where authors used the mean-square criterion for obtaining the coefficients d_{kj} and c_{kj}:

$$c_{kj} = \alpha_{kj} \, E\left[X^{6k} \dot{X}^{6j+2}\right] \Big/ E\left[X^{8k} \dot{X}^{8j+2}\right] \tag{5.27}$$

$$d_{kj} = \beta_{kj} \, E\left[X^{6k} \dot{X}^{6j+2}\right] \Big/ E\left[X^{8k} \dot{X}^{8j+2}\right] \tag{5.28}$$

Anh and Di Paola (1996) then replaced higher-order non-linear terms into the original non-linear terms

$$c_{kj} \overset{\bullet}{X}{}^{4k} X^{4j+1} \rightarrow q_{kj} \overset{\bullet}{X}{}^{4k} X^{4j+1} \tag{5.29}$$

$$d_{kj} X^{4k} \overset{\bullet}{X}{}^{4j+1} \rightarrow b_{kj} X^{2k} \overset{\bullet}{X}{}^{2j+1} \tag{5.30}$$

where

$$b_{kj} = d_{kj}\, E\left[X^{6k} \overset{\bullet}{X}{}^{6j+2}\right] \Big/ E\left[X^{4k} \overset{\bullet}{X}{}^{4j+2}\right] \tag{5.31}$$

$$q_{kj} = c_{kj}\, E\left[X^{6k} \overset{\bullet}{X}{}^{6j+2}\right] \Big/ E\left[X^{4k} \overset{\bullet}{X}{}^{4j+2}\right] \tag{5.32}$$

This step is followed by the conventional replacement

$$b_{kj} X^{2k} \overset{\bullet}{X}{}^{2j+1} \rightarrow h_{kj} \overset{\bullet}{X} \tag{5.33}$$

$$q_{kj} \overset{\bullet}{X}{}^{2k} X^{2j+1} \rightarrow l_{kj} X \tag{5.34}$$

where

$$h_{kj} = b_{kj}\, E\left[X^{6k} \overset{\bullet}{X}{}^{6j+2}\right] \Big/ E\left(\overset{\bullet}{X}{}^{2}\right) \tag{5.35}$$

$$l_{kj} = q_{kj}\, E\left[\overset{\bullet}{X}{}^{6k} X^{6j+2}\right] \Big/ E[X^2] \tag{5.36}$$

Let us show that the procedure by Anh and Di Paola (1996) can be directly obtained via modified stochastic linearization technique. Indeed, we demand statistical orthogonality of the difference of the left- and right-hand sides in eq. (4.139):

$$e_1 = \beta_{kj} X^{2k} \overset{\bullet}{X}{}^{2j+1} - d_{kj} X^{4k} \overset{\bullet}{X}{}^{4j+1} \tag{5.37}$$

with $\overset{\bullet}{X}{}^{4k} X^{4j+1}$, i.e., we require

$$\left(e_1, \overset{\bullet}{X}{}^{4k} X^{4j+1}\right) = 0 \tag{5.38}$$

where (., .) is the inner product defined as

$$(\phi, \psi) = E[\phi, \psi] \tag{5.39}$$

Thus, eq. (5.38) becomes:

$$E\left[\left(\beta_{kj} X^{2k} \overset{\bullet}{X}{}^{2j+1} - d_{kj} X^{4k} \overset{\bullet}{X}{}^{4j+1}\right) \overset{\bullet}{X}{}^{4k} X^{4j+1}\right] = 0 \tag{5.40}$$

yielding the expression (5.28) for d_{kj}. Analogously, the orthogonality requirement

$$\left(e_2, \overset{\bullet}{X}{}^{4k} X^{4j+1}\right) = 0 \tag{5.41}$$

where e_2 is the difference between the left- and right-hand sides in eq. (5.25)

$$e_2 = a_{kj} \dot{X}^{2k} X^{2j+1} - c_{kj} \dot{X}^{4k} X^{4j+1} \qquad (5.42)$$

yields eq. (5.28).

The results of the second step is likewise deducible from the requirements

$$\left(e_3, X^{2k} \dot{X}^{2j+1}\right) = 0 \qquad (5.43)$$

$$\left(e_4, \dot{X}^{2k} X^{2j+1}\right) = 0 \qquad (5.44)$$

where e_3 is the difference between the left and the right sides of eq. (5.30):

$$e_3 = d_{kj} X^{4k} \dot{X}^{4j+1} - b_{kj} X^{2k} \dot{X}^{2j+1} \qquad (5.45)$$

and e_4 is the difference between the left and the right-hand sides of eq. (4.142):

$$e_4 = c_{kj} \dot{X}^{4k} X^{4j+1} - q_{kj} \dot{X}^{2k} X^{2j+1} \qquad (5.46)$$

Equations (5.43) and (5.44) lead to eqs. (5.31) and (5.32), respectively. In perfect analogy, eqs. (5.35) and (5.30) are obtained by postulating following conditions

$$\left(e_5, \dot{X}\right) = 0 \qquad (5.47)$$

$$(e_6, X) = 0 \qquad (5.48)$$

where

$$e_5 = b_{kj} x^{2k} \dot{X}^{2j+1} - h_{kj} \dot{X} \qquad (5.49)$$

$$e_6 = q_{kj} \dot{X}^{2k} X^{2j+1} - l_{kj} X \qquad (5.50)$$

As is seen, eqs. (5.31), (5.32), (5.35) and (5.36) are obtained by stochastic Galerkin-type orthogonality conditions. As a result, the final, linear replacement takes place:

$$g\left(X, \dot{X}\right) = \sum_{k=0}^{N} \sum_{j=0}^{N} \left(h_{kj} \dot{X} + l_{kj} X\right) \qquad (5.51)$$

where

$$h_{kj} = \frac{E\left[X^{2k} \dot{X}^{2j+2}\right]}{E[X^2]} \frac{E\left[X^{6k} \dot{X}^{6j+2}\right]}{E\left[X^{4k} \dot{X}^{4j+2}\right]} \frac{E\left[X^{6k} \dot{X}^{6j+2}\right]}{E\left[X^{8k} \dot{X}^{8j+2}\right]} \beta_{kj} \qquad (5.52)$$

$$l_{kj} = \frac{E\left[\dot{X}^{2k} X^{2j+2}\right]}{E[X^2]} \frac{E\left[\dot{X}^{6k} X^{6j+2}\right]}{E\left[\dot{X}^{4k} X^{4j+2}\right]} \frac{E\left[\dot{X}^{6k} X^{6j+2}\right]}{E\left[\dot{X}^{8k} X^{8j+2}\right]} \alpha_{kj} \qquad (5.53)$$

Anh and Di Paola (1996) evaluated by their method several oscillators. For the Duffing oscillator in the investigated numerical range, the numerical results led to roughly half the percentagewise error of that resulted by the conventional stochastic linearization technique, i.e. without recourse to amending the original system by the higher non-linearity degree. As noted, Anh and Di Paola (1996) call their method as "a regulated Gaussian equivalent linearization (RGEL)." As is seen, RGEL can be interpreted as a multiple orthogonalization technique.

Consider the following nonlinear system, usually called as Atalik and Utku (1976) oscillator:

$$\ddot{X}(t) + \beta \dot{X}(t) + a X^3(t) = F(t) \tag{5.54}$$

where β is the damping constant, a is the nonlinear stiffness constant and $F(t)$ is a Gaussian white noise process with

$$E[F(t)] = 0, \quad E[F(t)F(t+\tau)] = 2d\beta\delta(\tau) \tag{5.55}$$

The exact stationary probability density function of the above system, obtained by the Fokker-Planck approach, is

$$p(x) = p_0 \exp\left(-\frac{a}{4d}x^4\right) \tag{5.56}$$

where p_0 is the normalization constant. To obtain the exact mean-square displacement,

$$\sigma_x^2 = E[X^2] = \int_{-\infty}^{+\infty} x^2 p(x) \, dx \tag{5.57}$$

we use the integration formula

$$\int_0^{+\infty} x^{s-1} \exp(-ax^h) \, dx = (h^{-1})\left(a^{-s/h}\right) \Gamma(s/h) \tag{5.58}$$

where $\Gamma(\bullet)$ is the Gamma function. The mean-square displacement becomes

$$\sigma_x^2 = \frac{(1/4)(a/4d)^{-3/4}\Gamma(3/4)}{(1/4)(a/4d)^{-1/4}\Gamma(3/4)} \approx 0.6760(a/d)^{1/2} \tag{5.59}$$

The equivalent linear system to eq. (5.44) can be written as

$$\ddot{X}(t) + \beta \dot{X}(t) + k_{eq} X(t) = F(t) \tag{5.60}$$

where the equivalent linear spring constant k_{eq} is found by processing the conventional linearization, as equal

$$k_{eq} = E\left[\frac{d}{dx}(\alpha X^3)\right] = 3\alpha E\left[X^2\right] \tag{5.61}$$

The mean-square value of the displacement of the linearized system is

$$E[X^2] = d/k \tag{5.62}$$

Thus, we obtain the approximate solution as

$$\sigma_{x_e}^2 = (d/3\alpha)^{1/2} \approx 0.5776\,(d/\alpha)^{1/2} \tag{5.63}$$

The percentagewise error committed by using the classical equivalent linearization technique in evaluating the mean-square displacement is

$$(0.6760-0.5776)/0.6760 = 14.6\% \tag{5.64}$$

Let us apply the RGEL method of Anh and Di Paola (1996). The scheme of the process can be read as follows

$$\alpha\,X^3(t) \rightarrow k_1\,X^5(t) \rightarrow k_2\,X^3(t) \rightarrow k_{eqI}\,X(t) \tag{5.65}$$

We can readily utilize the results obtained by Anh and Di Paola for the Duffing oscillator of which the Atalik and Utku oscillator is a particular case:

$$\alpha\,X^3(t) \rightarrow \frac{\alpha}{9E[X^2(t)]}\,X^5(t) \rightarrow \frac{7\alpha}{9}\,X^3(t) \rightarrow \frac{7\alpha}{3}\,X(t) \tag{5.66}$$

One gets the equivalent linearized equation

$$\ddot{X}(t) + \beta\dot{X}(t) + \frac{7}{3}\alpha E[X^2(t)]\,X(t) = F(t) \tag{5.67}$$

The mean-square value of $X(t)$ is evaluated by the following expression

$$E[X^2(t)] = \sqrt{\frac{3d}{7\alpha}} \approx 0.6546\left(\frac{d}{\alpha}\right)^{1/2} \tag{5.68}$$

Now, the percentagewise error found by using the RGEL linearization technique to calculate the mean-square displacement is

$$(0.6760-0.6546)/0.6760 = 3.17\% \tag{5.69}$$

We note a significant, over fourfold, improvement, which demonstrates the extreme efficiency of the RGEL method. Naturally, the question of continuing the process to greater order arises. However, calculation of such process beyond the first step in eq. (5.55), namely,

$$a\,X^3(t) \rightarrow \frac{a}{9E[X^2(t)]}\,X^5(t) \rightarrow \frac{a}{117\sigma_X^2 E[X^2(t)]}\,X^7(t)$$

$$\rightarrow \frac{11\,a}{117E[X^2(t)]}\,X^5(t) \rightarrow \frac{77\,a}{117}\,X^3(t) \rightarrow \frac{77\,a}{39}E[X^2(t)]\,X(t) \qquad (5.70)$$

leads to numerically worse results than the one previously found. Hence, for the Atalik and Utku oscillator, the optimum number of regulation step is unity. It is important to note that the evaluation of the two steps in Duffing oscillator leads to the same conclusion. The question arises if there is an oscillator where the optimum number of regulation steps is greater than one. The reply to this question is shown in the next section to be affirmative.

Consider now the nonlinear oscillator by Lutes and Sarkani (2004)

$$\dot{X}(t) + k\,X^a(t)\,\mathrm{sgn}[X(t)] = F(t) \qquad (5.71)$$

where a is a real number, $F(t)$ is a zero – mean, stationary Gaussian white noise with spectral density S_0. Lutes and Sarkani (2004) derive the probability density of the response

$$p_{X(t)} = A\exp\left[-\frac{k\,u^{a+1}}{(a+1)\,\pi\,S_0}\right] \qquad (5.72)$$

where

$$A = \left(\frac{k}{(a+1)\,\pi\,S_0}\right)^{\frac{1}{a+1}}\left(\frac{a+1}{2}\right)\left[\Gamma\left(\frac{1}{a+1}\right)\right]^{-1} \qquad (5.73)$$

The variance of the response

$$\sigma_X^2 = 2A\int_0^{\infty} u^2 \exp\left[-\frac{k\,u^{a+1}}{(a+1)\,\pi\,S_0}\right]\,du \qquad (5.74)$$

is obtained exactly

$$\sigma_{X,exact}^2 = \left(\frac{\pi\,S_0}{k}\right)^{\frac{2}{a+1}}(a+1)^{\frac{2}{a+1}}\,\Gamma\left(\frac{3}{a+1}\right)\left[\Gamma\left(\frac{1}{a+1}\right)\right]^{-1} \qquad (5.75)$$

Lutes and Sarkani (2004) also derived the approximate response via the classical stochastic linearization technique as follows:

$$\sigma_{X,approx}^2 = \left(\frac{\pi\,S_0}{k}\right)^{\frac{2}{a+1}}\left[\frac{(2\pi)^{1/2}}{2^{a/2}a\Gamma(a/2)}\right]^{\frac{2}{a+1}} \qquad (5.76)$$

The error η between exact and approximate solutions defined as

$$\eta = \frac{\left|\sigma^2_{X,exactx} - \sigma^2_{X,approx}\right|}{\sigma^2_{X,exactx}} \times 100\% \tag{5.77}$$

is shown in Fig. 5.1. Lutes and Sarkani (2004) concluded that "the statistical linearization gives a good approximation of the response variance only when a is relatively near unity." Indeed, for $a = 1$ the error equals zero. For $a = 2$, the error constitutes $\eta = 5.6\%$; for $a = 3$, the error equals $\eta = 14.6\%$; for $a = 4$, the error is $\eta = 22.8\%$; for $a = 5$, the error reaches $\eta = 29.9\%$. It appears to be of interest to investigate this oscillator via the modified Anh and Di Paola (1996) approach.

For simplicity we will limit ourselves by considering the case when a is a positive integer; then instead of $X^a(t)\text{sgn}[X(t)]$ we have $[X^a(t)]$. We intend to replace the power oscillator with nonlinear term X^a by a linear oscillator with the term $k_{eq} X$, the difference of powers being $a - 1$. During the "regulation" procedure by Anh and Di Paola (1996) we are recommended to increase the nonlinearity, i.e., power a by original power plus the increment $a - 1$, i.e., to use the new regulation power of $a + (a - 1) = 2a - 1$. Hence, the procedure can be represented schematically as follows:

$$k\, X^a(t) \rightarrow k_1 X^{2a-1}(t) \rightarrow k_2\, X^a(t) \rightarrow k_{eq}\, X(t) \tag{5.78}$$

We form a difference $k\, X^a(t) - k_1 X^{2a-1}(t)$ and demand the statistical orthogonality of this expression to $X^{2a-1}(t)$,

$$E\{\, [k\, X^a(t) - k_1 X^{2a-1}(t)]X^{2a-1}(t)\, \} = 0 \tag{5.79}$$

which leads to

$$k_1 = k\frac{E[X^{3a-1}(t)]}{E[X^{4a-2}(t)]} \tag{5.80}$$

The general expression for $E[X^a(t)]$ is

$$E[X^a(t)] = \sigma^a_X \int_{-\infty}^{\infty} \frac{\xi^a}{\sqrt{2\pi}} \exp\left(-\xi^2/2\right) d\xi \tag{5.81}$$

We obtain the first coefficient k_1 as follows:

$$k_1 = k\, \sigma^{a-1}_X \frac{\int_{-\infty}^{\infty} \xi^{3a-1} \exp\left(-\xi^2/2\right) d\xi}{\int_{-\infty}^{\infty} \xi^{4a-1} \exp\left(-\xi^2/2\right) d\xi} \tag{5.82}$$

Proceeding in perfect analogy, we obtain the following equivalent coefficients,

$$k_2 = k \frac{\left(\int\limits_{-\infty}^{\infty} \xi^{3a-1} \exp(-\xi^2/2) d\xi\right)^2}{\int\limits_{-\infty}^{\infty} \xi^{4a-1} \exp(-\xi^2/2) \, d\xi \int\limits_{-\infty}^{\infty} \xi^{2a} \exp(-\xi^2/2) \, d\xi} \tag{5.83}$$

$$k_{eq} = k \frac{\sigma_X^{a+1}}{E[X^2(t)]} \frac{1}{\sqrt{2\pi}} \frac{\left(\int\limits_{-\infty}^{\infty} \xi^{3a-1} \exp(-\xi^2/2) d\xi\right) \int\limits_{-\infty}^{\infty} \xi^{a+1} \exp(-\xi^2/2) d\xi}{\int\limits_{-\infty}^{\infty} \xi^{4a-1} \exp(-\xi^2/2) d\xi \int\limits_{-\infty}^{\infty} \xi^{2a} \exp(-\xi^2/2) d\xi} \tag{5.84}$$

$$= \frac{k \, E[X^2(t)]^{\frac{a-1}{2}}}{R} \tag{5.85}$$

where

$$R = \sqrt{2\pi} \frac{\int\limits_{-\infty}^{\infty} \xi^{4a-1} \exp(-\xi^2/2) \, d\xi \int\limits_{-\infty}^{\infty} \xi^{2a} \exp(-\xi^2/2) \, d\xi}{\left(\int\limits_{-\infty}^{\infty} \xi^{3a-1} \exp(-\xi^2/2) d\xi\right)^2 \int\limits_{-\infty}^{\infty} \xi^{a+1} \exp(-\xi^2/2) d\xi} \tag{5.86}$$

The equation of motion becomes

$$\dot{X}(t) + \frac{k \, E[X^2(t)]^{\frac{a-1}{2}}}{R} X(t) = F(t) \tag{5.87}$$

We deduce the mean-square displacement

$$E\left[X_{regulated}^2(t)\right]_I = \left(\frac{\pi S_0}{k}\right)^{\frac{2}{a+1}} R^{\frac{2}{a+1}} \tag{5.88}$$

In order to express R via Gamma function, note that

$$\int\limits_{-\infty}^{\infty} \xi^{2a} \exp(-\xi^2/2) \, d\xi = 2 \int\limits_{0}^{\infty} \xi^{2a} \exp(-\xi^2/2) \, d\xi \tag{5.89}$$

Then we make a change in the variable $\eta = \xi^2/2$, to get

$$\int\limits_{-\infty}^{\infty} \xi^{2a} \exp(-\xi^2/) \, d\xi = 2^{a+\frac{1}{2}} \int\limits_{0}^{\infty} \eta^{a-\frac{1}{2}} \exp(-\eta) \, d\eta \tag{5.90}$$

According to the definition of the Gamma function,

$$\Gamma(z) = \int\limits_{0}^{\infty} t^{z-1} \exp(-t) dt \tag{5.91}$$

we finally obtain,

$$\int_{-\infty}^{\infty} \xi^{2a} \exp(-\xi^2/2) \, d\xi = 2^{a+\frac{1}{2}} \Gamma\left(a + \frac{1}{2}\right)$$ (5.92)

By applying this process to the other integrals, we get

$$E\left[X^2_{regulated}(t)\right]_I = \left(\frac{\pi S_0}{k}\right)^{\frac{2}{a+1}} \left[\sqrt{\pi} 2^{-\frac{a+1}{2}} \frac{\Gamma\left(\frac{4a-1}{2}\right) \Gamma\left(\frac{2a+1}{2}\right)}{\Gamma\left(\frac{3a}{2}\right)^2 \Gamma\left(\frac{a+2}{2}\right)}\right]^{\frac{2}{a+1}}$$ (5.93)

Table 5.1 presents the percentagewise error due to approximate nature of the solutions (with both conventional linearization and RGEL method) in comparison to the exact solution provided by eq. (5.46), for different integer values of a.

Tab. 5.1: Error incurred by using a single step in the Anh and Di Paola regulation.

a	$\sigma^2_{X,exact}$	$\sigma^2_{classical\ X,approx}$	Error, %	$E[X^2_{regulated}(t)]_I$	Error, %
1	1	1.0000	0	1.0000	0
2	0.7765	0.7323	5.6877	0.7824	0.7713
3	0.6760	0.5774	14.5904	0.6547	3.1546
4	0.6175	0.4764	22.8490	0.5620	8.9861
5	0.5786	0.4055	29.9225	0.4917	15.0206
6	0.5505	0.3529	35.8981	0.4367	20.6846
7	0.5291	0.3124	40.9630	0.3925	25.8224

We can observe that there is an important improvement in the performance of the stochastic linearization when we utilize the RGEL method. That is, whereas for $a = 2$, the classical linearization is in error of about 5.69%, the regulated linearization has an error that is over 7 times less, namely 0.77%. For large value of a namely, $a = 5$, the regulated linearization has about half the error of that classical linearization namely 15% versus 29.9%. For even larger values of a, the error is much less than that in the classical scheme but still quite large: for $a = 7$, Anh and Di Paola approach leads to 25.8% error, whereas the classical approach is associated with error of about 41%. Still, the regulation reduces the error in this case by about 15%.

A natural question arises: What is the effect of additional steps in regulation? Anh and Di Paola (1996) considered only a single-step regulation. Here, the two-step regulation is performed, as illustrated schematically below:

$$k\,X^a(t) \rightarrow k_1\,X^{2a-1}(t) \rightarrow k_2\,X^{3a-1}(t)$$
$$\rightarrow k_3\,X^{2a-1}(t) \rightarrow k_4\,X^a(t) \rightarrow k_{eq,II}\,X(t)$$ (5.94)

Proceeding in perfect analogy with a single step procedure, we get

$$k_1 = k \, \sigma_X^{a-1} \frac{\int\limits_{-\infty}^{\infty} \xi^{3a-1} \exp(-\xi^2/2)d\xi}{\int\limits_{-\infty}^{\infty} \xi^{4a-1} \exp(-\xi^2/2)d\xi} \tag{5.95}$$

$$k_2 = k_1 \, \sigma_X^{1-a} \frac{\int\limits_{-\infty}^{\infty} \xi^{5a-3} \exp(-\xi^2/2)d\xi}{\int\limits_{-\infty}^{\infty} \xi^{6a-4} \exp(-\xi^2/2)d\xi} \tag{5.96}$$

$$k_3 = k_2 \, \sigma_X^{a-1} \frac{\int\limits_{-\infty}^{\infty} \xi^{5a-3} \exp(-\xi^2/2)d\xi}{\int\limits_{-\infty}^{\infty} \xi^{4a-2} \exp(-\xi^2/2)d\xi} \tag{5.97}$$

$$k_4 = k_3 \, \sigma_X^{a-1} \frac{\int\limits_{-\infty}^{\infty} \xi^{3a-1} \exp(-\xi^2/2)d\xi}{\int\limits_{-\infty}^{\infty} \xi^{2a} \exp(-\xi^2/2)d\xi} \tag{5.98}$$

$$k_{eq,II} = k_4 \, \sigma_X^{a+1} \frac{\int\limits_{-\infty}^{\infty} \xi^{a+1} \exp(-\xi^2/2)d\xi}{\sqrt{2\pi} \, E[X^2(t)]} \tag{5.99}$$

After expressing $k_{eq,II}$ via k, the initial equation of motion is replaced by

$$\dot{X}(t) + \frac{k \, E[X^2(t)]^{\frac{a-1}{2}}}{Q} X(t) = F(t) \tag{5.100}$$

where

$$Q = \sqrt{2\pi} \, \frac{\left(\int\limits_{-\infty}^{\infty} \xi^{4a-2} \exp(-\xi^2/2)d\xi\right)^2 \int\limits_{-\infty}^{\infty} \xi^{6a-4} \exp(-\xi^2/2)d\xi \int\limits_{-\infty}^{\infty} \xi^{2a} \exp(-\xi^2/2)d\xi}{\left(\int\limits_{-\infty}^{\infty} \xi^{5a-3} \exp(-\xi^2/2)d\xi\right)^2 \left(\int\limits_{-\infty}^{\infty} \xi^{3a-1} \exp(-\xi^2/2)d\xi\right)^2 \int\limits_{-\infty}^{\infty} \xi^{a+1} \exp(-\xi^2/2)d\xi} \tag{5.101}$$

We arrive at the mean-square displacement,

$$E\left[X_{regulated}^2(t)\right]_{II} = \left(\frac{\pi S_0}{k}\right)^{\frac{2}{a+1}} Q^{\frac{2}{a+1}} \tag{5.102}$$

or, via the Gamma functions

$$E[X_{regulated}^2(t)]_{II} = \left(\frac{\pi S_0}{k}\right)^{\frac{2}{a+1}} \left[\sqrt{\pi} 2^{\frac{a+1}{2}} \frac{\Gamma\left(\frac{6a-3}{2}\right) \Gamma\left(\frac{2a+1}{2}\right) \Gamma\left(\frac{4a-1}{2}\right)}{\Gamma\left(\frac{5a-2}{2}\right)^2 \Gamma\left(\frac{3a}{2}\right)^2 \Gamma\left(\frac{a+2}{2}\right)}\right]^{\frac{2}{a+1}} \tag{5.103}$$

The Roman subscript II indicates that the result is obtained in the second step of the regulation process. The Tab. 5.2 presents a comparison of the two-step procedure with the exact solution on one hand, a single-step procedure, and classical stochastic linearization.

Tab. 5.2: Error incurred by using an extended two-step regulation.

a	$\sigma^2_{X,exact}$	$E[X^2_{regulated}(t)]_I$	Error,%	$E[X^2_{regulated}(t)]_{II}$	Error, %
1	1	1.0000	0	1	0
2	0.7765	0.7824	0.7713	0.8205	5.6693
3	0.6760	0.6547	3.1546	0.7117	5.2820
4	0.6175	0.5620	8.9861	0.6251	1.2229
5	0.5786	0.4917	15.0206	0.5554	4.0131
6	0.5505	0.4367	20.6846	0.4988	9.4038
7	0.5291	0.3925	25.8224	0.4521	14.5548

We note that two-step regulation provides an additional improvement in comparison to the single-step regulation; for the moderate value of $a = 4$, the classical stochastic linearization is associated with the error of about 23%; the single-step regulation results in error of about 9% whereas the two-step regulation leads to the error of 1.23%. Thus, the error in two-step regulation is about 18 times less than in the classical scheme, and about 7 times less than in a single-step regulation. For larger values of a, though still much better than the classical of single-step regulation linearization, the two-step regulation reduces the error by about additional 10% in comparison with a single-step regulation: The classical linearization yields about 41% for $a = 7$; the single-

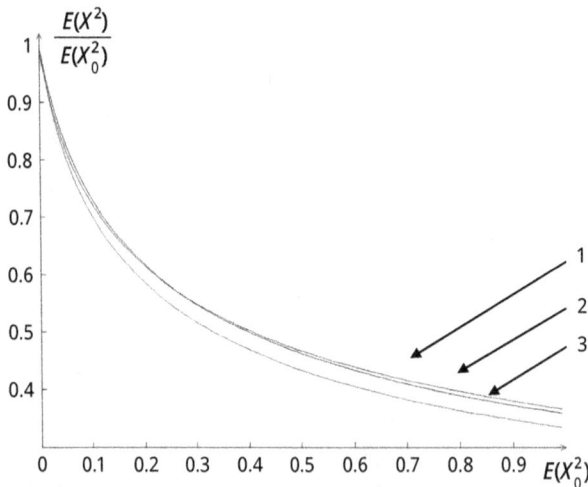

Fig. 5.1: Comparison of approximate solutions with exact results: (1) exact solution; (2) potential energy criterion in eq. (4.143); (3) classical stochastic linearization results.

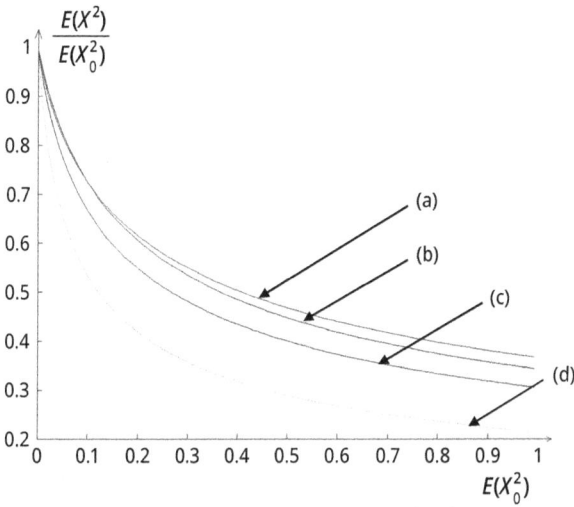

Fig. 5.2: Contrast of potential and complementary energy criteria with exact solution: (a) exact solution, (b) criterion based on equality of mean-squares of potential energies, in eq. (4.160), (c) criterion based on complementary energy in eq. (4.146), (d) criterion based on equality of mean-squares of complementary energies.

step regulation leads to 26%, whereas the two-step regulation results in 14.6% of error. We should note that the two-step regulation turns out to be the optimal one for the Lutes and Sarkani oscillator, since it turns out that the three-step regulation yields greater errors, than the two-step version. The optimum number of steps needed should be established for each oscillator at hand, (Fig. 5.2).

Note that in recent decades the interest in fractional calculus grew considerably in describing nonlinear materials and response (DiPaola *et al.*, 2022). They state: "The interest in fractional calculus is motivated by its unique capabilities in describing fractal or nonlocal media; it is also a valuable tool and well-established framework for capturing complex long-memory or multiscale phenomena in materials." In this respect, the works of Spanos and Evangelatos (2010), Spanos and Malara (2014) and Spanos *et al.* (2019) are of interest. The interested reader can consult with the review paper by Di Paola *et al.* (2013). Crandall (2006) reviewed the stochastic linearization's first fifty years. Elishakoff and Crandall (2017) discussed the progress of stochastic linearization during the past 60 years.

In this chapter we present a methodology for the simple exposition of the celebrated stochastic linearization technique. Moreover, this study presents two alternative derivations of the classical scheme of stochastic linearization technique that may prove useful in reinforcing its foundations. It should be stressed that Spanos and Zeldin (1994), Spanos *et al.* (2002), Failla *et al.* (2003), Ghanem and Spanos (1989), and Spanos *et al.* (2002) extensively applied the Galerkin method to various problems of nonlinear stochastic dynamics.

Chapter 6
Essay on the role of the Monte Carlo method in stochastic mechanics

6.1 Introduction

According to Hammersley and Handscomb (1964) "Monte Carlo methods comprise that branch of experimental mathematics which is concerned with experiments on random numbers. In the last decade they have found extensive use in the fields of operational research and nuclear physics, where there are a variety of problems beyond available resources of theoretical mathematics." Shinzuka's insistence that this technique is not the method of last resort, as some maintain, is well known. Neither is it a poor relative of analytical methods. Indeed, whereas analytical, closed-form or exact solutions are always welcome and preferable, there is only a small fraction of the problems that lend themselves to exact or closed-form solutions. Let us again quote Hammersley and Handscomb (1964): "One of the main strengths of theoretical mathematics is its concern with abstraction and generality: one can write down symbolic expressions or formal equations which abstract the essence of a problem and reveal its underlying structure. However, the same strength carries with it an inherent weakness: the more general and formal its language, the less is theory ready to prove a numerical solution

Fig. 6.1: Casino building in Monte Carlo (courtesy by Wikimedia Commons).

https://doi.org/10.1515/9783111354231-006

in a particular application. The idea behind Monte Carlo approach . . . is to exploit the strength of theoretical mathematics while avoiding its associated weakness by replacing theory by experiment whenever the former falters." In these circumstances approximate solutions play a central role both in the research and engineering application. Approximate methods consist of analytical or numerical methods. In most cases, the results of highly sophisticated approximate analytical tools are applied to quite simplified models. This is not to say that approximate analytical methods are without great merit. Yet, those analytical techniques that can be utilized to wide classes of problems and to multi-degree of freedom systems ought to be pursued. For example, from this point of view of simplicity, the method of stochastic linearization stands out as a technique that is applicable to a wide range of problems. Even then, the reliability estimation via the latter technique appears to be questionable. The only universal technique applicable to both simple and practical problems is the Monte Carlo method. Remarkably, authors, in order to substantiate their analytical findings via approximate techniques, conduct wide range Monte Carlo analyses.

Thus, Monte Carlo analyses are performed irrespective of the availability of the analytical method or not. As a result, the Monte Carlo method is invariably an integral part of any rigorous investigation. In these circumstances it is clear that much effort ought to be put into development of the Monte Carlo method in order to extend its applicability to widest possible range of problems. Yet this ought to be done not at the expense of analytical techniques that can serve as benchmark solutions. Closed-form solutions appear to be an indispensable part of a large program for its validation. Still, the proper balance ought to be found between the exact or approximate analytical and purely numerical techniques, which hopefully ought not be devoid of intellectual context. The physical insights that stem from benchmark closed-form solutions of simple problems on one hand, and the Monte Carlo-type solutions of complex problems

Fig. 6.2: Dilbert (courtesy of Wikimedia Commons).

ought to become a simultaneous emphasis of stochastic mechanics, especially in system's engineering, where only analytical bounds for the system's reliability are available. It appears that we need first to write the appropriate differential or integral equations; understand basic principles of the phenomenon analytically, both in deterministic and stochastic settings. Then, often, in order not to sacrifice the problem to small or large parameters, one would proceed to the Monte Carlo method. The results of the latter could guide us about which analytical techniques could support the Monte Carlo results, and in which ranges of the parameter variations. One can ask: "Can a number, say 9, appear, numerous times in the sequence of realization of the Monte Carlo Method?" The answer is provided by Samuel Kotz and Donna Stroup (1983): "consider a dramatic moment at the casino in Monte Carlo on August 18, 1913. A roulette wheel, when spun, has an equal chance of choosing a red and black number. On this evening, black came up 26 times in a row . . . A frantic rush to bet on red enriched the casino considerable."

In what follows some unorganized (but hopefully not disorganized) comments are made about this method. "Opinions," wrote the novelist George Eliot in 1859, "are a poor cement between human souls." If this chapter will contribute to better appreciation of the synchronized interplay of analytical and numerical methods in stochastic mechanics, it was no written in vain. The name "Monte Carlo Method" was introduced as a code name for secret work of neutron diffusion problem. The name itself was chosen because roulette – with which the casino town Monte Carlo (Fig. 6.1) is traditionally associated – is one of the simplest tools for generating random numbers (Fig. 6.2, Fig. 6.3).

6.2 Probability and computers

Nahin (2000) writes: "The concept of using probabilistic simulation to solve physical problems is generally credited to the brilliant Polish mathematician Stanislaw Ulam (1909–1984). Ulam was a key player in the American atomic bomb project (code-named the Manhattan District Project) during the war years 1943 to 1945 in Los Alamos, New Mexico. After the war, he continued to make seminal contributions to the development of the American hydrogen fusion bomb (the H-bomb). The A-bomb project was an effort that required the solution of an enormous number of extraordinarily difficult problems, many of which had an intense mathematical nature." The Monte Carlo method originated in 1953, from the article by Metropolis *et al.* titled "Equations of State Calculations by Fast Computing Machines." Thus, it is a hybrid of probability and computers, although even in the last century people would toss a coin and calculate the relative frequency of the outcomes. Louchard and Latouche (1983) comment: "Probability theory and computer science: the former is an ancient science, where names such as Fermat and Pascal play a pre-eminent role – the analysis of games of chance has progressively added to the modern concepts of the theory of probability – the latter led us in thirty years from the 30 tons and 18,000 tubes of ENIAC to the silicon chip and its tens of thousands of transistors per square millimeter.

The connections between the two sciences are ancient and multiple. Pascal himself, in 1642, invented one of the first mechanical adding machines: fifty were built, of which a few may be found in museums, alongside the prototypes of Babbage, and ENIAC itself."

Fig. 6.3: Random number generator (courtesy of Wikimedia Commons).

The Monte Carlo method is the natural output of the interaction between the probability and computers. Ulam (1976) describes the idea of simulation as follows: "The idea for what was later called the Monte Carlo method occurred to me when I was playing solitaire during [an] illness. I noticed that it may be much more practical to get an idea of the probability of the successful outcome of a solitaire game . . . by laying down the cards, or experimenting with the process and merely noticing what proportion comes out successfully, rather than to try to compute all the combinatorial possibilities . . . This is intellectually surprising, and if not exactly humiliating, it gives one a feeling of modesty about the limits of rational or traditional thinking. So, from a sick man playing a lonely game of cards came one of the great ideas of computational physics." "Moreover, "in a sufficiently complicated problem, actual sampling is better than an examination of all the chains of possibilities. It occurred to me then that this could be equally true of all processes involving branching of events, a sin the production and further multiplication of neutrons of some kind of material containing uranium or other fossil elements. At each stage of the process, there are many possibilities determining the fate of the neutron. It can scatter at one angle, change its velocity, be absorbed, or produce more neutrons by fission of the target nucleus, and so on. The elementary probabilities are individually known . . . But the problem is to know what a succession and branching of perhaps hundreds or thousands or millions will do . . . The [Monte Carlo method's] idea was to try out of thousands of such possibilities and, at each stage, to select by chance, by means of a "random number" with suitable probability, the fate [of a neutron]. After examining the possible histories of only a few thousand, one will have a good sample and an approximate answer to the problem." According to Nahin (2000) the following words of Ulam are "prophetic": "All one needed was to have the means of producing such sample histories. It so happened that commuting machines were coming into existence and here was something suitable for machine calculation."

Here a note is necessary about priority. Nahin (2000) stresses: "An historical purist might object to naming Ulam as the inventor of Monte Carlo [method], as one can find anecdotal use of random numbers in solving physical problems in much earlier literature. In particular, the great Scottish engineer/scientist William Thompson, later Lord Kelvin (1824–1907), used the drawing of numbers at random (written on slips of paper and pulled out of hat) in a paper published before 1900. But it was Ulam, fifty years later, who started the scientific study of the Monte Carlo idea itself to the point that it is today a recognized specialty of computational physics." History of the Monte Carlo Method is discussed by Metropolis (1985, 1987), Eckhardt (1987), Brown (2011), Benov (2016), Bielajw (2021) and Mascagni (2022).

Atanassov and Dimov (2008) write: "The year 1949 is generally regarded as the official birthday of the Monte Carlo method when the paper of Metropolis and Ulam (1949) was published, although some authors point to earlier dates. Ermakov (1985), for example, notes that a solution of a problem by the Monte Carlo method is contained in the Old Testament. In 1777 G. Compte de Buffon posed the following problem: suppose we have a floor made of parallel strips of wood, each the same width, and we drop a needle onto the floor. What is the probability that the needle will lie across a line between two strips? The problem in more mathematical terms is: given a needle of length l dropped on a plane ruled with parallel lines t units apart, what is the probability P that the needle will cross a line? He found that $P = 2\ l/(\pi t)$. In 1886, Marquis Pierre–Simon de Laplace showed that the number π can be approximated by repeatedly throwing a needle onto a lined sheet of paper and counting the number of intersected lines. The development and intensive applications of the method is connected with the names of John von Neumann, E. Fermi and G. Kahn, who worked at Los Alamos (USA) for 40 years for the Manhattan project."

Ferson (1996) and Atanassov and Dimov (2008) discuss what Monte Carlo methods can and cannot accomplish, whereas Kroese *et al.* (2014) and Borkar (2022a, 2022b) underline importance of the Monte Carlo method. Brown (2011) deals with the future prospects of the method.

As Shinozuka and Deodatis (1996) write, "Several methods are currently available to solve a large number of problems in mechanics involving uncertain quantities described by stochastic processes, fields, or waves. At this time, however, Monte Carlo simulation appears to be the only universal method that can provide accurate solutions for certain problems in stochastic mechanics, involving nonlinearity, system stochasticity, stochastic stability, parametric excitations, large variations of uncertain parameters, etc. and that can assess the accuracy of other approximate methods such as perturbation, statistical linearization, closure techniques, stochastic averaging, etc. . . ." As Eckhardt (1987) writes, "although computers cannot create randomness *de novo*, they can take a smidgen of *pseudorandom* numbers. As the name suggests, these numbers are not truly random, but they work well enough to fool most probabilistic algorithms (in other words, computers not only play dice, but they also cheat)." Apparently with the first paper by Shinozuka on this topic published in 1972, a computational mechanics' period began in stochastic mechanics.

6.3 Closed-form and numerical solutions: friends or foes?

The line of reasoning about numeric methods is not confined only to stochastic mechanics. This is how Zienkiewicz (1970), one of the pioneers of computational deterministic mechanics, justified the use of the finite element method:

> The engineer is in need of numbers with which his design process can be described is often impatient with the "properly formulated" problem for which complex equations exist but for which only comparatively trivial solutions can be achieved by classical mathematics. At such times, he often formulates ad hoc, crude models which, having served the purpose once, are later discarded . . . Its [finite element method's] popularity among engineers is assured and, at long last, an arousal of interest among applied mathematics promises at least a consideration of fundamentals.

Due to spectacular achievements in the deterministic FEM, one could anticipate that the analytical solutions and techniques would become less attractive for younger generations. Most unfortunately this phenomenon is taking place, and the development of numerical techniques takes place often in full divorce with analytical elegance, we are longing for hybrid methods. As Obraztsov (1975) maintains,

> the analytical methods have formulated and will formulate the way of thinking. The developmental speed in the area of numerical mathematics is defined by the level of the fundamental investigations. Therefore, only a synchronized development of the analytical and discrete methods secures a required progress.

Many specialized journals and conferences came into existence for computational mechanics. This is how one of the respectable journals declared its aims and scope:

> The trend in many scientific journals of a trivial "closed form" solution in preference to the more general if mathematically less elegant numerical process is reversed as a conscious policy of the journal.

It is pleasing that stochastic journals avoided such strong sentiments against "closed-form" or exact solutions, even along with the appreciation of the capabilities of the Monte Carlo techniques. This leads to the questions of mutual influence between approximate techniques, of which the Monte Carlo method is the prime technique, and analytical methods.

Here additional point must be mentioned, due to Hammersley and Handscomb (1964):

> "some people feel genuine distress at the idea of a mathematical result [provided by the Monte Carlo method-I. E.] which is necessarily not absolutely certain." In this connection the quote of John von Neumann (1951, 1963) appears to be relevant: "Anyone who considers arithmetical methods of producing random digits is, of course, in a state of sin."

Hammersley and Handscomb (1964) reply to such a possible objection as follows: "Applied mathematics is not in any case a black-and-white subject; even in theoretical applied mathematics there is always doubt whether the postulates are adequately valid for the situation under consideration . . ." Moreover, Robert Coveyou (1969) maintained (see also Ivars Peterson, 1998): "The generation of random numbers is too important to be left to chance," with addendum, as it were by Knuth, "Random numbers should not be generated with a method chosen at random."

6.4 Positive interplay between numeric analysis and analytics

Since the Monte Carlo method is essentially a multiple application of the deterministic method, it makes sense to recount some interesting instances of interaction between exact and approximate solutions in the deterministic mechanics. We will bring examples from structural dynamics.

Grossi and Bhat (1991) studied free vibrations of tapered beams using the characteristic orthogonal polynomials method, developed earlier by Bhat (1985). They compared their results with those reported by Goel (1976) obtained by the use of Bessel functions. Since Goel's (1976) derivations are associated with the exact solutions, while Grossi and Bhat's (1991) results were based on Rayleigh's method, the former results must be less than the latter ones. As Grossi and Bhat (1991) reported, for the tapered beams in question, "both Rayleigh-Schmidt and Rayleigh-Ritz methods yield upper bounds." They added that "unfortunately, several values reported by Goel (1976) are higher than the values obtained by the two above-mentioned methods." Auciello (1996) studied in detail this particular disagreement of the numerical results with general theorems. It was established that Goel's (1976) exact solution contained an error. It is remarkable that as an interesting byproduct of the approximate analyses, the revision of the exact solution became necessary. An analogous situation took place also in regard to the fundamental frequency of the pinned circular plate. Laura, Paloto, and Santos (1975) showed that for the Poisson's ratio $v = 0.3$, the non-dimensional fundamental frequency was $\Omega = 4.947$. It was found by the single polynomial comparison function via the Galerkin method. It turned out to be *lower* than the value of the exact frequency, which was reported in the literature to equal 4.977. Several years later it was shown that their approximate value was "almost perfect" (Laura, 2000). Indeed, as Leissa (2000) writes, "because of the inaccuracy and lack in detail imperviously appearing literature [about the exact values] Narita and Leissa (1980) published a paper in the 'Journal of Sound and Vibration' in 1980 devoted solely to the pinned circular plate. In Table 4 of that paper, it gives the exact fundamental frequency to six figures as 4.93515, when Poisson's ratio is 0.3." Thus, an approximate result reported by Laura, Paloto and Santos (1975) is 0.24% greater than the exact value. These are two examples of the positive interplay between exact and appropriate analyses, namely when the updating of the exact results turned out to be necessary as a result of numerical calculations. It appears worthwhile to corroborate exact solutions even when they are obtainable, with some approximate techniques, to distance oneself, as much as possible, from inaccuracies. This appears to be routinely done by stochastic analysts who invariably try to corroborate their analysis by the Monte Carlo method.

It is recalled that one of the authors' papers (Elishakoff, 1979) devoted to the Monte Carlo method inspired a development of one of the versions of the stochastic finite element method.

6.5 Do not overemphasize neither asymptotology nor simulation

Author recalls a story told some decades ago about a fresh and able applied mathematics graduate in Russia. After the successful interview in a company, he asked:

- "Does your company deal with problems with small parameters?"
- "No," was the answer.
- "Does your firm consider problems with the large parameters?" inquired the candidate. The reply was negative again.
- "I cannot possibly deal with problems that do not lend themselves to the asymptotic methods," said the applicant and refused to take a job.

Surely, some researchers can still confine themselves merely to the "asymptotology" (a term coined by Kruskall), study of problems with small or large parameters. Yet, well-rounded research cannot abandon an "electronic slave." Whereas analytical solutions, although confined to specific ranges of parameter variations may possess an elegance, they are only the first step in dealing the complex problems, in order to understand the physical phenomenon in the first approximation. Rigorous investigation of practical, real-world problems cannot limit itself with asymptotology alone. Since even for asymptotic solutions computers are used (Rand and Ambuster, 1987) why not utilize also fully computerized solutions, especially for engineering applications?

Yet there is a danger of overemphasizing the numerical solutions and the Monte Carlo method. Some researchers become overly zealous with simulations. Johnston (1961) notes:

> There are many advantages in simulated tests, carried out while with the aid of computer, in comparison with real tests in an actual testing machine. No machining is involved, no materials need to be acquired, and there is no scatter in the test results! Moreover, the precision of results, although based on a simulated and idealized material, permits a study of details of behavior that is not possible in ordinary laboratory tests. It would be impossible to completely duplicate the observations that may be made on the basis of the simulated tests.

Singer (1997) refutes this claim:

> It was forgotten that the simulation was so successful because the physical phenomena in this case were well known and had been extensively explored by very many real experiments. New phenomena have still to be found and properly understood in physical test, before even the powerful computers of today can give a reliable simulation and then extend the range of parameters.

On the other extreme, even if one would settle with the notion that the Monte Carlo method is the method of last resort, as some analytically minded researchers maintain, this still would have a positive connotation. At least, when all analytical approaches fail, there is then the method one can resort to. What would one do without such a method?

6.6 Stochasticity combined with asymptotology is not enough

Since for each realization of the stochastic variable, function, or field one has to deal with deterministic calculation, one immediately sees that deterministic mechanics is a cornerstone of the Monte Carlo method. Since one of the competing deterministic theories must be chosen by the stochasticity investigator, it is quite transparent to state that an excellent command of the deterministic mechanics is necessary for rigorous stochastic calculation. This implies that not confining one's research only to stochastic mechanics may be recommendable. Application of various theories in the stochastic context may lead to totally different stochastic responses. For example, in the Crandall's problem (1979), it turned out (Elishakoff and Lubliner, 1985) that use of the Timoshenko-Ehrenfest theory led to the mean-square values that were over 100% greater than those obtained with in the Bernoulli-Euler theory in some cases. This fact may even be more pronounced for determination of the probability of failure. Since this quantity must be extremely small, it may be tremendously influenced by the adopted deterministic theories.

As is seen, disadvantages of the employed deterministic theories fully manifest themselves in the stochastic solution of the problem. Likewise, advantages of the deterministic methods and solutions may be fully implemented in the stochastic analysis. Indeed, most advanced techniques of deterministic analysis can readily be implemented into the stochastic treatment of the problem.

The marriage of the finite element method and stochasticity is one magnificent example of incorporating the powerful deterministic technique into the stochastic analysis. In this regard, one should mention that presently the stochastic finite element method (SFEM) usually uses, either directly or indirectly, the perturbation method. This is thus an example of the hybrid use of "asymptotology" and stochasticity, since SFEM assumes that stochastic variations are small. Shinozuka and Yamazaki (1988) mention: "Monte Carlo simulation methods can provide useful alternative for the problems to which the perturbation method does not apply very well . . . [This becomes] more crucial as higher-order solutions are sought, as the degree of the material property variability becomes more pronounced . . ."

Marchante (1997) writes: "The Stochastic Finite Element Codes are, in general, limited to nonlinear problems with a reduced number of degrees of freedom and relatively small scatter."

Yet, one could visualize problems in which stochastic variability is large. Initial studies in this direction were performed by Elishakoff, Ren and Shinozuka (1996).

Still, this effort was confined to specific applications. Combining of the finite element and the Monte Carlo method was facilitated by Yamazaki *et al.* (1986).

6.7 About modeling

Bolotin (1984) stressed: "In principle such modeling [the Monte Carlo method] may give only that, which is put into the model."

Indeed, this reminds us of GIGO principle: "Garbage in, garbage out"; yet it could be also optimistically modified to mean "Good model in, good results out.." It is still surprising that single-degree-of-freedom models, or sometimes the half-degree-of-freedom models are utilized for stochastic modeling of real structures, like a bridge, or a wind excited structure, or an earthquake excited structure. Such a stochastic modeling forgets that even the deterministic calculations often involve FEM with tens of thousands of elements. Neyman (1945) expresses this in following manner:

> Every attempt to use mathematics to study some real phenomena must begin with building a mathematical model of these phenomena. Of necessity the model simplifies the matters to a greater or lesser extent and a number of details are ignored. Success depends on whether or not the details ignored are really unimportant in the development of the phenomena studied. The solution of the mathematical problem may be correct, and you may be in violent conflict with realities, simply because the original assumptions of the mathematical model diverge essentially from the conditions of the practical problem considered. Beforehand, it is impossible to predict with certainty whether or not a given mathematical model is adequate. To find this out, it is necessary to deduce a number of consequences of the model and to compare them with observation.

According to Høyland and Rausand (1984), a "pioneer in statistics, George E. P. Cox, repeatedly points out that 'no model is absolutely correct. In particular situations, however, some models are more useful than others'."

As Klejnem (1994) notes,

> All scientific methods are based on assumptions, which limit the applicability of these methods. These assumptions may be documented explicitly, or they may be left implicit. Many practitioners do not know *when* to use *what* method.

One of the goals of stochastic mechanics ought to be to explain which questions may be asked in practice, and which methods can answer these questions. In this context the recent study of Schuëller *et al.* (1998) appears to be of much interest. They conducted an important benchmark study in which eight examples of non-linear structural systems (both with single and multiple degrees of freedom) "have been solved independently by the participants by applying either their own method or the method of their preference." Authors arrived at the conclusion that for nonlinear MDOF-system "procedures as Equivalent Linearization (EQL) and Controlled Simulation (CST) appear to be more suitable."

6.8 About required precision: accountability

Popper (1982) begins his quest for honesty in scientific endeavors by noting that it is necessary to lay down certain ground rules on the precision required in scientific evaluations:

> The fundamental idea underlying scientific determinism is that the structure of the world is such that every future even can in principle be rationally calculated in advance, if only we know the laws of nature and the present state of the world. But if *every* event is to be predictable, it must be predictable *with any desired degree of precision*: for even the minutest difference in measurement may be claimed to distinguish between different events.

Moreover,

> Our theory will have to account for the imprecision of the prediction given the degree of precision which we require of the prediction; the theory will have to enable us to calculate the degree of precision in initial conditions that would suffice to give us a prediction of the required degree of precision. I call this demand the principle of accountability.

Tekkens (1994) further elaborates on this topic:

> Whenever they fail in their prediction, scientists tend to blame the poor accuracy of the observations, the lack of computer power and the inadequate parametrization in their numerical models, rather than their own lack of skill in computing the accuracy that can be obtained with present resources.

Tekkens' (1994) view is very strong:

> Popper holds researchers responsible for their own work . . . To put it bluntly, a calculation that does not include a calculation of its predictive skill is not a legitimate scientific product.

It is instructive to stress here a view by Ditlevsen (1982), according to whom "any reliability measure defined in connection with a limit state theory of a high reliability technological system is a purely formal comparative measure of safety. It only makes sense to make comparisons within classes of 'similar' technological systems which are all accessible to the same theory."

6.9 Generating randomness from chaos?

Recent chaos theory generated several thousands of papers up to now. There are numerous definitions of chaos that are commonly in use. Nine different definitions are critically reviewed by Brown and Chua (1996). Their conclusion is that "at present *chaos* is a philosophical term, not a rigorous mathematical term." Engineers are usually defining a chaotic dynamical system the one that exhibits a dramatic sensitive dependance on initial conditions. Another definition of chaos, in terms of stochasticity, was given by Iyengar (1993):

> Let the initial conditions of all the variables be specified as random variables with small variances. Let us consider the variance of the resulting response ensemble. We ask, in the limit [when] the initial variances go to zero, does the response variance in time also go to zero? If it does, we call the given nonlinear system to be deterministic, otherwise the system is chaotic. It may be noted that this definition characterizes stochastically the dependence of the system on the initial conditions.

Schiehlen and Bestle (1988) concluded in their interesting study that chaotic motions of nonlinear dynamical system can be modeled in the case of long periods as stochastic processes.

The Monte Carlo method involves the need of the pseudo-random number generators. Natural question arises: Since the sequences generated by the pseudo-number generators are repeatable, once the same seed number is used, how can we talk about the randomness of the generated numbers? The answer was furnished by Hammersley and Handscomb (1964)

> this should not affect the person who has to use them, since the question he should be asking is not 'Where did these numbers come from?' but 'Are these numbers correctly distributed?' and this question is answered by statistical tests on the numbers themselves.

Indeed, the sequence of pseudo-random numbers pass the statistical tests of randomness. Four tests of randomness have been developed. Hull and Dobell (1962) write:

> In this way [by using statistical tests] one may avoid becoming involved in any philosophical arguments about the meaning of randomness, arguments which, according to Kendall and Babington-Smith 'are of an abstract and metaphysical character bordering at times on the theological'.

As Abramowitz and Stegun (1965) note

> the use of random numbers in electronic computers has resulted in a need for random numbers to be generated in a completely deterministic way.

As Schiehlen (2000) writes ". . . consideration of chaos and randomness is most challenging. Probability theory usually starts with describing experiments. Throwing a well-designed die, you can get a uniform distribution for the six numbers of the die. However, a die is nothing else than a highly nonlinear mechanical system which may also be modeled by a very complex deterministic equations of motion. Randomness occurs due to the high sensitivity of the system with respect to the initial conditions which cannot be controlled properly. The probabilistic approach is an idealization of the real world to set mathematically consistent results. On the other hand, chaos is more difficult to define but it is closer to the real world."

Oishi and Inoue (1982) proposed a new method for designing pseudo-random number generators by making use of chaotic first-order nonlinear differential equations. Brown and Chua (1996) conclude that:

> Given any definition of chaos or pseudo-randomness, these will always be some "clearly chaotic/pseudo-random" dynamical system whose chaos/pseudo-randomness cannot be established by that definition.

This observation did not disturb Chua, Yao and Yang (1990) to publish a paper under the title "Generating Randomness from Chaos and Constructing Chaos with Desired Randomness." One may suggest that for *practical* purposes one can utilize the chaotic systems for pseudo-random generators. But a keen opponent may still ask: Is there a definition of practicality?

Here comes a thought-provoking article by Kalman (1994) who reminds us that according to De Finetti (1974), "Probabilities do not exist." Kalman (1994) claims that

> The majority of observed phenomena of randomness in Nature (always excluding games of chance) cannot be and should not be explained by (conventional) probability theory; there is little or no experimental evidence in favor of (conventional) probability but there is massive, accumulating evidence that explanations and even descriptions should be sought outside of the conventional framework.

It is strongly felt that the paper by Kalman (1994) ought to be read and re-read, digested and both critically and self-critically evaluated by the stochastic community. There are fundamental issues that need deep thought and introspection. Surely, anti-optimization techniques (also known in the literature as unknown-but-bounded, guaranteed, convex analyses or Wald's min-max) of design with uncertainty avoid notions of chaos and stochasticity. It is a pleasure to record Shinozuka's (1970) contribution in this subject some 30 years ago, as well as putting our heads together on the related topic recently (Li et al., 1996a, 1996b). Not all agree that this approach is any sounder than the probabilistic approach, or that it fundamentally differs from it.

6.10 An essential question

Marek (2000) writes: "While the scientific elite is approaching the problems related to the substance and applicability of this powerful tool 'from above'," our team is trying to bring a 'Monte Carlo message' in a simple form (limited to structures) to tens of thousands of designers, specification writing committees and others involved in structural design and in reliability assessment. Let me call this approach 'from below'." Best part of the designers of structures at present is unaware of all what Monte Carlo means."

Spanos and Zeldin (1998) state:

> It is suggested that with the advent of computational hardware, the probability community continues assessment the relative advantages of often intricate analytical or semi-analytical solutions over simple and versatile Monte Carlo analysis of stochastic mechanical systems.

It is hoped that consensus will be reached in the stochastic community on this important subject.

In one regard the present-day analytical solutions and the Monte Carlo method appear to be in perfect *tune* with each other: Both seem to *fail* to deal with extremely large structural systems, like a train, airplane, ocean liner, nuclear power plant *etc.* Writes the Nobel Prize Laureate K. Wilson:

> Computers certainly expand the possibility of theorists, but their results are practically bounded by a number of degrees of freedom. The numerical integration methods are not applicable, where a number of integrated values exceed 5–10, the partial differential equations become extremely complicated for n independent variables greater than 3. Although the Monte-Carlo and statistical averaging methods give a possibility to include the cases of thousands and millions of variables, but a slow convergence of these methods leads to a long computational time even for highly powerful computers.

Note that ironically, the analytical averaging methods (whose proponents may dislike Monte-Carlo solutions) and Monte-Carlo solutions (whose protagonists may maintain that the results of averaging methods' scientists are nearly immediately available on the computer) are put, as if were into the same basket!

For large complex system, a "crude" method, like a Statistical Energy Method (Lyon, 1975; Clarkson *et al.*, 1981) appear to be the only method that produces results (can we trust them?). Perhaps the methods of multi-body dynamics, or super-elements, will be a useful alternative.

Three decades ago, Zienkiewicz (1970) proclaimed in the reference to the finite element method, ". . . Perhaps without undue exaggeration, it may be said that with present day size of computers, solutions can be obtained to all solid mechanics/structural situations on a practical basis." An essential question begs to be asked: Can analogous statements be made as applied to stochastic mechanics? It humbly appears to this writer that the reply to this inquiry is not yet affirmative (see also comment of Prof. Z. Bažant on p. 26).

Finally, the multilevel Monte Carlo methods should be mentioned (Giles, 2008, 2012, 2015); Teckentrup, 2013; Butler *et al.*, 2015).

6.11 Conclusion

It appears appropriate to mention that in the pen and pencil Monte Carlo method one can use a fair dice, or better yet, a fair decahedron. It appears that we, as a stochastic community, need more fairness too. In a special issue of the "International Journal of Non-Linear Mechanics," that was devoted to Professor Shinozuka's 60th birthday, Professor Spanos (1991) beautifully mentioned the notion "above all, fairness." In order the uncertainty modeling to become a more attractive subject, in addition to professionalism, we all ought to be fair to the myriad of viewpoints of the members of this research community, to avoid creation of a very stochastic (or fuzzy or set-theoretic) "ghetto." Indeed, the Jewish sages of the second century of common era, in their collective tractates "*Mishna*" maintained, in the language spectacularly familiar to the Monte-Carlo analysts: ". . . whereas the man prints many coins from one die, each one is similar to the other, the Supreme King of Kings, the Holy One blessed be He stamped every man with the die of the first human and yet no one exactly resembles his fellow."

We need to find a proper balance in maintaining good human relations, and simultaneously avoiding "team errors," attendant to so called "groupthink." As Janis's (1972) analysis shows the interest of the group members shifted to maintain their good human relations, rather than to find the best decision to a given problem. Most unfortunately, they fell into a wrong decision (Sasou and Reason, 1999).

Only the very tip of the iceberg of questions has been discussed in this chapter. Many questions have been left out, due to the space limitation; partial list of these is as follows: Conditional simulations of random fields (see Kameda and Morikawa, 1992; Hoshiya, 1994; Ren *et al.*, 1995); homogenization approach and its correlation with the Monte Carlo methodology; modeling the human error; modeling the modeling errors (can we treat what we do not know as stochastic, like an unknown effect to be uncovered in the future?); modeling so called "information gaps," in the context of uncertainty modeling, of which the Monte Carlo method appears to be the universal vehicle. Objections of Finetti (1974), Kalman (1994) and thoughts by Brown and Chua (1996) will show in the future whether stochastic mechanics is a well-designed practical tool or it will collapse as an "unsinkable" Titanic did. Are the alternatives to stochasticity notion (fuzziness or anti-optimization) any sounder? Not everyone thinks so!

As Brown (2010) notes: "During 1947, von Neumann (Richtmyer and von Neumann, 1947) developed the first known computer program for performing Monte Carlo calculations. The description of the program logic is remarkably similar to what is still done today in modern Monte Carlo codes. (In fact, portions of MCNP coding appear to have originated from von Neumann's program, migrated through several generations of assembly language and Fortran.) In describing the program, von Neumann estimated that 100 neutrons with 100 collisions would take 5 h on the ENIAC (with its 18,000 vacuum tubes). Today that would take milliseconds. The first symposium on the Monte Carlo method was held in 1949." Moreover, Brown (2010) emphasizes "the maturity and rich list of features in today's most-used Monte Carlo codes were cited as distinct benefits of the past 63 years (prior to 2010 when Brown's paper was published–I.E.). of Monte Carlo development. Extending those Monte Carlo codes with ever more features over the years has a cost, however. As software packages become larger, the burden for maintenance, documentation, user support, and verification and validation grow excessively. After 20–30 years of adding new features, new development is impeded by older code structure and logic, obsolete data structures, and coding style that often resembles assembly language. The algorithms and data structures in many mature codes were first developed for problems with only a dozen or so regions and materials. While the codes have been (painfully) modified over the years to now handle many 1000s of materials and regions, extension to millions or billions of materials and regions is difficult."

Finally, it appears instructive to quote Sharon Bertsch McGrayne (2011): "The combination of Bayes and Markov Chain Monte Carlo has been called arguably the most powerful mechanism ever created for processing data and knowledge." For a definitive review of the Monte Carlo method, see also the papers by Hurtado and Barbat (1998) and Marek (2001).

Risk assessment via double Monte Carlo method was conducted by Ali *et al.* (2012). Using this method for system reliability was elucidated by Marseguerra and Zio (2002). Papadrakakis and Lagaros (2002) combined neural networks and Monte Carlo Simulation in the reliability-based optimization context.

Not everyone subscribes to the Monte Carlo method in particular, or probabilistic methodology in general. Here it appears instructive to quote from an otherwise definitive paper by Champneys *et al.* (2019), referring to the buckling of axially compressed shells. Authors note: "Much has been written on the classical problem of the buckling of an axially loaded cylindrical shell (see e.g., Hutchinson and Thompson, 2018; Hunt, 2011) and references therein. It carries all the hallmarks of a classical subcritical buckling problem, in particular its notorious sensitivity to imperfections. Its apparent simplicity, yet underlying complexity, has ignited significant academic dueling over the course of the last century (cf. different "resolutions" of the "paradox" by Zhu *et al.* (2002) and Elishakoff (2012), and caused design engineers many a headache." First of all, it is an honor for this writer to be listed in the company of emeritus faculty member of the Cambridge University, Professor Christopher Reuben Calladine, FRS FREng, for he is one of the authors of the paper (Zhu E. *et al.*, 2002).

Champneys et al. (2019) write: "One approach to deal with such imperfection-sensitivity [of shells] is through stochastic methods. Elishakoff, Arbocz, and others pioneered developments, such as the international databank of imperfections (see Arbocz and Hol (1991); Elishakoff, Li and Starnes (2001)) and references therein. While such methods appeal at one level, from a modeling point of view there is also significant sensitivity to the precision . . . the chosen numerical method, and useful analyses typically require many Monte Carlo realizations. Also, from a practical perspective, estimating a safety margin of a particular specimen would necessitate comprehensive imaging and analysis of all its imperfections. Unfortunately for modern, lightweight composite structures, such imperfections often occur beneath the surface layer and are hard to characterize in practice. There has therefore long been a search for a lower-bound criterion below which a violently subcritical structure, such as a cylindrical shell cannot buckle." Naturally, the lower-bound approach does not contradict the Monte Carlo analysis; these methods should be used in a careful and perhaps combined manner. The lower bound methodology can be contrasted with the first-order second moment method by Elishakoff, van Manen, Vermeulen and Arbocz (1987) or the papers by Ben-Haim and Elishakoff (1989) and Elseifi *et al.* (1999) where a non-probabilistic, lower-bound methodology was adopted, close by the spirit to that utilized by Champneys *et al.* (2019). The latter authors appear to criticize without checking the current state of the literature. The results of above papers were included in the book by Elishakoff, Li and Starnes (2001), that is cited by Champneys *et al.* (2019). We will take the liberty of citing one author about our work on imperfection sensitivity ". . . It was not until 1979, when Elishakoff published his reliability study . . . that a method has been proposed, which made it possible to introduce the results of imperfection surveys . . . into the analysis . . ." (Arbocz, 1991).

Apparently, Champneys, his coauthors, and many other readers would benefit from reading the insightful paper by Kroese et al. (2014) tellingly titled *Why the Monte Carlo Method Is So Important Today.* An interesting account of using the Monte Carlo method for solution of linear elasticity problem was provided by Shia and Hui (2000), as an alternative to applying the finite element method or boundary element method. Monte Carlo simulation and kriging were combined in the study by Echard *et al.* (2011). Au and Wang (2014) developed a class of powerful simulation techniques called Markov Chain Monte Carlo method (MCMC), an important machinery behind Subset Simulation that allows one to generate samples for investigating rare scenarios in a probabilistically consistent manner. Papaioannou *et al.* (2013) considered reliability sensitivity analysis with Monte Carlo Methods. We will reproduce only one praise of Monte Carlo method, belonging to Doucet *et al.* (2001):

> Monte Carlo methods are revolutionizing the on-line analysis of data in fields as diverse as financial modelling, target tracking and computer vision. These methods, appearing under the names of bootstrap filters, condensation, optimal Monte Carlo filters, particle filters and survival of the fittest, have made it possible to solve numerically many complex non-standard problems that were previously intractable.

The reviews of the Monte Carlo method were furnished by Spanos and Zeldin (1998), Hurtado and Barbat (1998), Marek (2001), Marek *et al.* (2001), Elishakoff (2003), and Schuëller (2009). The Monte Carlo method can be successfully applied also to interval and fuzzy sets-based analyses (Kreinovich *et al.*, 2007; Zhang H. *et al.*, 2010; Jahani *et al.*, 2014; Vieira *et al.*, 2021).

Review of improved Monte Carlo approaches was given by Hu *et al.* (2016).

It should be noted that the alternative to Monte Carlo simulation was given by Jie Li and his collaborators (Li, 2016; Li and Chen, 2004, 2009). According to Li (2016), "The principle of probability preservation can be stated as follows: If the random factors involved in a stochastic system are retained, then the probability will be preserved in the state evolution process of the system." Moreover, "The probability density evolution method, as a new perspective to stochastic systems, provides a powerful tool for the analysis, control, and critical performance assessment of stochastic systems. Derived from the principle of probability preservation, the generalized probability density evolution equation incorporated with the uncoupled physical equations, takes significant advantages over traditional probability density evolution equations in understanding the essential characteristics of engineering structures and systems, especially the randomness propagation in nonlinear dynamical systems even in general physical systems" (Li, 2016). An intuitive basis for justification of this method was given in the paper by Ang (2017).

A broader perspective on the Monte Carlo method was furnished by Spanos and Zeldin (1998). Uncertainty quantification through Monte Carlo Method in a cloud computing setting was provided by Cunha *et al.* (2014). Handbooks on Monte Carlo simulation were composed by Kroese et al. (2014), Brandimarte (2014). Handbook on Markov chain Monte Carlo method was authored by Brooks *et al.* (2011). Recent review papers on the Monte Carlo Method include those by Zhang (2021), Świechowski *et al.* (2023), Wills and Schön (2023), Georgescu (2023), and Song and Kawai (2023).

Chapter 7
Fuzzy sets–based interpretation of the safety factor

> Is uncertainty the same as randomness? If we are not sure about something, is it only up to chance? Do the notions of likelihood and probability exhaust our notions of uncertainty? Many people trained in probability and statistics believe so. Some even say so and say so loudly. Bart Kosko (1990)

Safety factor is a universally utilized concept in several branches of engineering. On one hand, most engineers, as it were, neglect uncertainty, but on the other hand, the allowable stress level was introduced long time ago as a ratio of the yield stress to the so-called safety factor to provide the region for the safe utilization of the structure. Thus, the uncertainty is introduced into practice by the "back door." This observation led to a considerable literature dedicated to the probabilistic interpretation of the safety factor. The present chapter deals with the novel aspect of elucidation of the concept of safety factor through the theory of fuzzy sets. The aim of the chapter is to present the safety factor that is uniformly employed by engineers, but in a new light. The safety factor in the fuzzy setting is introduced. The ideas are illustrated on two strengths of materials problems; simple examples are chosen so as to allow for clearer illustration of ideas.

7.1 Introduction

Safety factors are integral part of any meaningful design in structural engineering for over two centuries. It appears useful to make pertinent comments on this concept as it is utilized in the deterministic context. The material properties are reflected in the stress-strain curves; the stress is defined as the ratio of the load applied to the specimen to the cross-sectional area; the strain is defined as the relative change of its length under the load. In some materials the stress–strain curve exhibits so called yield stress σ_y, corresponding to the increase of the strain without increase of the stress. This level, naturally, is a dangerous level. Engineers demand the stress to be below this level, i.e., $\sigma < \sigma_y$. Moreover, engineers introduced the 'fence' against it, by requiring it to be sufficiently away, not just below the yield stress. Instead of the inequality $\sigma < \sigma_y$, engineers demand that the following inequality to hold $\sigma \leq \sigma_{all}$, where σ_{all} is the allowable stress, in contradistinction with σ_y which is an unallowed level. The allowable stress is expressed as a fraction of the yield stress, $\sigma_{all} = \sigma_y / s_{req}$ where s_{req} is the required safety factor. Is it obvious that the required factor of safety ought to exceed unity, so that the stress remains below the yield level. The question is, "How do the engineers get the value of s_{req}?"

https://doi.org/10.1515/9783111354231-007

In deterministic setting it is obtained in an *ad hoc* manner, namely, by considerations of the accumulated experience and the level of responsibility of the structure; more specifically, no analytical techniques are used to justify the numerical level of the safety factor. Its value, be it 1.2, 1.5, 2.3, or even 3 (the latter value is often used in civil engineering), for decades was assigned at the *will* of the engineer; presently it is *codified*. Usually, the allowable stress is introduced by the relationship $\sigma_{all} = \sigma_y / s_{req}$, where σ_y is the material yield stress and s_{req} is the required safety factor. Numerous authors argue (for detailed discussion and pertinent quotes the reader can consult the recent monograph by Elishakoff (2004)) that there are various unforeseen uncertainties that should be considered in some global manner. It is no surprise, therefore, that the safety factors often are referred to as *"factors of ignorance."*

Although the safety factor is introduced in an *ad hoc* manner, it permits us to solve three following problems in strength of materials. The *first problem* may be referred to as *checking safety*. In this case both the loads and the geometrical dimensions are known. Engineers are required to check if the safety of an existing structure is maintained. One proceeds with finding the fields of displacements, strains, and stresses in the structure. Once the maximum stress level σ_{max} is evaluated, one calculates the actual safety factor (see Elishakoff, 2004) $s_{act} = \sigma_y / \sigma_{max}$. If the inequality $s_{act} \geq s_{req}$ is fulfilled, i.e., if the actual safety factor is greater than or equal to the required safety factor, the structure is declared to be safe. Otherwise, it is considered unsafe. The *second problem* deals with determining of the maximum allowable level of the load $P_{max,all}$ so that the equality $\sigma_{max} = \sigma_{all}$ is satisfied. The *third problem* is that of design. The minimum required geometrical dimension $a_{min,req}$ is determined so that equality $\sigma_{max} = \sigma_{all}$ is satisfied. Most of the structural codes in aerospace, mechanical, civil, and naval engineering are the above format for the design of the structure in consideration.

Despite the spectacular success of this concept for ensuring engineering safety, the lack of any *analytical* technique to justify safety factors was deemed unsatisfactory by many engineering mechanicals. It was their desire to attempt to directly connect safety factors with non-deterministic analysis, rather than acknowledge its origin due to uncertainty but prescribe it without *any* analysis. Such as non-deterministic, namely, stochastic analysis was pioneered by Freudenthal (1938) and Rzhanitsyn (1947), apparently independently. The monograph by Elishakoff (2004) extensively exposes this concept in probabilistic setting. The concept of reliability turns out to be an effective tool while calibrating safety factors associated with different failure modes as shown by Verhaeghe and Elishakoff (2013). Wang X-J, Shi Q. *et al.* (2019) contrasted the reliability-based and safety factor-based analyses.

Here we address the following question that, most unfortunately, was not posed before: Is it possible to put the safety factor in the realm of uncertainty analysis as exemplified by the *fuzzy sets*? The reasonableness of this question stems from the recognition of the fact that often times fuzzy-sets-based approach may prove to be advantageous over the purely probabilistic methodology. It is shown in this study (Sections 7.4–7.8) that the reply to the above question is *affirmative*.

It appears instructive, before proceeding with the concept of the fuzzy safety factor, to first present its probabilistic counterpart, in order to be in position to compare the two methodologies. In this section, we first reiterate (in Section 7.2) the connection that exists between the safety factor and probabilistic notion of reliability, although on examples that were not studied before. In Section 7.3 it is shown that these two concepts are *compatible* with each other on the example of triangularly distributed random stress. In Section 7.4 we introduce the safety factor in the framework of fuzzy sets theory and dub it as the *central fuzzy safety factor*. To the best of our knowledge, this task has not been dealt with before. In Section 7.5 we give some strength of materials applications. In particular, fuzzy numbers with either triangular or quadratic membership function are considered in detail for the *possibilistic* description of the stress. Remarkable relationships are established between the possibility and necessity measures and the respective central fuzzy safety factors.

An apparent skeptic may emphasize that researchers were *long* interested in re-interpretation of (probabilistic) reliability in the context of fuzzy sets. Indeed, there are several references that utilize the notion that is referred to as *fuzzy reliability* (see the papers by Wang C. and Wang W., 1986; Cai, 1996; Utkin and Eurov, 1996; Wu, 1997; Guo and Lu, 2003) or *fuzzy probabilistic method* (Wang J.H. et al., 2007) or *random-fuzzy analysis* (Guo *et al.*, 2001a, 2001b; Guo and Lu, 2003; Haldar and Reddy, 1992, Möller and Beer, 2004). Likewise, in the context of another non-deterministic analysis, namely, *convex modeling*, Ben-Haim (1994) discusses the so called *non-probabilistic reliability* or *robust reliability* (Ben-Haim, 1996) (see also Ben-Haim's (1996) paper with a telling title "Must Reliability Be Probabilistic?"). This suggestion did not meet a universal acceptance: Good (1995) suggests that the non-probabilistic reliability is an "oxymoron." We do not share such a radical sentiment against either "non-probabilistic reliability" or "fuzzy reliability." On the other hand, since the engineers almost uniformly use the safety factor, it is *more* pertinent to express safety factor either in direct probabilistic or direct fuzzy environment rather than to extend notion of reliability to fuzziness or to convexity. This chapter attempts to speak, as it were, the language utilized by the engineering profession, via a "translator" which understands the fuzzy sets theory. It is hoped that the existing significant gap in the non-deterministic analyses of structures is filled by this study. In the next section, we start with *probabilistic* interpretation, our main goal in mind, to present a *fuzzy safety factor*.

7.2 Probabilistic interpretation of the safety factor

Herewith, we again quote Freudenthal (1968), ". . . It seems absurd to strive for more and more refinement of methods of stress-analysis if, in order to determine the dimensions of the structural elements, its results are subsequently compared with so called working stress, derived in a rather crude manner by dividing the values of somewhat dubious material parameters in conventional materials tests by still more dubious empirical numbers called safety factors." This pioneering consideration led to the probabilistic interpretation of the safety factor. Several probabilistic approaches to it have been presented in the literature. Freudenthal (1968) apparently was the first one to introduce the central safety factor

$$s = E\left(\sum_g\right)/E\left(\sum\right)$$

(7.1)

where the stress $\sum$ and the yield stress $\sum_y$ are treated as random variables, denoted by capital letters; for their possible values the lower-case letters are used, $E(\cdot)$ means mathematical expectation.

In probabilistic context the inequality $\sigma \leq \sigma_{all}$ is replaced by the probabilistic counterpart

$$R = Prob\left(\sum \leq \sigma_y\right)$$

(7.2)

where R is reliability, the probability that the stress will not exceed the yield stress. Let us consider the simple case in which the yield stress is deterministic, but the stress is a random variable. In these circumstances

$$R = F_\Sigma\left(\sigma_y\right)$$

(7.3)

where $F_\Sigma(\cdot)$ equals cumulative distribution function of the stress:

$$F_\Sigma(y) = Prob\left(\sum \leq y\right)$$

(7.4)

where y is a possible value, the stress may take. In probabilistic analysis safety requirement lies in the demand that the reliability is not less than a required one r

$$R \geq r$$

(7.5)

In view of eqs. (7.3) and (7.5) safety required reads:

$$F_\Sigma\left(\sigma_y\right) \geq r$$

(7.6)

In the design problem the required values of the system, i.e., the minimum required geometrical dimension or the maximum allowable level of the load, are obtained from the equality

$$F_\Sigma\left(\sigma_y\right) = r$$

(7.7)

This topic is considered in much detail in monograph of Elishakoff (2004).

7.3 Concepts of reliability and safety factor are compatible

In this section we illustrate the direct connection that exists between the reliability and the safety factor in probabilistic setting. Let us first review the results pertaining to the case when the stress has a uniform probability density function (Fig. 7.1). The results are given in (Elishakoff, 2004). The safety factor compatible with required reliability r is expressed in an analytical form

$$s = 1 + v_\Sigma \sqrt{3}(2r - 1) \tag{7.8}$$

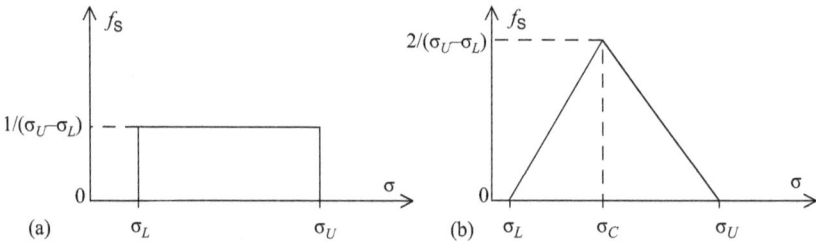

Fig. 7.1: Probability density of the stress: (a) uniform density, (b) triangular density function of the random variable Σ.

where v_Σ is the conventional coefficient of variation $v_\Sigma = \sqrt{Var(\Sigma)}\big/E(\Sigma)$, $Var(\Sigma)$ is the stress variance.

Now we consider a case in which the stress Σ is a random variable with the triangular probability density function (see Fig. 7.1b)

$$f_\Sigma(\sigma) = \begin{cases} \dfrac{2}{(\sigma_U - \sigma_L)(\sigma_C - \sigma_L)} \dfrac{\sigma - \sigma_L}{}, & \text{for } \sigma_L < \sigma < \sigma_C \\[3mm] \dfrac{2}{(\sigma_U - \sigma_L)(\sigma_U - \sigma_C)} \dfrac{\sigma_U - \sigma}{}, & \text{for } \sigma_C < \sigma < \sigma_U \\[3mm] 0, & \text{otherwise} \end{cases} \tag{7.9}$$

where σ_L and σ_U are the lowest and the greatest values that the stress may take, whereas the value σ_C of stress corresponds to the maximum value of density function. In order not to complicate the analysis, we treat the yield stress σ_y to be a deterministic quantity, i.e., to take a single value σ_y with unity probability. The probability distribution function

$$F_\Sigma(\sigma) = Prob(\Sigma \le \sigma) = \int_{-\infty}^{\sigma} f_\Sigma(t)dt \tag{7.10}$$

reads

$$F_\Sigma(\sigma) = \begin{cases} 0, & \text{for} \quad \sigma < \sigma_L \\ \dfrac{(\sigma - \sigma_L)^2}{(\sigma_U - \sigma_L)(\sigma_C - \sigma_L)}, & \text{for} \quad \sigma_L \le \sigma < \sigma_C \\ 1 - \dfrac{(\sigma_U - \sigma)^2}{(\sigma_U - \sigma_L)(\sigma_U - \sigma_C)}, & \text{for} \quad \sigma_C \le \sigma < \sigma_U \\ 1, & \text{for} \quad \sigma_U \le \sigma \end{cases} \tag{7.11}$$

Bearing in mind eqs. (77.3) and (7.11), we get the reliability

$$R = \begin{cases} 0, & \text{for} \quad \sigma_y < \sigma_L \\ \dfrac{(\sigma_y - \sigma_L)^2}{(\sigma_U - \sigma_L)(\sigma_C - \sigma_L)}, & \text{for} \quad \sigma_L \le \sigma_y < \sigma_C \\ 1 - \dfrac{(\sigma_U - \sigma_y)^2}{(\sigma_U - \sigma_L)(\sigma_U - \sigma_C)}, & \text{for} \quad \le \sigma_y < \sigma_U \\ 1, & \text{for} \quad \sigma_U \le \sigma_y \end{cases} \tag{7.12}$$

We note that the mean value of the stress is

$$E(\Sigma) = (\sigma_L + \sigma_C + \sigma_U)/3 \tag{7.13}$$

whereas the variance of the stress equals

$$Var(\Sigma) = (\sigma_U^2 + \sigma_C^2 + \sigma_L^2 - \sigma_U \sigma_C - \sigma_U \sigma_L - \sigma_C \sigma_L)/18 \tag{7.14}$$

For the sake of simplicity, we will concentrate on the particular case in which $\sigma_C = (\sigma_L + \sigma_U)/2$. Equations (7.12) and (7.14) become

$$E(\Sigma) = (\sigma_L + \sigma_U)/2 \quad Var(\Sigma) = (\sigma_U - \sigma_L)^2/24 \tag{7.15}$$

From these two equations we express σ_L and σ_U as the function of the mean and variance

$$\sigma_L = E\left(\sum\right) - \sqrt{6Var\left(\sum\right)}, \quad \sigma_U = E\left(\sum\right) + \sqrt{6Var\left(\sum\right)} \tag{7.16}$$

Let $\sigma_C \le \sigma_y < \sigma_U$, in accordance with eq. (7.12), we have

$$R = 1 - \frac{2[E(\sum) + \sqrt{6Var(\sum)} - \sigma_y]^2}{24Var(\sum)} \tag{7.17}$$

By substituting eq. (7.16) and dividing both the numerator and the denominator by the mean value of the actual stress squared $[E(\sum)]^2$, and introducing the coefficient of variation v_Σ of the actual stress, we get, instead of eq. (7.17)

$$R = 1 - \frac{2(1+\sqrt{6}v_\Sigma - s)^2}{24v_\Sigma^2}, \quad \text{for} \quad \sigma_C \le \sigma_y \le \sigma_U \qquad (7.18)$$

As is seen the reliability R is *directly* expressed in terms of the central safety factor s and the coefficient of variation of the involved random variable v_Σ. Thus, the reliability methods enable to *introduce safety factors rigorously, rather than arbitrarily.* When we seek the reliability evaluation, these parameters s and v_Σ are independent: the knowledge of both is needed for the reliability evaluation. If, however, the reliability level is set constant r-- called a required reliability--due to the code requirement, the coefficient of variation v_Σ is treated as an independent variable, whereas the safety factor is considered as its function, given as follows

$$s = 1 + v_\Sigma \sqrt{6}\left(1 - \sqrt{2}\sqrt{1-r}\right), \quad \text{for} \quad \sigma_C \le \sigma_y \le \sigma_U \qquad (7.19)$$

This relationship is depicted in Fig. 7.2. Maximum value of the safety factor

$$s_{\max} = 1 + v_\Sigma \sqrt{6} \qquad (7.20)$$

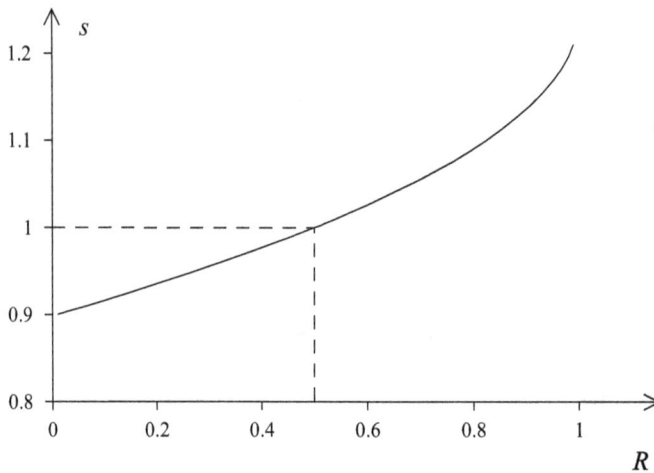

Fig. 7.2: Triangular probability density function: relation between safety factor and reliability for $v_\Sigma = 0.1$.

is achieved when reliability tends to one from below. For example, for the coefficient of variation 0.05 the maximum safety factor assumes the value 1.12; for the coefficient of variation 0.1 the maximum safety factor equals 1.24; for coefficient of variation 0.15 it takes a value 1.37 etc. We conclude that with greater variation of the involved random variable, the safety factor must be *increased.* It is seen that the reliability context allows one to make *quantitative* judgments in terms of the *required reliability* and the *coefficient of variation.*

7.4 Fuzzy safety factor

In the past century, non-probabilistic approaches have been proposed in structural engineering for problems involving uncertain variables such as fuzzy sets developed by Black (1937) and Zadeh (1965, 1973, 1975a, 1975b, 1975c, 1994, 2008a, 2008b), convex models (Ben-Haim, 1985;1996a, 1996b; 2001;2006; Ben-Haim and Elishakoff, 1990; Elishakoff, Lin and Zhu, 1994; Elishakoff and Ohsaki, 2010) random sets (Goodman, 1982; Kendall, 1974; Dubois and Prade,1991; Matheron, 1995; Robbins, 1994, 1995; Bernardini, 1999, 2000; Tonon, 2004a, 2004b; Tonon and Bernardini, 1999; Tonon *et al.*, 1999, 2001; Molchanov, 2005; Fetz and Oberguggenberger, 2016), possibility theory (Yager, 1982, 1986, 1987, 1991, 1993; Stroud *et al.*, 2002; Bae *et al.*, 2003, 2004a, 2004b; Penmesta and Grandhi, 2003), Dempster-Shafer theory (Dempster, 1968; Yager and Liu, 2008), etc. Main motivation for adopting non-probabilistic approaches is the high sensitivity of failure reliability to the tails of probability distributions of the random variables involved in the analysis (Ben-Haim and Elishakoff, 1990; Elishakoff, 1999a, 2000a).We quote here Corotis (2015) noting: "Possibility theory and evidence theory both differ from probability theory in the modeling of ignorance. Probabilistically, ignorance is represented by uniformly distributing the total probability (i.e., unity) to individual elements of the universe (Ross, 2010). It follows then that the choice of a uniform distribution is founded upon no evidential basis (Ross, 2010). Both possibility theory and evidence theory recognize ignorance explicitly, including total ignorance. Such a scenario is defined by maximally imprecise probabilities (or possibilities), i.e., the set of all probability distributions on X. These are referred to as vacuous probabilities (Shafer, 1976)" (see also Ballent *et al.*, 2019). Langley (2000) states: "Various methods exist for assessing the safety of a structural or mechanical system that has uncertain parameters. These methods are either statistical (probabilistic), in which case the probability of failure is sought, or deterministic (possibilistic), in which case bounds on the response are sought. Well-known statistical methods include the first-order reliability method (FORM) and the second-order reliability method (SORM), while deterministic methods include interval analysis, convex modeling, and fuzzy set theory (although the categorization of the latter approach as a deterministic method is debatable). The development of probabilistic and possibilistic methods has tended to occur independently, with specialized algorithms being developed for the implementation of each technique. It is shown here that a wide range of probabilistic and possibilistic methods can be encompassed by a single mathematical algorithm, so that, for example, existing codes for FORM and SORM can potentially be employed for other methods, thus allowing the designer to readily choose the method most suited to the available data."

In the present study, the safety factor approach is established in the framework of fuzzy numbers theory. In the field of structural engineering, often there are cases when load can be treated as a fuzzy number. Hence, during the design process of a structural member, it is useful to define stress $\sum$ through membership function μ_Σ, while the strength of the beam σ_y may still be considered as a deterministic quantity, in the simplest approximation. Since $\sum$ is a fuzzy number, so is the ratio:

$$S = \sigma_y / \sum \qquad (7.21)$$

which is called a fuzzy safety factor. In Fig. 7.3 a generic membership function of fuzzy stress is shown. Hereinafter σ_L will represent the lower bound of the support of the membership function, while the stress σ_U will be associated with the upper bound of the support of the membership function. The value σ_C of stress corresponds to $\mu_\Sigma = 1$.

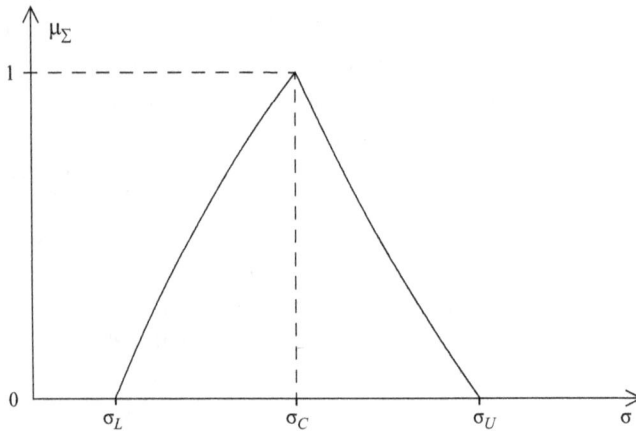

Fig. 7.3: Membership function of fuzzy stress $\sum$.

Note that S constitutes a fuzzy variable, unless the fuzzy variables σ and σ_y are proportional. One can envision use of the fuzzy safety factor in the codes. However, presently, the codes make use of deterministic numbers as safety factors. Therefore, it makes sense to resort to defuzzification (Zhao and Goving, 1991). In this study defuzzification is performed via centroid method (Möller and Beer, 2004).

We define *central fuzzy safety factor* as equal to the ratio of the abscissas of center of gravity of the yield stress and of center of gravity of the stress, respectively,

$$s = G\left(\sum_g\right) / G\left(\sum\right) \qquad (7.22)$$

where $G(\cdot)$ means the abscissa of the center of gravity of a fuzzy number:

$$G\left(\sum\right) = \frac{\int_{\sigma_L}^{\sigma_U} \sigma \mu_\Sigma(\sigma)\, d\sigma}{\int_{\sigma_L}^{\sigma_U} \mu_\Sigma(\sigma)\, d\sigma} \qquad (7.23)$$

Since the yield stress is treated here to be a deterministic value, eq.(7.22) becomes:

$$s = \sigma_y / G\left(\sum\right) \tag{7.24}$$

We also evaluate the lower and upper bounds of the fuzzy central safety factor s. To do this, we need some preliminary considerations involving the concepts of possibility and necessity and their relationship with fuzzy set theory, as defined in the monograph by Dubois and Prade (1980) [see also the paper by Savoia (2002)]. We are interested with the possibility and necessity associated with the event A that the fuzzy stress satisfies the inequality $\sum \leq \sigma_y$, where σ_y is a deterministic quantity:

$$Possibility(A) = \prod\left(\sum \leq \sigma_y\right) = \sup_{\sigma \leq \sigma_y} \mu_\Sigma(\sigma) \tag{7.25}$$

$$Necessity(A) = N\left(\sum \leq \sigma_y\right) = 1 - \prod(\overline{A}) = 1 - \prod\left(\sum > \sigma_y\right) = 1 - \sup_{\sigma > \sigma_y} \mu_\Sigma(\sigma) \tag{7.26}$$

As Dubois and Prade (1980) demonstrate, the necessity and possibility represent the respective lower and upper bounds of the probability of the event A:

$$N(A) \leq P(A) \leq \Pi(A) \tag{7.27}$$

where the necessity and possibility are defined, respectively, as:

$$N(A) = P_*(A), \qquad \Pi(A) = P^*(A) \tag{7.28}$$

More explicitly, since the probability that the stress does not exceed the yield stress σ_y is nothing else but the reliability [see eq. (7.2)], we get:

$$N\left(\sum \leq \sigma_y\right) \leq R \leq \prod\left(\sum \leq \sigma_y\right) \tag{7.29}$$

To reiterate, possibility and necessity of fulfilling the inequality $\sigma \leq \sigma_y$ are upper and lower bounds of reliability, respectively. We can therefore use an alternative notation. The lower bound of reliability will be denoted thereinafter as R_*, whereas its upper bound will be denoted as R^*. Hence

$$R_* = N\left(\sum \leq \sigma_y\right), \quad R^* = \Pi\left(\sum \leq \sigma_y\right) \tag{7.30}$$

It is of interest to note that Cai K.Y. (1996) uses possibility of successful event as his definition of the fuzzy reliability, dubbed by him as *posbist reliability*.

Being inspired by the monograph by Kaufmann and Gupta (1985) we also define the left and right values of abscissas of centers of gravity pertaining to the possibility and the necessity measures, $\tilde{G}_{pos}$ and $\tilde{G}_{nec}$, respectively. These are evaluated as the static moments of the hatched areas in Fig. 7.8b–c divided by respective areas. Note that Fig. 7.8a represents the triangular membership function; y_Π denotes the vertical axis passing through the center of gravity of the hatched area of possibility function; y_N indicates the vertical axis passing through the center of gravity of the hatched area of necessity function. These axes will be needed further to calculate the second moments of the respective areas. For possibility function the abscissa of center of gravity is written as:

$$\tilde{G}_{pos} = \frac{\int\limits_{\sigma_L}^{\sigma_C} \sigma\Pi(\sum \leq \sigma)d\sigma}{\int\limits_{\sigma_L}^{\sigma_C} \Pi(\sum \leq \sigma)d\sigma} \tag{7.31}$$

The abscissa of center of gravity of the hatched area in the necessity function $\tilde{G}_{nec}$ is written as

$$\tilde{G}_{nec} = \frac{\int\limits_{\sigma_C}^{\sigma_U} \sigma\Pi(\sum \leq \sigma)d\sigma}{\int\limits_{\sigma_C}^{\sigma_U} \Pi(\sum \leq \sigma)d\sigma} \tag{7.32}$$

where the domain of integration is $\sigma_C \leq \sigma \leq \sigma_U$. The abscissa of center of gravity $\tilde{G}_{pos}$ is smaller than $\tilde{G}_{nec}$. Moreover, the following relation holds

$$\tilde{G}_{pos} \leq G\left(\sum\right) \leq \tilde{G}_{nec} \tag{7.33}$$

Instead of the abscissas of the centers of gravity of lower and upper bounds of probability, Dempster (1967) and Dubois and Prade (1985) define the lower and upper mathematical expectations as Lebesgue-Stieltjes integrals:

$$E_*\left(\sum\right) = \int\limits_{-\infty}^{\infty} \sigma d\Pi\left(\sum \leq \sigma\right), \quad E^*\left(\sum\right) = \int\limits_{-\infty}^{\infty} \sigma dN\left(\sum \leq \sigma\right) \tag{7.34}$$

Thus, they define the first moments of possibility and necessity functions in a way analogous to the probability theory, dealing with $N(\sum \leq \sigma)$ and $\Pi(\sum \leq \sigma)$ as analogues of the cumulative distribution function.

It is also instructive to deal with the lower and upper bounds of the central fuzzy safety factor s. In view of the definition (7.24) and inequality (7.33) we get the lower and upper bounds, respectively,

$$s_{nec} = s_* = \sigma_y/\tilde{G}_{nec}, \quad s_{pos} = s^* = \sigma_y/\tilde{G}_{pos} \tag{7.35}$$

Hence, we arrive at the inequality for the central fuzzy safety factor s

$$s_{nec} \leq s \leq s_{pos} \tag{7.36}$$

At first glance the eq. (7.36) is not of much use, because the central fuzzy safety factor can be directly calculated from eq. (7.24). Yet it will be shown that the definition of lower and upper bounds of s is important in order to find a relationship between the central fuzzy safety factor and bounds on reliability.

7.5 Bounds on reliability in a basic problem of the strength of materials

We study a basic problem in which an element (Fig. 7.4) is subjected to a fuzzy stress σ, with a membership function $\mu_\Sigma(\sigma)$, while the yield stress is assumed to be a deterministic quantity σ_y.

Based on the value of the strength σ_y we can have four possible cases:

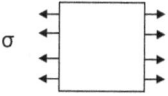

Fig. 7.4: Element subjected to a stress.

7.5.1 Case 1: $\sigma_y \leq \sigma_L$

The first case is the one where the strength is smaller than the lower bound of the support of μ_Σ (see Fig. 7.5a),

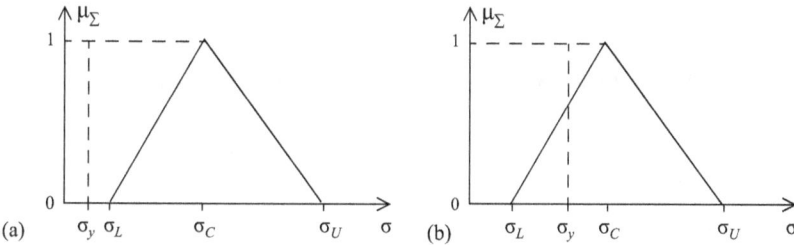

Fig. 7.5: Interrelation of fuzzy stress $\sum$ and deterministic yield stress σ_y: (a) $\sigma_y \leq \sigma_L$, (b) $\sigma_L \leq \sigma_y \leq \sigma_C$.

i.e., $\sigma_y \leq \sigma_L$. The possibility and necessity of the event A become, respectively,

$$R_* = N\left(\sum \leq \sigma_y\right) = 1 - \prod\left(\sum \geq \sigma_y\right) = 1 - \sup_{\sigma \geq \sigma_y} \mu_\Sigma(\sigma) = 1 - 1 = 0$$

$$R^* = \prod\left(\sum \leq \sigma_y\right) = \sup_{\sigma \leq \sigma_y} \mu_\Sigma(\sigma) = 0 \qquad (7.37)$$

We reiterate that in according to eq. (7.29) the reliability takes values between possibility and necessity of fulfilling a mission specified in the event A. Hence, in this case

$$0 = N\left(\sum \le \sigma_y\right) \le R \le \prod\left(\sum \le \sigma_y\right) = 0 \tag{7.38}$$

leading to the conclusion that in this case reliability vanishes identically, as it should be.

7.5.2 Case 2: $\sigma_L \le \sigma_y \le \sigma_C$

We consider now the second case in which σ_y lies between the lower bound of the support σ_L and σ_C the value corresponds to $\mu_\Sigma = 1$ (see Fig. 7.5b). We have

$$R_* = N\left(\sum \le \sigma_y\right) = 1 - \prod\left(\sum \ge \sigma_y\right) = 1 - \sup_{\sigma \ge \sigma_y} \mu_\Sigma(\sigma) = 1 - 1 = 0$$

$$R^* = \prod\left(\sum \le \sigma_y\right) = \sup_{\sigma \le \sigma_y} \mu_\Sigma(\sigma) = \mu_\Sigma(\sigma_y) \tag{7.39}$$

Thus, reliability is bounded as follows

$$0 = N\left(\sum \le \sigma_y\right) \le R \le \prod\left(\sum \le \sigma_y\right) = \mu_\Sigma(\sigma_y) \tag{7.40}$$

implying that reliability is between zero and the values of membership function evaluated at σ_y.

7.5.3 Case 3: $\sigma_C \le \sigma_y \le \sigma_U$

Let us deal now with the third case of interest $\sigma_C \le \sigma_y \le \sigma_U$ (see Fig. 7.6a).

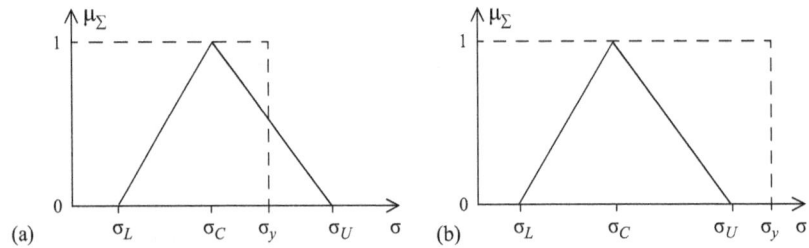

Fig. 7.6: Interrelation of fuzzy stress $\sum$ and deterministic yield stress σ_y: (a) $\sigma_C \le \sigma_y \le \sigma_U$, (b) $\sigma_y > \sigma_U$.

We get,

$$R_* = N\left(\sum \le \sigma_y\right) = 1 - \prod\left(\sum \ge \sigma_y\right) = 1 - \sup_{\sigma \ge \sigma_y} \mu_\Sigma(\sigma) = 1 - \mu_\Sigma(\sigma_y)$$

$$R^* = \prod\left(\sum \le \sigma_y\right) = \sup_{\sigma \le \sigma_y} \mu_\Sigma(\sigma) = 1 \tag{7.41}$$

The reliability is bounded, therefore, as follows:

$$1 - \mu_\Sigma(\sigma_y) = N\left(\sum \le \sigma_y\right) \le R \le \prod\left(\sum \le \sigma_y\right) = 1 \tag{7.42}$$

7.5.4 Case 4: $\sigma_y > \sigma_U$

In the fourth case we confine ourselves to the case $\sigma_y > \sigma_U$ (see Fig. 7.6b). We get,

$$R_* = N\left(\sum \le \sigma_y\right) = 1 - \prod\left(\sum \ge \sigma_y\right) = 1 - \sup_{\sigma \ge \sigma_y} \mu_\Sigma(\sigma) = 1 - 0 = 1$$

$$R^* = \prod\left(\sum \le \sigma_y\right) = \sup_{\sigma \le \sigma_y} \mu_\Sigma(\sigma) = 1 \tag{7.43}$$

The reliability is bounded, therefore as follows,

$$1 = N\left(\sum \le \sigma_y\right) \le R \le \prod\left(\sum \le \sigma_y\right) = 1 \tag{7.44}$$

leading to the conclusion that in this case reliability takes the maximum value 1, as it should be.

We can summarize the four cases in the following two expressions, where we write the lower and upper bounds of reliability for different values of yield stress σ_y

$$R_* = R_{nec} = \begin{cases} 0, & for \quad \sigma_y < \sigma_C \\ 1 - \mu_\Sigma(\sigma_y), & for \quad \sigma_C \le \sigma_y < \sigma_U \\ 1, & for \quad \sigma_U \le \sigma_y \end{cases} \tag{7.45}$$

$$R^* = R_{pos} = \begin{cases} 0, & for \quad \sigma_y < \sigma_L \\ \mu_\Sigma(\sigma_y), & for \quad \sigma_L \le \sigma_y < \sigma_C \\ 1, & for \quad \sigma_C \le \sigma_y \end{cases} \tag{7.46}$$

To be on the safe side, we choose *design reliability* $R_{design}(\sigma_y)$ as the lower bound for the reliability

$$R_{design}(\sigma_y) = R_* = N\left(\sum \le \sigma_y\right) \tag{7.47}$$

Note that it is takes values different from zero for $\sigma_y > \sigma_C$. In this way we use only the right branch of membership function of fuzzy stress μ_Σ and we neglect, as it were, in order to be on the safer side. In the following two sections we present two specific cases, exemplified with triangular and quadratic fuzzy stresses, respectively.

7.6 Triangular fuzzy stress

In their paper, Kosheleva *et al.* (2019) explained why triangular or trapezoidal membership functions are working extremely well in fuzzy sets-based analysis. Therefore, we let the stress σ to be a fuzzy number with a triangular membership function that is represented as:

$$\mu_\Sigma(\sigma) = \begin{cases} 0, & \text{for} \quad \sigma \leq \sigma_L \\ \dfrac{\sigma - \sigma_L}{\sigma_C - \sigma_L} & \text{for} \quad \sigma_L \leq \sigma \leq \sigma_C \\ \dfrac{\sigma_U - \sigma}{\sigma_U - \sigma_C} & \text{for} \quad \sigma_C \leq \sigma \leq \sigma_U \\ 0, & \text{for} \quad \sigma \geq \sigma_U \end{cases} \tag{7.48}$$

Let us consider the simple case of a symmetric fuzzy number $\sigma_C = (\sigma_L + \sigma_U)/2$. By introducing in eqs. (7.45), (7.46) the triangular membership function (eq. (7.48)), the necessity and possibility measures of event A, that the inequality $\sum \leq \sigma_y$ is fulfilled, expressed as the functions of the yield stress σ_y are:

$$R_* = N\left(\sum \leq \sigma_y\right) = \begin{cases} 0, & \text{for} \quad \sigma_y < \sigma_C \\ 1 - 2\dfrac{\sigma_U - \sigma_y}{\sigma_U - \sigma_L}, & \text{for} \quad \sigma_C \leq \sigma_y < \sigma_U \\ 1, & \text{for} \quad \sigma_U \leq \sigma_y \end{cases} \tag{7.49}$$

$$R^* = \Pi\left(\sum \leq \sigma_y\right) = \begin{cases} 0, & \text{for} \quad \sigma_y < \sigma_L \\ 2\dfrac{\sigma_y - \sigma_L}{\sigma_U - \sigma_L}, & \text{for} \quad \sigma_L \leq \sigma_y < \sigma_C \\ 1, & \text{for} \quad \sigma_C \leq \sigma_y \end{cases} \tag{7.50}$$

Figure 7.7 depicts the possibility and necessity functions for this case. In light of eqs. (7.31) and (7.32) we can calculate the abscissas of centers of gravity for necessity and possibility function

$$\tilde{G}_{nec} = (\sigma_L + 5\sigma_U)/6, \qquad \tilde{G}_{pos} = (2\sigma_L + \sigma_U)/3 \tag{7.51}$$

It is instructive to evaluate how much far spread are the values of the possibility and the necessity functions from their centers of gravity $\tilde{G}$. To do this we define the moments of inertia of the hatched areas of possibility and necessity functions with respect of their vertical central axes y_Π and y_N, respectively, as shown in Figs. 7.7, 7.8b–c:

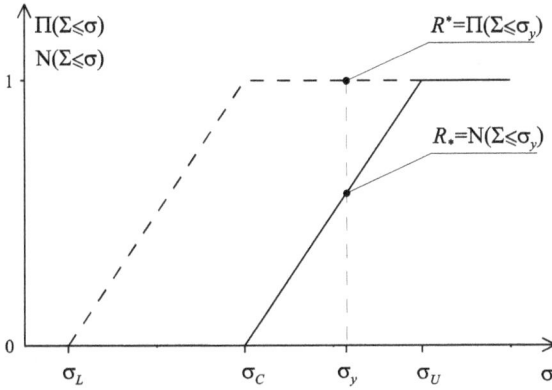

Fig. 7.7: Triangular membership function of stress $\sum$: possibility $\Pi(\sum \leq \sigma)$ and necessity $N(\sum \leq \sigma)$ of fulfilling the inequality $\sum \leq \sigma$.

$$\tilde{I}_{pos} = \int_{\sigma_L}^{\sigma_C} [\sigma - \tilde{G}_{pos}]^2 \Pi\left(\sum \leq \sigma\right) d\sigma \tag{7.52}$$

$$\tilde{I}_{nec} = \int_{\sigma_C}^{\sigma_U} [\sigma - \tilde{G}_{nec}]^2 N\left(\sum \leq \sigma\right) d\sigma \tag{7.53}$$

In the case of triangular membership function μ_Σ the two central moments of inertia read:

$$\tilde{I}_{nec} = (\sigma_C - \sigma_L)^3/36, \quad \tilde{I}_{pos} = (\sigma_U - \sigma_C)^3/36 \tag{7.54}$$

When $\sigma_C - \sigma_L = \sigma_U - \sigma_C$, or σ_C is placed in the middle of σ_L and σ_U we have

$$\tilde{I} = \tilde{I}_{nec} = \tilde{I}_{pos} = (\sigma_U - \sigma_L)^3/288 \tag{7.55}$$

Note that in contrast to probabilistic case the second central moments include the third power of the stress. This is because we calculate the moments of inertia of possibility and necessity functions that are somewhat analogous to the probabilistic cumulative distribution function or associated reliability and rather than to the probabilistic density function. For other approaches to the concepts of fuzzy variance the reader can consult the monograph by and Möller and Beer (2004).

We also introduce the coefficient of variation associated with possibility and necessity, respectively, as

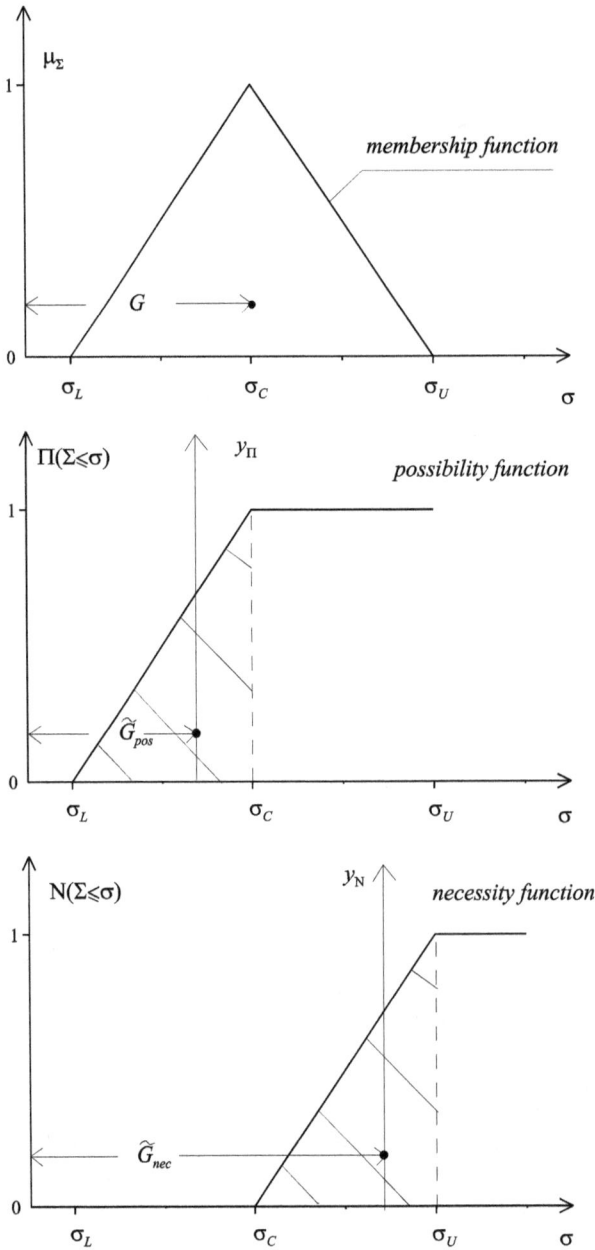

Fig. 7.8: (a) Triangular membership function of fuzzy stress $\sum$, (b) Possibility of fulfilling a specified mission $\sum \leq \sigma$, (c) Necessity of fulfilling a specified mission $\sum \leq \sigma$.

$$v_{nec} = \sqrt[3]{\tilde{I}_{nec}/G_{nec}}; \quad v_{pos} = \sqrt[3]{\tilde{I}_{pos}/G_{pos}} \tag{7.56}$$

From eq. (7.55) we first express the denominator in eq. (7.49)

$$\sigma_U - \sigma_L = 2\sqrt[3]{36\tilde{I}} \tag{7.57}$$

Let us express σ_L and σ_U via necessity measure. Then from eqs. (7.51) and (7.55) we get

$$\sigma_L = \tilde{G}_{nec} - (5/3)\sqrt[3]{36\tilde{I}}, \quad \sigma_U = \tilde{G}_{nec} + (1/3)\sqrt[3]{36\tilde{I}} \tag{7.58}$$

We also let $\sigma_C = G(\sum) \le \sigma_y < \sigma_U$. In accordance with eq. (7.49), we obtain

$$R_{nec} = 1 - \frac{\tilde{G}_{nec} + 1/3\sqrt[3]{36\tilde{I}} - \sigma_y}{\sqrt[3]{36\tilde{I}}}, \quad for \quad \sigma_C \le \sigma_y \le \sigma_U \tag{7.59}$$

By dividing both numerator and denominator by the value $\tilde{G}_{nec}$, and introducing the coefficient of variation v_{nec} of the necessity, we get, instead of eq. (7.59)

$$R_* = R_{nec} = \frac{2}{3} + \frac{S_{nec} - 1}{v_{nec}\sqrt[3]{36}}, \quad for \quad \sigma_C \le \sigma_y \le \sigma_U \tag{7.60}$$

In eq. (7.60) the parameters s_{nec} and v_{nec} are independent of each other. As is seen the lower bound of reliability R_{nec} is *directly* expressed in terms of the lower bound of central fuzzy safety factor s_{nec} and the coefficient of variation v_{nec} of the involved fuzzy number. The lower bound of the central fuzzy safety factor s_{nec} corresponding to the required reliability r_{nec} is obtained from eq. (7.60) by putting $R_{nec} = r_{nec}$:

$$s_* = s_{nec} = 1 + v_{nec}\sqrt[3]{36}[r_{nec} - (2/3)], \quad for \quad \sigma_C \le \sigma_y \le \sigma_U \tag{7.61}$$

When the lower bound of reliability satisfies the inequality $(2/3) \le R_* \le 1$, the lower safety factor exceeds unity, and its maximum value

$$s_{* \, max} = 1 + v_{nec}\sqrt[3]{4/3} \tag{7.62}$$

is achieved when the lower bound of reliability tends to unity, from below. For example, for coefficient of variation 0.05 the maximum value of the lower bound of safety factor assumes the value 1.055; for the coefficient of variation 0.1 the maximum value of s_* equals 1.11; for coefficient of variation 0.15 it takes a value 1.165 etc.

In the same manner we find a relationship between the upper bound of reliability R^* and the upper bounds of the central fuzzy safety factor s^*. From eqs. (7.51) and (7.55) we have

$$\sigma_L = \tilde{G}_{pos} - (2/3)\sqrt[3]{36\tilde{I}} \quad \sigma_U = \tilde{G}_{pos} + (4/3)\sqrt[3]{36\tilde{I}} \tag{7.63}$$

Let $\sigma_L \leq \sigma_y \leq \sigma_C$, in accordance with eq. (7.50), we arrive at

$$R_{pos} = \frac{\sigma_y - \tilde{G}_{pos} + 2/3\sqrt[3]{36\tilde{I}}}{\sqrt[3]{36\tilde{I}}}, \quad for \quad \sigma_L \leq \sigma_y \leq \sigma_C \tag{7.64}$$

By introducing the coefficient of variation v_{pos} of the actual stress, we get, instead of eq. (7.64)

$$R^* = R_{pos} = \frac{2}{3} + \frac{S_{pos} - 1}{v_{pos}\sqrt[3]{36}}, \quad for \quad \sigma_L \leq \sigma_y \leq \sigma_C \tag{7.65}$$

Note that s_{pos} and v_{pos} in eq. (7.65) are treated as independent of each other. As is seen the upper bound of reliability R_{pos} is *directly* expressed in terms of the upper bound of central fuzzy safety factor s_{pos} and the coefficient of variation of possibility function v_{pos}. The lower bound of the central fuzzy safety factor s_{pos} corresponding to the required reliability r_{pos} is expressed as follows

$$s^* = s_{pos} = 1 + v_{pos}\sqrt[3]{36}\left[r_{pos} - (2/3)\right], \quad for \quad \sigma_L \leq \sigma_y \leq \sigma_C \tag{7.66}$$

Obviously, the relationships in eqs. (7.61) and (7.66) are mathematically analogous, but they hold for different ranges of the strength σ_y. For sake of additional safety, we use the lower bound structure's reliability in the design process (see Fig. 7.9).

It appears instructive to compare the results for the safety factor obtained in the probabilistic setting first for the case of uniform density function, i.e., a linear cumulative distribution function with the result obtained within the fuzzy set theory for the case of triangular membership function. In both cases we established a remarkable relationship expressing the central safety factor as a linear function of reliability and an appropriately chosen coefficient of variation.

The structure of the formulas appears to be similar: The safety factor depends on the required reliability and the coefficient of variation in the probabilistic setting; in the fuzzy setting the safety factor depends on the value of the required level of necessity and coefficient of variation. For the probabilistic case we use the result Elishakoff (2004, eq. (7.8)) that for reliability greater than half we get a central safety factor greater than unity. For fuzzy sets-based treatment the lower bound of central safety factor is greater than unity only if the lower bound of reliability is bigger than a value $r_{nec} = 2/3$. This value is obtained from eq. (7.60) by setting $s_* = 1$. This value is independent of coefficient of variation v_{nec}. Thus, in order to have a safety factor larger than unity the required level of necessity has to be greater than its probabilistic counterpart, i.e., required reliability.

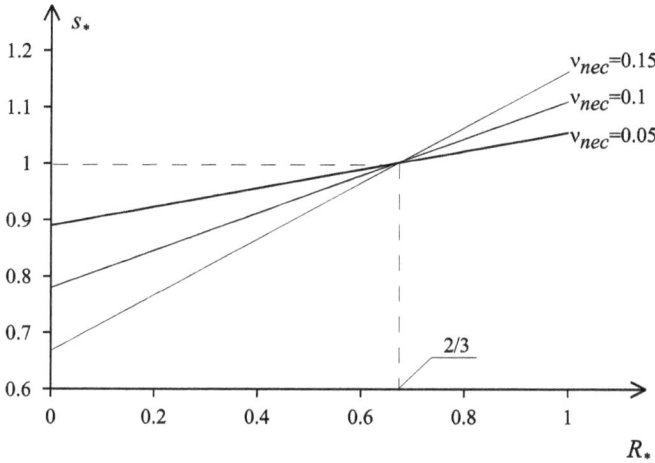

Fig. 7.9: Triangular membership function: relationship between lower bound of reliability R_* and the lower bound of central fuzzy safety factor s_* for different values of coefficient of variation $\Box_{nec}$.

7.7 Quadratic fuzzy stress

Let now us use another fuzzy number with the following membership function of the actual stress (see Fig. 7.10a)

$$
\mu_\Sigma(\sigma) = \begin{cases}
0, & \text{for} \quad \sigma < \sigma_L \\[2mm]
\dfrac{(\sigma - \sigma_L)^2}{(\sigma_U - \sigma_L)(\sigma_C - \sigma_L)}, & \text{for} \quad \sigma_L \le \sigma < \sigma_C \\[2mm]
\dfrac{(\sigma_U - \sigma)^2}{(\sigma_U - \sigma_L)(\sigma_U - \sigma_C)}, & \text{for} \quad \sigma_C \le \sigma < \sigma_U \\[2mm]
0, & \text{for} \quad \sigma_U \le \sigma
\end{cases}
\tag{7.67}
$$

For the sake of simplicity, we will concentrate on the particular case in which $\sigma_C = (\sigma_L + \sigma_U)/2$. We note that according to eq. (7.23) the abscissa of center of gravity of the stress equals

$$
G = (\sigma_L + \sigma_U)/2 = \sigma_C
\tag{7.68}
$$

In view of eqs. (7.5) and (7.6) we can write the lower and upper bounds of reliability as

$$
R_* = N\left(\sum \le \sigma_y\right) = \begin{cases}
0, & \text{for} \quad \sigma_y < \sigma_C \\[2mm]
1 - \dfrac{(\sigma_U - \sigma_y)^2}{2(\sigma_U - \sigma_L)^2}, & \text{for} \quad \sigma_C \le \sigma_y < \sigma_U \\[2mm]
1, & \text{for} \quad \sigma_U \le \sigma_y
\end{cases}
\tag{7.69}
$$

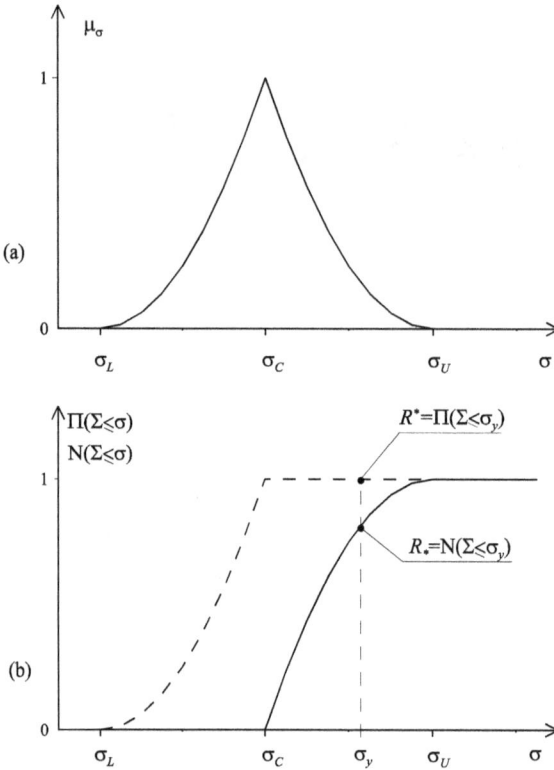

Fig. 7.10: (a) Quadratic membership function of stress $\sum$, (b) possibility $\Pi(\sum \leq \sigma)$ and necessity $N(\sum \leq \sigma)$ of fulfilling the inequality $\sum \leq \sigma$.

$$R^* = \Pi\left(\sum \leq \sigma_y\right) = \begin{cases} 0, & \text{for} \quad \sigma_y < \sigma_L \\ \dfrac{(\sigma_y - \sigma_L)^2}{2(\sigma_U - \sigma_L)^2}, & \text{for} \quad \sigma_L \leq \sigma_y < \sigma_C \\ 1, & \text{for} \quad \sigma_C \leq \sigma_y \end{cases} \tag{7.70}$$

Figure 7.10 depicts the necessity and possibility that inequality $\sum \leq \sigma$ is fulfilled as functions of σ; for $\sigma = \sigma_y$ necessity and possibility represent the lower and upper bounds of reliability.

From eqs. (7.31) and (7.32) we calculate the abscissas of centers of gravity of hatched areas of necessity (see Fig. 7.11) and possibility functions:

$$\tilde{G}_{nec} = \frac{9\sigma_C + 11\sigma_U}{20} = \frac{9\sigma_L + 31\sigma_U}{40}$$

$$\tilde{G}_{pos} = \frac{\sigma_L + 3\sigma_C}{4} = \frac{5\sigma_L + 3\sigma_U}{8} \tag{7.71}$$

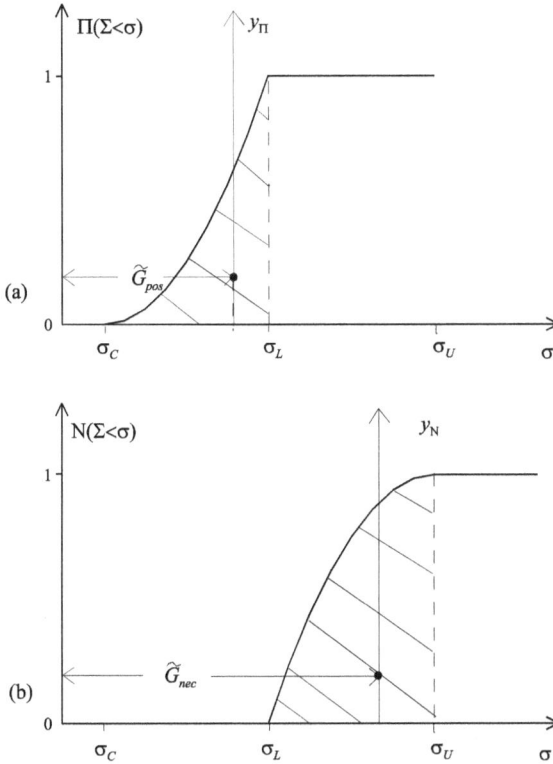

(a)

(b)

Fig. 7.11: Quadratic membership function: (a) possibility of event that the inequality $\sum \le \sigma$ is fulfilled, (b) necessity of event that the inequality $\sum \le \sigma$ is fulfilled.

The moments of inertia of the hatched areas of necessity and possibility functions respect to their vertical central axes y_N and y_Π as shown in Fig. 7.11 are:

$$\tilde{I}_{nec} = \frac{31}{3840}(\sigma_U - \sigma_L)^3, \quad \tilde{I}_{pos} = \frac{1}{1280}(\sigma_U - \sigma_L)^3 \tag{7.72}$$

Let us express the lower and upper bound of the support of the fuzzy stress in terms of necessity function. Then from eqs. (7.71) and (7.72) and we obtain,

$$\sigma_L = \tilde{G}_{nec} - \frac{31}{10}\sqrt[3]{\frac{60}{31}\tilde{I}_{nec}}, \quad \sigma_U = \tilde{G}_{nec} + \frac{9}{10}\sqrt[3]{\frac{60}{31}\tilde{I}_{nec}} \tag{7.73}$$

Let $\sigma_C = G(\sigma) \le \sigma_y < \sigma_U$. In accordance with eq. (7.69), we arrive at

$$R_{nec} = 1 - \frac{(\tilde{G}_{nec} + 9/10 \sqrt[3]{60/31 \tilde{I}_{nec}} - \sigma_y)^2}{32(60/31 \tilde{I}_{nec})^{2/3}}, \quad \text{for} \quad \sigma_C \le \sigma_y \le \sigma_U \qquad (7.74)$$

By dividing both the numerator and denominator by $\tilde{G}_{nec}$, and introducing the coefficient of variation v_{nec} of the necessity, we get, instead of eq. (7.74)

$$R_* = R_{nec} = 1 - \frac{(1 + 9/10 \sqrt[3]{60/31 v_{nec}} - s_{nec})^2}{v_{nec}^2 32(60/31)^{2/3}}, \quad \text{for} \quad \sigma_C \le \sigma_y \le \sigma_U \qquad (7.75)$$

It is remarkable that as in the case of the triangular membership function, here too the lower bound of reliability R_{nec} is *directly* expressed in terms of the lower bound of central fuzzy safety factor s_{nec} and the coefficient of variation of the involved fuzzy number v_{nec}. The lower bound of the central fuzzy safety factor s_{nec} corresponding to the required reliability $r_{nec.}$ is expressed as follows

$$s_* = s_{nec} = 1 + v_{nec} \sqrt[3]{60/31} \left[(9/10) - 4\sqrt{2}\sqrt{1 - r_{nec}} \right], \quad \text{for} \quad \sigma_C \le \sigma_y \le \sigma_U \qquad (7.76)$$

This relationship is depicted in Fig. 7.12. It is interesting to note that the maximum value of lower bound of central fuzzy safety factor is achieved when the lower bound of reliability equals unity.

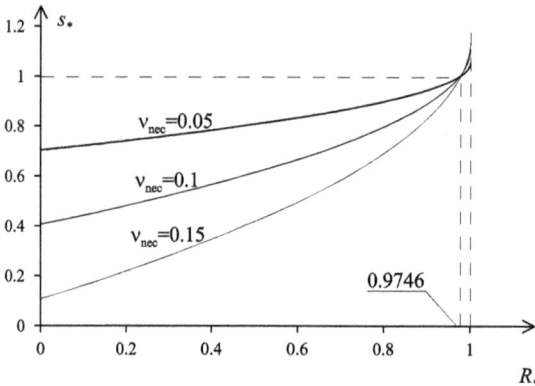

Fig. 7.12: Quadratic membership function: the lower bound of safety factor vs the lower bound of reliability for different values of v_{nec}.

Analogously, we find the relationship between the upper bound of central fuzzy safety factor and the upper bound of reliability

$$s^* = s_{pos} = 1 + v_{pos} \sqrt[3]{20} \left[4\sqrt{2r_{pos}} - 1.5 \right], \quad \text{for} \quad \sigma_L \le \sigma_y \le \sigma_C \qquad (7.77)$$

We compare the results obtained by using triangular density function, i.e., a quadratic cumulative distribution function with the result obtained with the fuzzy set approach for the case of quadratic membership function. In both cases we establish a *remarkable* relationship expressing the central safety factor as function of reliability and an appropriately chosen coefficient of variation.

For the probabilistic case of triangular density we establish in eq. (7.19) that for reliability greater than half we get a central safety factor greater than unity (see Fig. 7.2). For fuzzy sets based treatment the lower bound of central safety factor is greater than unity only if the lower bound of reliability is bigger than a value $r_{nec} = 0.9746$, obtainable from eq. (7.61) by setting $s_* = 1$. This value is independent of coefficient of variation v_{nec}. In a very narrow region for the lower bound of reliability $0.9746 \le R_* \le 1$, the lower safety factor exceeds unity, and attains a maximum value which equals

$$s_{*max} = 1 + (9/10)\sqrt[3]{60/31}\,v_{nec}. \qquad (7.78)$$

7.8 Conclusion

In this chapter, following the studies of Elishakoff and Ferracuta (2006a, 2006b), we defined the central *fuzzy safety factor* and introduce its lower and upper bounds. This is done in order to establish a relationship between bounds of probabilistic reliability and bounds of fuzzy central safety factor; moreover, we obtain the lower and upper bounds of probabilistic reliability as specified either for triangular or quadratic membership functions of the stress using possibility and necessity measures.

Some reasonable questions arise as a result of this study: *What to do if the possibility distributions differ from triangular and quadratic ones?* Indeed, such a question was posed by a colleague. Naturally, in order to come up with fuzzy safety factor one needs to have the information on the membership functions of input quantities. If the information differs from being a triangular or a quadratic function, then by following the general methodology, outlined in the present study, one can establish appropriate fuzzy safety factor. Therefore, the considered triangular and quadratic membership functions should be considered as methodical examples.

It is shown in this study that instead of *ad hoc* assignment of the safety factors one can analytically obtain its value once the probabilistic or fuzzy performance criteria are formulated. Further investigations appear to be needed to illustrate the superiority of the proposed approach to the present practice of *more or less arbitrary* assignment of safety factors.

Despite numerous papers that apply fuzzy sets to engineering problems, the apparently most important subject of the *fuzzy safety factor* has never been elucidated before papers by Elishakoff and Ferracuti (2006a, 2006b) and Savoia *et al.* (2005). It should be noted that many researchers who embrace fuzzy sets-based analysis criticize – sometimes harshly – probabilistic methods; but then they utilize probabilistic approaches to construct membership function(s)! Interested readers can consult paper by Civanlar and Trussell (1986) on "constructing membership functions using statistical data," published in the prestigious journal *Fuzzy Sets and Systems*. The interested reader can also consult the papers by Hasuike *et al.* (2015) and Gao *et al.* (2021), as representative examples. One would have preferred to read about constructing the membership functions from the raw data.

More general treatment of uncertainties is given by random sets (Goodman, 1982; Kendall, 1974; Dubois and Prade,1991; Matheron, 1995; Robbins, 1994, 1995; Bernardini, 1999, 2000; Tonon, 2004a, 2004b; Tonon and Bernardini, 1999; Tonon *et al.*, 1999, 2001; Molchanov, 2005; Fetz and Oberguggenberger, 2016). For structural engineering applications the interested reader can consult with definitive monograph of Bernardini and Tonon (2010). One can argue that it would be of interest to recast the safety factor notion also in terms of random sets. The same applies to the interesting possibility theory (Yager, 1982, 1986, 1987, 1991, 1993; Stroud *et al.*, 2002; Bae *et al.*, 2003, 2004a, 2004b; Penmesta and Grandhi, 2003), evidence theory (Soundappan *et al.*, 2004; Oberkampf and Helton, 2005) and the theory of grey numbers (Rao and Liu, 2017; Liu and Rao, 2018; Yang X. *et al.*, 2018). We would like to bring to the attention of the readers a definitive paper by Weichselberger (2000) provocatively titled "The Theory of Interval-Probability as a Unifying Concept for Uncertainty." It resonates well with the papers by Elishakoff and Colombi (1993a, 1993c) and by Zhu and Elishakoff (1996a, 1996b). It appears instructive to cite Zhang and Jiang (2021): "Evidence-theory-based structural reliability analysis (ETRA) has been gaining increasing attentions in practical engineering problems, with the continuous increase in complexity of modern engineering structures and our design requirements for reliability. This chapter reviews several main directions of evidence-theory-based reliability analysis, and it can be found that after decades of development, remarkable achievements have been obtained in areas of efficient reliability analysis, correlated reliability analysis, hybrid reliability analysis, and reliability-based design optimization." Interested reader can also consult with paper by Jiang *et al.* (2013b).

Finally, we would like to quote from Ross (2010, p.7): "In the coming years it will be the consequence of this isomorphism that will make fuzzy systems more and more popular as solution schemes, and it will make fuzzy systems theory a routine offering in the classroom as opposed to its previous status as a 'new, but curious technology.' Fuzzy systems, or whatever label scientists eventually come to call it in the future, will be a standard course in any science or engineering curriculum. It has all of what algebra has to offer, plus more, because it can handle all kinds of information not just numerical quantities."

Chapter 8
Possible limitations of probabilistic methods in engineering

Probability is the only satisfactory description of uncertainty. D.V. Lindley (1987)

My thesis, paradoxically, and a little provocatively, but nonetheless genuinely, is simply this:
PROBABILITY DOES NOT EXIST.
B. De Finetti (1974)

8.1 The idea of the poll

Denis Lindley's statement is shared by numerous researchers in uncertainty analysis of structures and not only structures. Still, the in words of Robert F. Nau (2001), "it is strange that the summary of a lifetime of work on the theory of X should begin by declaring that X does not exist, but so begins De Finetti's Theory of Probability (1970/ 1974)" with the above quote. Nau's (2001) conclusion is in the title of his paper, "De Finetti was right: probability does not exist."

During the years 1988–1996, the present writer conducted a survey among the professional colleagues on the titled topic. A questionnaire was distributed to numerous researchers around the world about the possible limitations of probabilistic methods and their applications in engineering. The questionnaire contained a single question requesting a response about possible, if any, limitations of probabilistic methods. Many of our colleagues replied, whereas some decided to decline the invitation to express their opinion on this topic. Since several years of the timespan are involved in conducting the survey, it is quite possible that some of the respondents have altered or changed their positions. Therefore, it was felt that the responses will be reproduced but the respondents' names will be listed separately, not immediately following the respective quotation. The main matter is the thoughts on the possible limitations of the probabilistic analysis. Yet some authors can be easily "detected" due to some reference(s), contents, or style. A total of 42 responses are reproduced here. It is hoped that these views will have a definitive impact on both the research and practice in the engineering field. Engineers and researchers who did not have a chance to participate in this survey, or in the past have declined the invitation to try to influence opinions of their peers, are kindly invited to try to influence opinions of their peers, are kindly invited to communicate their views, on this paramount topic in order to have the sequel of this overview to appear in several years.

https://doi.org/10.1515/9783111354231-008

8.2 Responses of the researchers

1

The question you have been bringing forward is constantly in my mind for a longer time already, perhaps in somewhat different shape. Obviously, the answer cannot be straightforward since it is necessary to consider the complexity of the problem. I don't believe that any single person can answer your question in full; therefore, your intentional enquiry is extremely valuable because it surely will collect many different, maybe even contradictory views. I would greatly appreciate getting some more detailed information about the results later. I suppose that the assembly of answers would be a first-class guideline for scholars and, simultaneously, an excellent opportunity for exchanging ideas. Well, here are my views:

(a) Firstly, we have to ask whether there are any limitations in the structural reliability methods (SRM's, I prefer this term over "probabilistic methods in structures," it is more general)? I think definitely yes.

(b) The SRM's form a close and limited set of methods and concepts, without any doubt, similarly as in other branches of technology (electrical, mechanical engineering, etc.), or in the very disciplines of the structural engineering (e.g., in structural mechanics).

(c) The limits of the SRM's have not yet been reached, however. In my opinion, two limits can be identified: (i) limit of the theoretical apparatus, outward limit, and (ii) limit of practical applications, inward limit.

(d) As far as the outward limit is concerned, we are now very close to it. The available SRM's are able to solve, with certain concessions to accuracy, any problem set by the current construction process. Of course, different methods used in specific cases may supply solutions which may differ in quantitative or even qualitative conclusions, but nevertheless, such solutions are now possible. That was not true before, say, 1980.

(e) On the other hand, we are still not yet able to describe the reliability state of the structure-load-environment reliability system in detail. We can establish failure probability of simple systems, or simplified systems. Board-minded simplifications are necessary, and in effect, the results obtained must be treated broad-mindedly. What we wait for, is something I call Finite Reliability Elements Method, FREM, which would make possible to describe the "reliability continuum," time behavior included, similarly as the FEM can describe the mechanical continuum (some work having been already under way). Powerful super-high-speed computers will be needed, such as are now occurring ("connection machine"). I don't believe there is anything fundamentally new which might be expected beyond this supposed outward limit.

(f) The inner space of the SRM's set is, however, very little explored and filled in, at the present time. I believe that only about 10% of possibilities available are materialized in practical applications and that a lot of space is still being left for the future. The main task here is the implementation of the accumulated theoretical knowledge. There is a large gap between the actual state of the theory and the current use of its conclusions. The values of various proportioning parameters (reliability coefficients, design strength, action combination coefficients, importance coefficients) are still empirical values, though some improvements have been achieved in the last decade. Testing and inspection are based on primitive approaches. The prime values of structural reliability (target values of failure probability, lifetime expectancy) are not yet established. Etc.

(g) To overcome the gap and to increase the implementation rate we need data (particularly on loads and environment), and active education of engineers in SRM's. If the situation in these areas improves, the inward limit will get closer.

(h) However, I do not believe in the current use of sophisticated design methods based on SRM's. The human mind has its limitations, too, and consequently, the direct application of SRM's in everyday practice is to be doubted. In civil engineering so many basically new technologies are arising every year that a simple following of the development becomes more and more difficult. How to ask, under such conditions, the Profession to accept SRM's as a direct tool? That is hardly possible. Therefore, any implementation efforts should concentrate on creating an acceptable "transfer function" between SRM's and current construction process (i.e., design, execution, testing, inspection, maintenance, and use).

Let me summarize:
(i) There are limitations to the structural reliability methods.
(ii) (ii) Outward and inward limits can be distinguished.
(iii) The SRM's are on the verge of arising at the outward limits.
(iv) Inward limits are still far from being reached in the nearest future.

2

A good way to visualize the rapid thinning out of the tails of the normal distribution is to consider a stationary normal process with zero mean. Consider the fraction of time such a process spends above various thresholds. For example, it is an easy exercise to show that the process spends:

16% of its time above the one-sigma level
2.3% of its time above the two-sigma level
0.13% of its time above the three-sigma level

The fraction of time above the threshold decreases dramatically as the threshold is raised further. My favorite example is for a threshold level slightly higher than ten-sigma. The fraction of time the process spends above this level is represented by the fraction below:

This shows that no one will ever live long enough to verify experimentally that any given random process is truly normal.

3

Concerning your question on the limits put on the use of probabilistic models in engineering, I would like to say, leaving aside all trivial and common-sense considerations, the following:

Engineers, especially those on the top, do not trust statements and "predictions" of probabilistic character. They must make decisions on important projects, large structures, and big investments, and they prefer to be completely sure that their decisions are true. Of course, a proper education in the theory of probability and mathematical statistics helps here, but there is a more profound, maybe subconscious case of such an attitude.

Most engineers are happier to look at the samples of the behavior of a system, at its time signatures than at probability density functions, and cumulative distribution functions that all seem alike to be practicing engineers. In addition, they become suspicious when a probabilistic talks about probabilities of failure, say and all that.

In my own experience, it is expedient to show to an engineer a set of samples of the system behavior, in particular the "worst" sample, the "best" one and an "average" or a typical one. If you tell the engineer that these samples are chosen, e.g., from 100 trials, he or she almost surely will say to you: "As I can judge, there is one chance from a hundred that my system will behave in such a bad manner." Moreover, if your prediction of sample behavior is reasonable, the engineer begins to believe you more and, maybe will associate "one chance from a hundred" with an estimate of the probability. In general, in my opinion, a representative set of time histories obtained with numerical simulation of the probabilistic model presents to engineers more information than the final results of statistical treatment.

4

The probabilistic approach demands a lot of input data that must be evaluated by statistical methods. The approach is quite clear for scientists and, therefore, they do not want to follow this long, tedious and miserable way. On the other hand, the practicing engineers demand that complete data from the theory. The final result is that the probabilistic methods have not yet been supported by sufficient experimental input data.

A classic example in this field is the theory of limit state for the design of engineering structures where some of the input coefficients are simply appreciated using an engineering feeling and not based on proper statistical observations and-or experiments.

Nevertheless, I hope that after evaluating all the answers of other colleagues your conclusions will not be too pessimistic.

5

There is no doubt about the power and beauty of the probabilistic approach to design . . . One of the main difficulties, however, has been the computations involved in conjunction with the extensive data required which has made the use of this approach rather unwarranted as compared to the simpler and cruder methods. These disadvantages are of course being reduced with the increasing use of digital computation in design and the progressive accumulation of relevant design data.

There is also another important problem I discussed in my paper to the 6th International Conference on Fracture in New Delhi in 1984 (reprint enclosed). This relates to not only the inevitable uncertainties in the design assumptions but to the occurrence of design errors (human error really). As shown in the paper there has been an improvement in this aspect over the years but there is major design discrepancies even in the more recent data from the aeronautical field where design calculations are very intensive. While this can be taken account of by using an appropriate probability distribution of design strength, the dispersion of results is so great that the resulting calculations by the reliability method show no advantage over the simpler design methods.

This led me to conclude in that paper that in view of the spasticity of the extensive data required and the probability of occurrence of design errors, limit state design (really a transition phase to the fully probabilistic approach) is a simpler and appropriate design procedure except in the aeronautical engineering field. In that field the overriding requirement for minimum weight is such that a great deal of analysis and computation is used in design and this trend is becoming rapidly more pronounced. My colleague Mr. Grandage and myself are both of the same view that in aeronautical engineering design the fully probabilistic approach should be progressively adopted.

These views were proposed in a paper I presented to the I.U.T.A.M. conference in Stockholm in 1984, at which I know you yourself presented a paper. In fact, in a meeting of the Scientific committee for that Conference a motion was passed that I.C.A.O. should be advised of the large number papers presented on Reliability and offering the co-operation of the Scientific Committee in testing a Reliability Approach, in parallel with existing methods, to the Airworthiness of oncoming aircraft types (see enclosure). Unfortunately, no positive action was taken by I.C.A.O. who rather side-stepped the issue by saying that the research sub-committee who would normally handle the matter had temporarily lapsed.

6

I represent the point of view that neither the probability of failure nor the reliability index β are adequate safety measures for structural design, because of the deterministic inference of inspection authorities. I have suggested two/or/three/separate safety indices: for resistance and for actions/also for model uncertainties. I called it "Conditional Certainty," i.e., the structural safety will be certain if the control is perfect. However, there are some problems when action combination rules are applied according to the level 2 design. Particular actions are controlled separately and the very rare case that their design values occur simultaneously is little probable but possible. On the other hand, the characteristic values accepted as the control values would give too conservative solutions. I try to give answer to this question in the paper presented during the Symposium in Lausanne, Switzerland, in July. Enclosed you will find a copy. Have I understood your question well? I am curious about the results of your enquiry.

7

The question you pose concerning possible limitations of probabilistic methods is very stimulating. I understand by limitations either (a) absolute impossibility to obtain satisfactory approximations to the solution; (b) temporary impossibility of this, considering present computational capabilities, or temporary disadvantage relative to at least one deterministic method. Let us take these three meanings one at a time.

Impossibility is relevant only insofar as the problem at hand could in principle be solved in a deterministic context but not when recognizing our uncertainties. (By " in principle" I mean given no constraints on computational capabilities.) This cannot happen. If we can solve a problem deterministically, we can always obtain a better probabilistic answer by using a Latin hypercube or simulation.

Using the same argument, we can equally discard the cases of temporal impossibility.

We are left with cases in which it is preferable at present to use a deterministic approach than a probabilistic one. By "preferable" we mean all implications considered including the (generalized costs of analysis and implementation, quality of approximations, and consequences thereof. (I say "at present" because future developments in computational capacity can favor the probabilistic approach in some instances). Here the answer is yes, indeed, probabilistic analysis can be excessively complicated to justify the increased accuracy or to make this type of analysis salable. I am assuming, of course, that when we talk of probabilistic methods, we have Bayesian methods in mind. We are then always open to objections in our choice of prior probabilities. I think, however, that such objections have been adequately answered by Bayesians and I find that, as a rule, there is no difficulty in convincing engineers of the validity of Bayesian statistics.

8

Fundamentally the final reliability estimate comes from a CDF function not too well known. Not only is a confidence level in the reliability desirable, but even then, it doesn't tell it all.

There are powerful stochastic influences at work "driving" additive combinations toward normality, and multiplicative combinations toward lognormality. How far the process has progressed and how close an additive or multiplicative association is to reality is another concern.

Fitting asymptotic distributions to stochastic samples is another source of error. Just because classical distributions are not truncated, doesn't mean a real phenomenon isn't.

A stream of small parts from an automatic screw machine will exhibit a diametral growth due to tool wear is nearly linear with part sequence number. Mix the parts and a claim for uniform random distribution can be argued. If the parts weigh thirty pounds then any mixing is less than thorough. What do you have compared to what you'd like to believe?

Data have been taken with bias, analyzed with bias, and used by others believing otherwise. There are continued presumptions of normality, even after transformations, which cloud data analysis results. There are undetected correlations of which the designer is unaware, and he acts as if independence is the rule of the day.

Engineers often dislike statistics because it can warn them not to believe what they really want to be so. The usual statistical remedy "take more data" is unwelcome, resisted, rationalized way.

9

Very briefly, I felt there are two areas of limitation for the application of probabilistic methods to structures, and one serious aspect of deficiency. One area of limitation is that failures generally occur due to some aspect of human error, beyond the normal realm of probabilistic modeling. A second limitation is the interdiction of human modification. By this I mean the control of live loads (on both buildings and structures), the tendency to reconfigure buildings, etc. The deficiency to which I refer is the lack of sufficient data in the extremes of occurrence that control reliability.

Of course, one could argue that the probabilistic framework, within its Bayesian and stochastic capabilities, could handle all of the above problems, as well as such other computationally intensive considerations as load path dependency and structural deterioration.

10

The limitations stem from lack of physical data and lack of understanding by the profession.

11

I have long been a skeptic of using probabilistic methods for structural design (not analysis) simply because of the lack of information regarding the probabilistic models and values of the parameters. As far as analysis is concerned, the lack of information of course stills our ability to predict as well. The argument given was the same as when Dar Kiureghian said is commenting on your remarks. That is design is a decision process which must be made even in the face of uncertainty and lack of information. If you look at it this way, it sounds reasonable and whether one is or not, one is forced to make the decision, right or wrong "rational" or "irrational" and the argument that probabilistic is better than deterministic or not can go on and on forever. In a nutshell, I remain a skeptic and pessimist and at a loss.

12

In regard to your recent questions on applications of probabilistic methods and their limitations, I wish to respond based on recent work done for the offshore oil industry on platforms and also highway bridge structures. A proposed code we developed for reliability based offshore platforms is approaching adoption. Code for reliability-based strength evaluation of existing bridges and also for safe remaining life (fatigue) are also nearing acceptance.

As a very quick reply to your question on limitations, I would indicate the following:
(a) The probabilistic approach may be more profitable initially for evaluating existing structures rather than new designs.
(b) Limited data makes engineers hesitant to adopt probabilistic methods when they may have difficulty defending new design concepts.

(c) Lack of knowledge of modern probability-based design approaches by engineers is of major importance.
(d) Lack of incentives for better utilization of resources in design, inspection, evaluation, etc. is important.

Overall, I think progress is being made – recent adoption of probability in buildings, bridges, offshore structures, etc. – but naturally, it is slower than expected.

13

All processes occurring in mechanical systems-as in inanimate Nature in general involve accidental deviations. In these circumstances a deterministic approach, appropriate for mean values, is insufficient for an exact description.

The difficulties inherent in applying a statistical approach consist in the following:

(a) Examined quantities specifically loads may be variously spread over different intervals. A distribution of peak values may differ from that of its frequently occurring (close to the mean) counterparts. The two distributions are necessarily related, but the actual relationship is not known because of the shortage of experimental peak data, peaks being infrequent by definition.
(b) In general, the volume of experimental information is patently inadequate. Available publications are basically concerned with theoretical problems, and in the absence of experimental grounding many of them lack any practical significance.
(c) Stochastic processes may change with time, and the variation pattern of their parameters is often unknown. Thus, for example, a fast railroad in Japan exhibited rapid increase of carloads, which in turn raised doubts regarding the bridges which had been designed for smaller loads. The same may occur with cranes, where the activity served by them may be intensified after they have been put into operations with the attendant changes in mean values and spreads.
(d) The principal shortcoming of modern statistical analysis of mechanical systems is its total disregard of practical applicability. The engineer is urged to replace the arbitrary safety factor by the concept of breakdown (or failure) probability, without any clue as to what values should be assumed for the latter.
(e) An answer to this question can only be given on the basis of the economic consequences of a breakdown or failure. This is especially important with new types of structures like nuclear power plants and offshore oil rigs, where an intuitive approach is useless in the absence of experience.
(f) An economically optimal solution necessitates knowledge of the weight and cast functions of the structure, as well as of data on exploitation expenses. Regrettably there has been no follow-up on this aspect over the last hundred years.

Accidental deviations can be minimized with the aid of active control, but that is a different aspect.

14

There can be no doubt that probabilistic methods must play a very important role in the future in assessing structures of all kinds. In fact, whether one considers wind or tides, or earthquakes or man-made loadings imposed on structures, in the end a failure is a matter of probability, and perhaps the most important engineering question to be asked and answered is "What is the probability that a given structure exposed to a given environment will survive for a given length of time?."

15

I have always been somewhat skeptical about the serviceability of the probabilistic approach, not because I doubt the reality of the basic principles and methods but because I believe the numerical values of many parameters, that have to be introduced into the calculations, are not reliable enough. Reliability will probably increase as time goes by, when the parameters, or rather their numerical values, become gradually better known. Admittedly, I am not very familiar personally with probabilistic methods and my relative skepticism may be partly due to that lack of conversancy.

16

Since I myself didn't work in structural design I have only indirect experience in this field. Generally speaking, in the problems I had to solve, the stochastic models were quite satisfactory, with one exception: the generation of a non-stationary random process in order to stimulate ground motion. I felt that in this direction (stochastic modeling) has not yielded everything.

17

As the mathematical treatment of structural reliability has evolved up to a very advanced state, this could give the wrong impression that we master safety within very narrow limits. However, as the evaluation of the obtained results concerns, one has to be aware of the inherent limitations by which these methods are conditioned. These limitations follow from:

- The degree of sophistication which can be achieved in the elaboration of the physical model and the probabilistic formulation.
- The often-considerable statistical uncertainties related to parameter estimation, due to the limited number of real-life data.
- Human errors, which actually are one of the main causes of structural failures, hence the importance of quality assurance in building practice.

As concrete strength occurs, I have made some contributions to the elaboration of more realistic probabilistic models in my doctoral thesis. Particular emphasis was given to the influence of autocorrelation on compliance control.

18

My personal view is that there are no real "limitations." It is always better to make a decision based on a probabilistic approach than on a deterministic approach. But of course, there are some areas where further research is needed, in order to improve the quality of the probabilistic approach. Some of these areas are
1. statistical data, especially on correlation,
2. computing time for complex problems,
3. assessing monetary values to intangibilities like human lives, natural and cultural values, and so on.

19

As I see it, there are four principal limitations to the power of probabilistic methods in structures (i.e., in the design of successful, economical structures and in the analysis of the risk of malperformance):

(a) Probabilistic theory applies to only one kind of uncertainty: Statistical. There are three other kinds, important in structures: Vagueness (e.g., what is "unacceptable vibration"?), lack of data, and conflict (e.g., deliberate attack). I place them all four at the corners of a tetrahedron; fuzzy methods, Bayesian methods, classical statistics, and conflict analysis each are disciplines that apply to solve problems purely in one corner of the tetrahedron. Real world problems, needless to say, lie somewhere inside the tetrahedron, and we have no integrated rationale for their analysis.

These four theories are complementary, but each is limited to a specific pure task. Moreover, each discipline has its own orientation that may not be relevant to the task at hand. For example, conflict analysis does not permit us to predict (i.e., assign a measure to) the event that reactor containment vessel will be breached in its lifetime by sabotage.

(b) Probabilistic theory in its present form does not account for the most complex components of the system: the humans who design, build, monitor, use, and maintain the structure, constantly interacting with it. Five out of six failures escape the trap of probabilistic theory (which may well be an optimal proportion).

(c) Probabilistic theory is intrinsically incomplete and must remain so. It cannot account for the failure modes that we cannot imagine Tacoma Narrows provides a familiar example.

(d) Probabilistic theory is just a minor refinement of a minor subprocess of structural production. All it does is guide the lily of the safety of structures which is a parenthetical attribute. "The safety of structures is essential and otherwise unimportant" was how Hardy Cross put it. Even if probabilistic theory were perfect, even if we could predict which structures were going to fail, we would not have a marked influence on public safety or on the cost of structures. Humility is called for, in the extreme.

20

I think the major effort in probabilistic methods is directed towards:

(a) A better knowledge of load and resistance parameters (material properties, load surveys, tests).
(b) Reliability analysis methods to calculate reliability for given parameters and limit states.

But the limits are due to:
(a) Human errors, the variation because *HE* exceeds the variation due to other factors. Because of *HE* there is a limit on maximum reliability. Models for *HE*, in design and construction, use of past experience, expert systems, and how to incorporate those models in design codes; these are issues for the next decade of research.
(b) Definition of limit states, in particular serviceability limit states. Experts do not agree as to what is acceptable and what is not, and limit state functions are the basis for reliability analysis. This is very clear in the development of reliability models for bridges.

Otherwise, the present theory of reliability offers very efficient procedures.

21

Based on my experience in the problem, I can inform you as follows:
(a) The universally accepted and semi-probabilistic approach in structural design of buildings, in which the probabilistic aspects are treated in defining the characteristic values of the loads and material strengths, and these are in turn associated with partial factors of safety behavior, accuracy, etc. does not make full allowance for the probabilistic nature of reliability of many structures. Under the traditional approach no stochastic functions of structural failure are considered, and only the overall variation of material strengths and of some loads (but not their interactions) are partially considered.
(b) Under the traditional approaches to design, it is impossible to obtain a structure with predetermined reliability, or different structures with the same reliability. At the same time, direct use of stochastic functions for evaluating structure reliability at high levels is impracticable not only because of uncertain data regarding small probabilities of the initial parameters. Safety factors should cover the possible consequences of failure caused by non-predictable systematic and, in particular, gross errors in design and construction. Gross errors have hitherto not been amenable to statistical description, and all decision-making in the problem is inevitably based on practical construction experience.
(c) The probabilistic approach to analysis of structural safety may serve for general or specific scientific purposes, or for some characteristic estimations of structural reliability, but all these are subject to available reliable initial data.
(d) Probability serviceability analysis of building structures may be applied for many cases in practice but calls for supplementation of the codes by suitable statistical characteristics of loads and materials and by probabilistic restrictions.

22

The probabilistic approach is a step forward from the deterministic approach, which ignored the possible variations in the values of the parameters used. The deterministic approach assumed all parameters to be fixed at its mean value and tried to cover for any variation by using the factor of safety approach.

In the probabilistic approach the variations in the parameters are recognized and used as statistical description in the study. The complete description would require the join and cross parameters involved, of all possible hierarchy levels, or some other equivalent characteristics. Generating this information would usually mean immense effort in terms of experimentation and testing of large numbers of samples and may prove to be prohibitive for most problems. We are able to use realistic assumption-like independence of many parameters to somewhat reduce the effort. We also make some other assumptions like ergodicity of the process; it is fitting the Gaussian, Poisson or some such other known description and behaving as a white noise, Markoff process, etc. for further simplifications.

Most of the time these simplifications are a great help in reducing the efforts in terms of calculations inputs to obtain a usable result. But in many cases, some of these assumptions may not be very accurate. This would lead to inaccurate results. In my opinion the source of error in this approach is the inaccurate probabilistic description of the parameters used.

The designer and the analyst usually have no means at hand to accurately establish the probabilistic description of any parameter. They have to depend mostly on published literature for this information. It would be a great help if there is a statistical evaluation and lasting cell to supply this information to them with proper verification.

23

It seems to me that probabilistic methods can be applied to the analysis of any static/ dynamic problem provided that probabilities are interpreted correctly, e.g., Bayesian or frequentist. Claims that one needs to use fuzzy sets to solve some cases have not been substantiated. I do not know of any example in which probability theory cannot be applied and fuzzy set theory is successful.

24

I am still continuing studies on the theory of stochastic dynamical systems and its application to engineering problems.

As far as the application of probabilistic methods to structure design, the meaningful situation is to take into account the following two phases:
(a) Random input-random response
(b) Nonlinearities
   (a) At the present stage, the input statistic is assumed to be the stationary Gaussian process. In the very near future, and extension of studies to the non-stationary random process is strongly expected.
   (b) A nonlinear characteristic exhibited in structure dynamics is limited to zero-memory type. However, bearing applications to structural systems in mind, hysteresis should be considered, which depends on the velocity. As far as nonlinearities of zero-memory type, the stochastic linearization technique is quite useful.

25

The "power and beauty of the probabilistic approach" is not under discussion: the problem is the key for entering this beautiful model, the data. In some fields of application, this lack of data leads to manipulation of the input. The output can then be forced to meet the assigned constraints and the procedure loses objectivity. In these fields alternative approaches are still to be investigated.

26

I can only comment that probabilistic methods in structural engineering will be restricted by the requirements that such methods be simple enough to be practical and that they not be based on models containing parameters whose values are not obtainable.

The introduction of load factors and strength reduction factors based on probabilistic notions into both concrete and steel codes has been the major contributions so far. Any additional applications will have to be comparable simplicity. In this regard, my ideas have not changed since my participation with [Fred] Freudenthal in the work of the Committee on Factors of Safety of ASCE about 40 years ago. That, of course, may be more a reflection of my senility than the soundness of my views.

27

Your question on possible limitations of probabilistic structural analysis is difficult to answer. One limitation, I think, stems from the fact that many practicing and design engineers are not well versed in probabilistic methods, which makes it difficult to be translated into practice. Another, I believe, has to do with the increasing mathematical sophistication we find in the publications. This only makes the gap wider. As I mentioned in my talk at Urbana last November, it is our responsibility to bridge this gap between theory and practice if we want to see a majority of our probabilistic results actually used in practice.

28

Yes, probabilistic methods have their limitations, just as do deterministic methods. These limitations arise from the extremely limited tools with which we must work, including the human brain.

At the moment, a recasting of the mathematical theory of probability is essential since the current theory does not contain a satisfactory paradigm of reality. And, oddly enough, today's problem arises from non-essential theoretical abstractions that, for example, conflict with Bayesian ideas for no good reason. And Bayesian concepts as well contain extraneous limitations.

So, to me, the problem of the moment is to return to the absolutely essential ideas that allow probabilistic ideas to be described and manipulated. Then, recast the entire theory without restriction. With this start most of the limitations will vanish, at least for our times.

29

I think I am not ready to reply to your question about "limitations." The question is either too shallow (e.g., as compared to what alternative?) or too deep (what is probability?) to be answered lightly, quickly, or even at all, at least without a much more specific context. What do you have in mind?

30

The true advantages and limitations of the probabilistic methods in structural mechanics need serious reflection.

In spite of the questions of purely practical nature (e.g., proper interpretation of events of very small probability, the value of probabilistic reasoning in some static problems etc.) there are more fundamental questions associated with the relationship between real world and the language of probability theory.

Probability theory deals with mathematical models of randomness. But the first question which arises is: what does it mean to be random? Most likely, phenomena or events do not have probabilities; only people speak about probabilities. From this reason a probability is also deemed as a measure of someone's degree of belief in an outcome. Of course, stating such questions we are entering into philosophy. But this is inevitable.

31

I personally do not see any contradiction between probabilistic and deterministic analysis, i.e., modeling. It is just a matter of information processing.

32

(a) The strongest limitation, in my opinion, is the problem that one wants to extend the method to extreme case events. Gauss which is very good for events with random error up to 1 to 2 or physical limitations for force β. What is the physical upper limit?

(b) The problem of associating with a socially acceptable (for decision means or) value system. Engineers in this sense are persons who have to accept the responsibility for their work: for them is not necessary, a safety factor.

33

First, I did write a paper in the ACI Journal entitled "Limitations in Application of Probabilistic Concepts" (Sexsmith and Nelson, ACI Journal, October 1969, No. 10, Proceedings, Vol. 66). On rereading it, it's amazing to me that not much has changed. I therefore refer you to that article for most of my comments.

My thoughts on current limitations/difficulties are:

(a) How to use the results. We need better understanding of consequences and decisions under certainty. It is not only a matter of rational decisions, in fact it often reduces to political or societal decisions, and they (society) are not well informed of comparative risk, thus they don't act rationally; yet as engineers we can't play God and usurp society decisions. There are a lot of interesting problems now regarding how much seismic retrofit we should have, how much pollution should be reduced etc.'

(b) There are still a lot of modeling problems and math problems, but generally the math problems are starting to be solvable in realistic models.'

(c) Our field needs some better unity in philosophy. Many mathematical exercises are done by people who have little interest in the philosophy of engineering design and construction thus they solve problems that don't need solving, while there are many real problems.

34

I am now responding to your enquiry about limitations. My first reaction is to see your enquiry as impossibly wide: probabilistic theory is useful in so many ways that it is useful, rather than to catalogue the many more ways that it isn't. I shall respond therefore only in this way, and only in regard to my own area of interest.

My particular interests have for a long time been concerned with reasonably detailed dynamic response statistics, for which the establishing of spectra forms an ideal basis. The application of spectral theory to the response of linear systems under stationary Gaussian random excitations is at the root of the whole problem, and within these restrictions it is resoundingly successful. Of course, no system is precisely linear, and no random process is precisely stationary or Gaussian, and you may regard this as a limitation. I prefer to assume that engineers have their own way of deciding whether a real system is linear or not: if you assume that it is, and get reasonable results, then it is. The situation that arises when such idealizations are not suitable must be dealt with according to the particular need.

Extension to multivariate problems proves surprisingly simple. Matrix algebra makes light of analytical treatment, and even in practical computation the need to compute a host of direct-and cross-spectral densities is balanced by the relative efficiency of computing spectra in mass. But of course, if there are other complications also, the existence of many variables makes matters much worse.

For nonlinear systems spectral descriptions are less attractive. Because responses cannot be expected to be Gaussian, spectral densities no longer provide a complete description of a response, and in any case, spectra are much more difficult to predict. Mean square responses come easily through Fokker-Planck, but even these can sometimes give a misleading impression of what is happening. Jump phenomena can occur even with random excitation and the Fokker-Planck result gives no inkling of this. Higher order spectra are sometimes useful, but the computing requirements are very demanding, and the spectra are difficult to interpret. Non- Gaussian processes can occur in the absence of explicit nonlinearity and the same difficulties of description prevail.

With nonstationary processes I have found that progress can only be made where the non-stationarity takes a reasonably simple form. Response descriptions tend to be incomplete, and computing facilities need to be extensive.

Finally, but exceedingly important, there is a fundamental difficulty which arises wherever practical data is to be turned into spectral descriptions. An ideal mathematical model can usefully represent a real system only over a limited range of values, and over a limited period of time. To compute a reliable spectral density even for a near-ideal random process requires a very large number of sample realizations to enable adequate smoothing to be applied. Even where a large number of samples are available, they may not be consistent enough to justify the assumption of stationarity. And a sufficiently large number of records may simply not be available. In such cases the usefulness of the spectra obtained will depend on the nature of the information required. Certainly, a spectrum based on a single realization is likely to be quite useless.

To me one of the attractions of the study of random vibration is its affinity with real life. I expect the response to your inquiry will be truly random process, and correspondingly difficult to analyze.

35

Probability measures of the phenomena require the definition of a sample description space, the set of all possible outcomes. In engineering design, or more generally to forecast the outcome of phenomena modeled by deterministic physical laws but dependent by variables with uncertain values, the application of the theory of random variables and stochastic processes leads to a measure of probability. This measure can be employed as a guideline in the design choices, on the basis of the frequentist interpretation of the probability (the risk is assumed equal to the consequences of the outcome multiplied by its probability measure).

In fact, in the field of complex phenomena, such as the overcoming of ultimate or serviceability limit states of structural systems, the integral application of this methodology is rare (however the probabilistic methods are frequently used to justify or test design rules or codes based on the systematic use of global or partial safety factors).

The main difficulties to more extensive applications derive from the lack of reliable statistical data to evaluate the joint probability distributions of the variables, and moreover from the weight of the numerical or analytical procedures in the case of a large number of variables and/or limit states to be considered.

But a more drastic limitation seems to me deriving in some cases, when the limit state criterion or the measure of the performance of the system cannot be described as a crisp set of elementary, well scattered events.

As an example, I mention the case of seismic vulnerability analyses or of the damage forecasting of building structural systems. The levels of damage are more related to some particular patterns in the cracks or yielding distributions of the different structural components than to a well-defined set of response parameters.

A posteriori, in the post-earthquake survey of the buildings, the classifications of damage level should be performed by expert people using also qualitative and subjective judgments (the classification is generally expressed by a linguistic variable or a conventionally correspondent value of a discrete variable from 0 to 1).

A priori, in the forecasting the expected damages of future earthquakes, it seems to be meaningless a measure of probability of the different levels of damage, also in the case when some of the involved variables (for example the ground motion) could realistically be modeled by random variables or processes.

36

To your question concerning possible limitations of the probabilistic methods in structural analysis in brief,

– see argumentation with respect to fuzzy sets,
– how to determine joint pdf's if the sample size in practice is very small and if there is not sufficient *a priori* knowledge.

37

As far as I understood your letter, you'd like us to mention something about we think the possible limitations of the probabilistic approach to the design of structures may be.

Since this is a very complex subject I hesitate to reply in short form. On the other hand, I am not regarding myself competent enough to cover the question of your letter (though nowhere explicitly posed) in full sense. But because you seem to appreciate any "input" let me tell you in brief these are my personal opinion on some aspects of the subject that I have gained through the last years' experience in the reliability assessment of nuclear power plant components.

In principle there are no limitations to the probabilistic approach but only to the deterministic approach.

The probabilistic approach should be beneficial for almost every structure, but its application is necessary only for high-risk structures if we take safety as a primary objective function.

Regarding such structures we have two problems: the adequacy of the mathematical models for the complex reality and the lack of appropriate statistical input data in most cases (which you know probably better than I do). This data lack is often virtually not overcome.

Consequently, reliability figures are only valid as the underlying models and data are. I admit that this is a triviality although perhaps still worthwhile to mention with respect to some overly enthusiastic engineers. There is still a danger of discrediting the probabilistic approach by insensible usage.

Due to the often statistically poor input data one cannot hope for absolute reliability figures but instead at least gain important insight about the interdependence of design, operating and maintenance parameters in a more than qualitative sense. Careful sensitivity studies may permit the ranking among alternatives.

With respect to the design process and complex structures I am afraid the state of the art in the probabilistic approach is not so far developed that it could be used efficiently in an optimization process with the costs as the primary objective function.

The situation concerning most of these limitations will improve perhaps slowly but steadily due to the common research activities of a lot of scientists all over the world. There is still a need for refined methods and especially for sufficient data.

38

Quite likely it concerned value theory since structural safety involves a choice among alternatives which must involve value. Probabilistic methods have not been routinely applied in structural practice as a result of this problem. Codes are legal documents that do not include the theory of values.

39

For your consideration and possible use, I am listing several limitations of probabilistic method in structural engineering as follows:
1. The event of structural failure or damage is not clearly defined (except for total collapse).
2. Usually, the "exact" mathematical representations of the limit states of complex structural systems are not known.
3. The tail ends of distribution functions cannot be ascertained statistically (not practical to have sufficiently large samples). The probability of failure is highly sensitive to probability distributions.

40

In itself the concept of probabilistic reliability theory is value free in its being a rational (that is, logically appreciative) argumentative model for analyzing structural safety problems. Therefore, the question of advantages or drawbacks cannot be answered as seen for the theory itself. The theory provides a tool for rational conglomeration of the information available for structural decision making. Thus, the safety measure defined and calculated by use of concepts and methods from the mathematical probability theory is essentially a measure of the information about the relevant structural situation with respect to reliability, and not a pure physical measure that exists objectively and independently of the information situation.

According to Hasofer (1984) (A.M. Hasofer in Symposium on Structural Technology and Risk (2nd, 1983, Univ. of Waterloo), M. Grigoriu, editor: Risk, Structural Engineering and Human Error, Univ. of Waterloo Press, Waterloo, Ontario, Canada, 1984) the reliability measure is a metaphysical quantity because it cannot be measured by any physical device. However, the measure makes sense as derived by anticipatory logical modeling on the basis of the rules and laws of geometry, physics, and probabilistic theory. As demonstrated by Hasofer (1984), the probability of failure can in principle be eliminated from the final decision formula such that it contains solely measured quantities, these measurements being actually made by a physical device or just professionally assessed by judgment. Therefore, there is no need for an interpretation except that probabilities are weights that by anticipatory rational modeling are put on the different possibilities to reflect the information about the likelihood of their relative occurrence within the set of relevant possibilities.

Most importantly, the theory of probability describes the laws of large numbers (in terms of concepts of convergence of sequences of random variables) and thereby it provides a help to a mental appreciation of the concept of probability. This also makes up the basis for computerized simulation studies.

To answer the posed question, one therefore must look at the theory from the outside. From an application point of view as seen by the day-to-day practicing engineer, the theory as a practical tool has the obvious drawback relative to simpler minded and less structured considerations that it requires more learning and thinking to appreciate the meaning of it. The practical calculations, however, should be a minor obstacle with the generally available program packages nowadays. Also, the theory can be used, and has been used, as the basis for making operational codes of practice and for setting the parameter values in such codes (safety factors, etc.)

What concerns comparisons with alternative frameworks for reliability assessment such as exemplified by the fuzzy set concept, the question can hardly be answered by anything else than it is a matter of taste and what can be accepted with respect to reasonability. What is considered as a satisfying information basis for a decision maker is up to the decision maker himself. If he thinks he gets rational and genuine information out of fuzzy set manipulations, then he is entitled under responsibility to make decisions on this basis. Fuzzy set theory may in the mind of the decision maker solve information processing problems. He may, however, be in a situation where he is forced to explain his decisions and their basis. Questions about the predictive power of the results of fuzzy set manipulation outside the set of problems to which they have been adapted (perhaps, adapted with purpose to give those answers that are wanted or are convenient) may then be fatal for the respectability of the reliability analysis, even though it is claimed to be based on scientific principles. The sense of explaining, for example, the laws of large numbers (prudently admitting that this proposition is to be the best knowledge of the writer, who honestly cannot claim to be familiar with more than the basic notions of fuzzy set theory which, by the way, also seems to be the case for those engineers that publish about applications of fuzzy set concepts to structural safety analysis). It seems that the fuzzy set concept and its associate, fuzzy logic, merely acts as an adaptive language that in some technical matters such as controllers and operators can be quite useful and effective.

The conclusion drawn by the writer is that reliability theory based on the probability concept (interpreted as a mathematical set measure supplemented by the concept of dependency and independency) is by far superior to the fuzzy set-based reliability theory both with respect to richness in structure and with respect to prediction abilities.

Another known alternative for reliability analysis is advocated and developed by the questioner. It is the deterministic convex modeling principle. Actually, this principle can be interpreted as a rather special case of probabilistic modeling. Assuming bounded convex supports of all the considered probability distributions and seeking optimal structural dimensions under the constraint of having zero failure probability in fact defines the problem solved by applying convey analysis. To the writer's opinion the application problem here is the arbitrariness of choosing the convex supports, then probabilistic assessments of this uncertainty should be introduced. Then the constraint of zero with respect to the failure event is maintained. This is not to say that convex modeling cannot be a useful tool in certain types of structural problems. For example, the application of convex modeling can be a way to come close to an optimally designed structure in a first calculation. Thereupon more refined probabilistic analysis might be relevant.

From a philosophical point of view the writer sees no drawback of probabilistic reliability analysis as a tool to rationalize the information treatment and to mentally interpret the results.

41

Most ill point of the probabilistic methods is the notion of white noise, which never occurs in practice. It is nice to compute with it, but it is dangerous to use it.

42

As you know, I am fully interested in probabilistic methods and probabilistic modeling, even if I focus more on the material scale than on the structural one. Of course, if materials and structures are often distinguished, the reason is more the field of abilities of researchers than scientific reasons. Some work on structures while others work on materials. In reality, the two fields deeply interact since a structure is made of materials and a material behavior is identified from a specimen whose mechanical response is itself that of a structure. These general comments are not new, but I think they are of specific interest in the probabilistic frame.

My research [is] devoted to the probabilistic modeling of materials with two aims, namely
–   to apply the results for predicting structural reliability,
–   to see how and when the materials response depends on the heterogeneities it contains.

It appears that in heterogeneous materials (concrete, ceramics, rocks, soils, woods), the heterogeneities play a great role on (and even drive) the response in the nonlinear domain and that accounting for their presence and influence may be a requirement when one wants to really understand and predict the material (thus the structural) response. Among the consequences, we have:
–   When one considers the structural response (we have recent interesting results on size-effects in glulam structural elements), modeling the local distribution of heterogeneities may be useful or even obligatory. We recently obtained some experimental results on the tensile failure of fiber reinforced concrete that contradict intuitive thinking and can be understood only after accounting for the material heterogeneity.

– The local material heterogeneity must be measured and modeled (for future simulations), and interactions between numerical model and material intrinsic model have to be studied (some problems of objectivity and mesh sensitivity are not straightforward).

Considering your question about the limitations of probabilistic models in structures, I can give from this material point of view. If the structural response appears as being "random" the cause is often the material heterogeneity (I do not consider external causes like stochastic loadings). In this case correct modeling must be done. This means

(a) to identify the scale of the relevant heterogeneities (some of them can have no influence on the macroscopic response),

(b) to measure and model them (PDF, spatial correlation functions),

(c) to simulate the heterogeneous media such as to obtain the structural response.

Many limits can be encountered during these steps:

(a) difficult to identify the relevant scale and parameters (f.i. if various scales seem to interact),

(b) difficult to measure the micro-characteristics (microstructural investigation, availability of data),

(c) difficult to simulate (computation cost, discretization problems).

Another (general) problem with probabilistic models is that we must be careful with predictions. Description is not prediction. A model can be called predictive only when it: (a) gives the expected response and (b) all its parameters are objectively identified from available data. As probabilistic models generally have a large number of parameters (compared to deterministic ones), their descriptive ability is very wide. Unfortunately, many of these parameters are often identified by reverse engineering techniques or fitted (from a posteriori results). For myself, I see here the main limitation of probabilistic models. But this is not specific to probabilistic models, it is just more visible in this case than for deterministic models (for which one knows that curve-fitting is widely used also).

8.3 Concluding remarks

As far as the authors' stand is concerned, it is reflected in the monograph by Ben-Haim and Elishakoff (1990), as well as in an extremely thought-provoking paper was written by Kalman (1994), which reminded the present writers that the results of the above-described questionnaire were not yet fully communicated to the engineering public.

We will be extremely indebted to hear from both researchers and practitioners in the field, with a view to preparing a sequel of this overview. One should stress again that since the present survey necessitated several years, it is quite plausible that the respondents have changed their position, partly because of the active and/or relevant research that took place during this time span, or other reasons. Yet it is hoped that this overview will be useful for both researchers and funding agencies, which should decide which aspects of probabilistic modeling should be strengthened. It appears instructive here to cite from the letter by Professor Mervin King (2020), one of the authors of the book *Radical Uncertainty* (Kay and King, 2020): " I have now had a chance to read your paper. What is fascinating about it is the similarity of the responses of so many of the 42 interviewees to our reactions to the work done by economists, especially in the field of macroeconomics. Your conclusion "we do not know" is full of wisdom. Ambiguity in the dictionary sense is part of all of the major practical problems that both engineering and economics confront. The questions are often not well defined. Hence it was interesting for me to see how much attention was focused on the engineering question of "what is really meant by being safe?" One of the quotations summarized the essential implication of radical uncertainty: "We have to consider that a single neglected or unrecognized risk can invalidate all the reliability calculations, which are based on known risks!" And I enjoyed the quotation from Kalman: " . . . Probability is an intellectual construct. It does not exist in the real world. It is not something quantifiable, measurable, and concrete. It is not of scientific concern today because it exists only in the self-interest of gamblers or the imagination of statisticians, or the mind-reading of philosophers. There was and is no such thing as a probabilistic revolution in science although there does seem to be something going on. I hope it is not a social Alzheimer's. It is the ultimate hypotheses Jingo . . ."

Many of the factors that we identify as problems in economics and business – the lack of adequate data to identify the tales of a distribution, the non–station narrative the of the processes generating outcomes, and the fact that the behavior of most economic variables depends on what we believe about the underlying process generating them – seem to appear in many of the comments of your respondents.

Finally, I was intrigued to read the reaction of one of the people to whom you sent the results of the survey. In the following quotation he could have been writing about macroeconomics: "If we were more collegial, then the field would flourish. Rather, we are all parochial, and we think that we are . . . creating a new existence by our little models. To some extent our community is incestuous. I have dropped out because I do not like the interaction, I had in the past with many of my colleagues. This may be true to some extent in all intellectual circles, but I believe that the stochastics community is far above average in the negative traits. After all, we just took the models of physics community and applied them to structural oscillators (linear or nonlinear) for almost forty years. What exactly have we accomplished except play fancy games?" In his paper Breitung (1992) notes that "The usual statistical models, classical and Bayesian, are not well suited for problems of structural reliability, since here we have rarely as assumed for these models, a sequence of identical experiments (structures) . . . If we have a probabilistic model in structural reliability, it is often not possible to validate the model in any way, i.e., to show that it describes the reality correct[ly] . . . For statistical inference in structural reliability not the statistical standard models should be used. It appears to be more useful to adopt empirical Bayes and predictivity models." The interested reader can also consult with the paper by Rocchetta *et al.* (2018) tellingly titled "Do We Have Enough Data? Robust Reliability via Uncertainty Quantification."

Likewise, Klir and Yuan (1995, p.419) write: "Civil engineering, when compared with other engineering disciplines, is fundamentally different in the sense that available theories never fully fit the actual design problem. This is because each civil engineering project is, by and large, unique; hence, there is almost never a chance to test a prototype, as in other engineering disciplines. The designer has to make decisions in spite of the high uncertainty he or she faces."

Corotis (2015, p.040801–9) notes: "The limitations of probability theory extend to the probability-based tools most prevalent in the literature. As discussed by Maes and Faber (2004), research suggests that limitations exist in utility modeling and rational decision-making, such as the Allais paradox and Bergen paradox."

8.4 The list of poll participants

The above replies to the questionnaire were kindly provided by numerous researchers around the world. These are (in alphabetical order):
-   Dr. J. R. Benjamin of the Jack Benjamin & Associates, Inc., U.S.A
-   Professor A. Bernardini of the University of Padua, Italy.
-   Dr. F. Bljuger of the National Building Institute, Israel.
-   Professor V. V. Bolotin of the Moscow Power Engineering Institute and State University, Russia.
-   Professor D. Breyse of the Universite Deaux 1, France.
-   Professor F. Casciati of the University of Pavia, Italy.
-   Professor C. A. Cornell of Stanford University, U.S.A.
-   Professor R. Corotis of the University of Colorado, U.S.A.
-   Professor S. H. Crandall of the M.I.T., U.S.A.
-   Professor F. L. DiMaggio of the Columbia University in the City of New York, U.S.A.
-   Dr. F. Dinca of the IFTM, Romania.
-   Professor O. Ditlevesen of the Technical University of Denmark, Denmark.
-   Professor B. Etkin of the Institute for Aerospace Studies, University of Toronto, Canada.
-   Doc. Ing. L. Fryba of the Institute of Theoretical and Applied Mechanics, Prague, Chech Republic.
-   Professor M. Grigoriu of Cornell University, U.S.A.
-   Professor J. Kogan of the Technion-Israel Institute of Technology, Israel (deceased).
-   Professor N. C. Lind of the University of Victoria, Canada.

- Professor C. R. Mischke of the Iowa State University, U.S.A.
- Professor F. Moses of the University of Pittsburgh, U.S.A.
- Prof. Dr. Inz. J. Murzewski of the Politechnika Krakowska, Poland.
- Professor Dr. Rer. Nat. H. G. Natke of the University of Hannover, Federal Republic of Germany.
- Prof. M. Novak of the University of Western Ontario, Canada (deceased).
- Professor A. Nowak of the University of Michigan, U.S.A.
- Professor A. O. Payne of the Royal Melbourne Institute of Technology, Australia.
- Professor E. Plate of the Karbsruhe University, Federal Republic of Germany.
- Professor J. D. Robson of the University of Glasgow, U.K.
- Professor E. Rosenblueth of the Instituto de Ingenieria, Ciudad Universitaria, Mexico (deceased).
- Professor P. Sagirow of the Stuttgart University, Federal Republic of Germany.
- Dr. Ing. Thomas Schmidt of the Universitat der Bundeswehr, Hamburg, Federal Republic of Germany.
- Professor G. Schuëller of the University of Innsbruck, Austria.
- Dr. R. G. Sexsmith of the R. G. Sexsmith Ltd, Canada.
- Professor K. Sobczyk of the Polish Academy of Science, Poland.
- Professor T. T. Soong of the University of Buffalo, U.S.A.
- Professor Y. Sunahara of the Kyoto Institute of Technology, Japan.
- Professor Ir. D. Vandepitte of the Rijksuniversiteit- Gent, Belgium.
- Professor T. Vrouwenvelder of the Delft University of Technology, The Netherlands.
- Professor M. Tichy of the Building Research Institute, Prague, Chech Republic.
- Dr. Ir. L. Taerwe of the Rijksuniversiteit-Gent, Belgium.
- Professor D. C. C. Tung of the North Carolina State University, U.S.A.
- Professor D. Yadav of the Indian Institute of Technology Kanpur, India.
- Professor M. T. Yao of the Texas A & M University, U.S.A.

Chapter 9
What may go wrong with probabilistic methods

> By contrast to the well-known probabilistic approach, the guaranteed approach does not require precious knowledge of probability distributions (which are seldom available in practical problem) and yields reliable estimates for the system behavior (Felix L. Chernousko, 1999)

This chapter is directed to a single objective: to illustrate the possible error associated with the effect of a small perturbation in the probability density on the structural reliability. This perturbation is associated with the interpretation of the experimental data, which lies as a basis of the probabilistic model involved. Moreover, the small perturbation in the probability density is still associated with the same probabilistic moments possessed by an unperturbed density.

9.1 Introduction

It appears that the investigation of a possible error associated with the probabilistic analysis must become an integral part of a meaningful probabilistic analysis. The usual stress requirement reads:

$$\sigma \leq \sigma_y \qquad (9.1)$$

where σ is an actual stress, whereas σ_y is a yield stress. But then one introduces uncertainty considerations into the picture by stating that we may know the loads precisely, or there may be an imprecision in measuring geometric parameters of the cross-sectional area, or we may have built an imperfect mechanical model to describe the behavior of the structure. Thus, the required safety factor k_{req} is introduced, and eq. (9.2) is replaced by

$$\sigma \leq \frac{\sigma_y}{k_{req}} \qquad (9.2)$$

Once the structure is designed, one can introduce the actual safety factor

$$k_{act} = \frac{\sigma_y}{\sigma_{max}} \qquad (9.3)$$

where σ_{max} is the maximum actual stress occurring in the structure. Thus, the design requirement can be formulated as follows

$$k_{act} \geq k_{req} \qquad (9.4)$$

In other words, the actual safety factor should not be less than the required one.

https://doi.org/10.1515/9783111354231-009

Can one quantify the actual safety factor of the uncertainty that is not hidden but directly introduced into the scene? Let us attempt to answer this question. Let the force P acting on a tension-compression element with cross-sectional area have a Weibull distribution with the following probability distribution

$$F_p(p) = 1 - exp\left[-\left(\frac{p - p_0}{w - p_0}\right)^k\right], \quad (k > 0, w > p_0, p \geq p_0) \tag{9.5}$$

For $p < p_0, F_p(p) \equiv 0$. We are interested in the reliability of the structure, i.e. the probability that the structure will perform its intended mission satisfactorily. Such a performance is identified with holding relationship (9.1) true. Thus, reliability becomes:

$$R = Prob\left(\sum \leq \sigma_y\right) = Prob\left(\frac{P}{b} \leq \sigma_y\right)$$

$$= Prob\left(P \leq \sigma_y b\right) = F_P(\sigma_y b) \tag{9.6}$$

Thus,

$$R = 1 - exp\left[-\left(\frac{\sigma_{yb} - P_o}{w - P_o}\right)^k\right] \tag{9.7}$$

How to define the safety factor in the context of probabilistic design? One natural way is to relate it to some characteristic load, say the average load. The average load equals

$$E(P) = p_0 + (w - p_0)\Gamma\left(1 + \frac{1}{k}\right) \tag{9.8}$$

where $\Gamma(x)$ is the Gamma function. Variance of the load is

$$V(p) = (w - p_0)\left[\Gamma\left(1 + \frac{2}{k}\right) - \Gamma^2\left(1 + \frac{1}{k}\right)\right] \tag{9.9}$$

The central safety factor is defined as the ratio of yield stress over the average stress $E(P)/b$:

$$k_c = \frac{\sigma_y b}{E(P)} = \frac{\sigma_y b}{P_o + (w - p_o)\Gamma\left(1 + \frac{1}{k}\right)} \tag{9.10}$$

Let us design the structure probabilistically. Probabilistic design requires that the reliability be not less than a codified value r:

$$R \geq r \tag{9.11}$$

Thus, in view of eq. (9.7) we get

$$1 - exp\left[-\left(\frac{\sigma_y b - p_0}{w - p_0}\right)^k\right] \geq r \tag{9.12}$$

The design value of the cross-sectional area b_{design} is found from an equality $R = r$ and becomes:

$$b_{design} = \frac{P_0 + (w - P_0)\left[\frac{\ell n1}{1-r}\right]^{1/k}}{\sigma_y} \tag{9.13}$$

Substitution of this expression in eq. (9.10) allows us to explicitly write the central safety factor in terms of the codified required reliability r:

$$k_c = \frac{P_0 + (w - P_0)\left[\frac{\ell n1}{1-r}\right]^{1/k}}{P_0 + (w - P_0)\Gamma\left(1 + \frac{1}{k}\right)} \tag{9.14}$$

Consider as an example the following set of parameters:

$$w = 3p_0, \quad k = 4 \tag{9.15}$$

Then,

$$k_c = \frac{1 + 2\ell n\left[\frac{1}{1-r}\right]^{1/k}}{1 + 2\Gamma(1.25)} \tag{9.16}$$

For $r = 0.9$, $k_c = 1.2314$; for $r = 0.99$, $k_c = 1.3971$; for $r = 0.999$, $k_c = 1.5082$; for $r = 0.9999$, $k_c = 1.5942$; for $r = 0.99999$, $k_c = 1.6653$; for $r = 0.999999$, $k_c = 1.7263$, etc. As we see, probabilistic models allow associate the required reliability directly with the safety factor. We learn several lessons from this simplest example:

(1) The concept of the safety factor, which is so often criticized by practitioners and researchers alike, is, in actuality, a powerful concept, which could be given a probabilistic interpretation.

(2) Within probabilistic interpretation, safety factor is not a mysterious quantity determined by the will of the designer, for whom this may a "personal factor of safety." If one can quantify, say through legislation, the required reliability, one can then also quantify the celebrated safety factors.

9.2 How reliable are reliability calculations?

Let us investigate now how a small error may affect the reliability calculations. Let us assume that instead of parameters w, p_o and k we have measured other values, w_1, p_1 and k_1, respectively. The analysis, however, was performed for values w, p_o and k. We may ask ourselves: What is the actual reliability R_{act}, or actual probability of failure $P_{f,act}$?

$$P_{f,act} = 1 - R_{act} \tag{9.17}$$

Actual reliability is given by eq. (7.7) with substituted actual values of the parameters

$$R_{act} = 1 - exp\left[-\left(\frac{\sigma_y b - p_1}{w_1 - p_1}\right)^{k_1}\right] \tag{9.18}$$

actual probability of failure being

$$P_{f,act} = exp\left[-\left(\frac{\sigma_y b - p_1}{w_1 - p_1}\right)^{k_1}\right] \tag{9.19}$$

Design of the structure has been performed using the values w, p_o and k, respectively. The appropriate value of the cross-sectional area is given by eq. (9.13). To calculate the actual probability of failure, corresponding to the design value, b_{design}, we substitute the expression b_{design} into eq. (9.19) to result in

$$P_{f,act} = exp\left[-\left(\frac{\sigma_y b_{design} - p_1}{w_1 - p_1}\right)^{k_1}\right] \tag{9.20}$$

or

$$P_{f,act} = exp\left[-\left(\frac{p_0 - p_1 + (w - p_0)\left[\frac{\ell n 1}{1-r}\right]^{1/k}}{w_1 - p_1}\right)^{k_1}\right] \tag{9.21}$$

Let us investigate some particular cases. In the simplest case $p_1 = p_0$, $w = w_1$, but $k_1 \neq k$. Then

$$P_{f,act} = exp\left[-\left(\ell n\frac{1}{1-r}\right)^{\frac{k_1}{k}}\right] \tag{9.22}$$

Since r is the required reliability, $1 - r$ is recognized in
Equation 9.22 as the allowed probability of failure $P_{f,all}$. Thus, eq. (9.22) can be rewritten as:

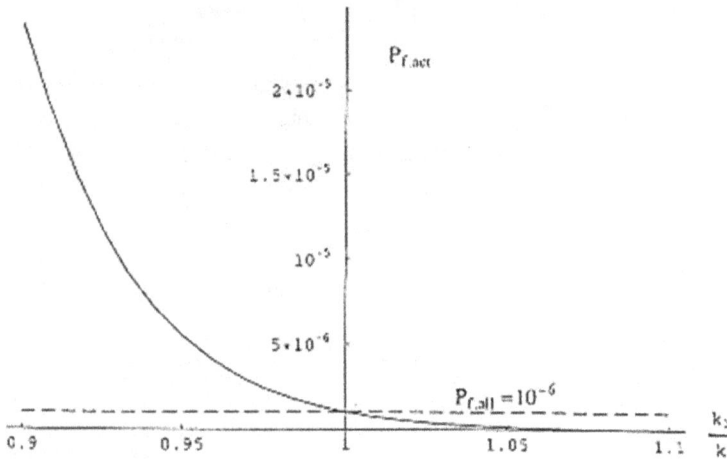

Fig. 9.1: Actual probability of failure as a function of the ratio k_1/k: for $k_1 = k$ the actual probability of failure coincides with the required one; for $\frac{k_1}{k} > 1$, the actual probability of failure is less than the required one, whereas for $\frac{k_1}{k} < 1$, the actual probability of failure may well exceed the allowed value, resulting in a detrimental state.

$$P_{f,act} = exp\left[-\left(\ell n\frac{1}{P_{f,all}}\right)^{\frac{k_1}{k}}\right]$$

(9.23)

Let $P_{f,all} = 10^{-6}$. Then

$$P_{f,act} = exp\left[-13.81551056^{k_1/k}\right]$$

(9.24)

This can be viewed as a function of the ratio k_1/k. This function is depicted in Fig. 9.1. As is seen, when

$$\frac{k_1}{k} = 1, \quad P_{f,act} = P_{f,all} = 10^{-6}$$

Yet, if $k_1 \neq k$, the actual probability of failure may differ from the allowed one. It is remarkable that there could be a serendipitous situation, i.e., when an error in measurement of k_1 may be a "favorable" nature: for $\frac{k_1}{k} > 1$, the actual probability of failure is less than the allowable one. Yet, when $\frac{k_1}{k} < 1$, the effect of a small error in evaluating k may be detrimental. If $\frac{k_1}{k} = 0.95$, the actual probability of $\frac{k_1}{k} = 0.93$, the actual probability of failure is approximately 10 times larger than the one which was permitted! We conclude that the probability of failure is too sensitive a parameter to allow for an imprecise input characteristic. We conclude that the accurate determination of the input's probabilistic characteristics must become an integral part of the rigorous probabilistic analysis. Paradoxically, probabilistic engineers advocate a detachment from experiments, since they neither try to validate their assumptions via experiments, nor introduce the study of the errors in their analyses. Yet, mathematicians do not advocate such a detachment; as Richard Bellman (1962) mentions

> We postulate then that one of the basic responsibilities of the mathematician is to examine the structure of the problems of society and to provide mathematical formulations which are easily susceptible to numerical solution in terms of the current technology. This means not only an examination of the computational aspects, but also of the experimental aspects. What information is required for which formulation, what sensing devices are available, what accuracy do they possess, what should be measured when, and so on?

Engineers dealing with uncertainty are well advised to follow these recommendations by a mathematician.

9.3 Effect of a small deviation in the probability density

We consider a fundamental problem in structural reliability, namely a bar with a cross-sectional area b and subjected to a load N, that is a random variable with the probability density $f_p(p)$, where p is a possible value that the load may take on. The following assumption is made for facilitating a simple analysis: The bar's material is perfectly elastic in compression but has a yield stress in tension σ_y. We also assume that the bar cannot lose its stability or have any other form of failure. The reliability of the structure is determined as the probability that the stress $\sum = P/b$ does not exceed the value of the yield stress:

$$R = prob\left(\sum \le \sigma_y\right) \tag{9.25}$$

In other words, let us visualize the situation in which the analyst has assumed the probability density of the load as that of the log-normal variable (Elishakoff, 1983):

$$f_p(p) = \begin{cases} \dfrac{1}{p\sigma_p\sqrt{2\pi}}\, exp\left[\dfrac{-(\ell np - a)^2}{2\sigma_p^2}\right], & \text{for } p > 0 \\ \{0, & otherwise \end{cases} \tag{9.26}$$

The parameters a, and σ_F characterize the probability density, namely, the mean value $E(P)$ and the variance $V(P)$ are expressed as:

$$E(P) = \exp\left(a + \frac{1}{2}\sigma_p^2\right) \tag{9.27}$$

$$V(P) = \exp\left(2a + \sigma_p^2\right)\left[\exp\left(\sigma_p^2\right) - 1\right]$$

We find the reliability of the structure:

$$R = Prob\left(P \leq \sigma_y b\right) = Prob\left(\ell nP \leq \ell n\sigma_y b\right)$$

$$= \frac{1}{2} + erf\left(\frac{\ell n\sigma_y b - a}{\sigma_P}\right) \tag{9.28}$$

$$erf(x) = \frac{1}{\sqrt{2\pi}} \int_0^x \exp\left(-\frac{t^2}{2}\right) dt$$

since ℓnP is a normal variable with mean "a," and variance σ^2.

Let us visualize now that the actual probability density slightly differs from the true one and reads, for $p > 0$:

$$f_P^{(\varepsilon)}(p) = \frac{C}{p\sigma\sqrt{2\pi}} \exp\left[-\frac{\ell np - a^2}{2\sigma_P^2}\right], C = \{1 + \varepsilon \sin[2\pi(\ell np - a)] \tag{9.29}$$

where ε is a constant, belonging to an interval $[-1, 1]$. For $p < 0$, the probability density vanishes. It can be shown that $f_P^{(\varepsilon)}(p)$ is indeed a probability density. It is non-negative and satisfies the equality

$$\int_{-\infty}^{\infty} f_P^{(\varepsilon)}(p) dp = 1 \tag{9.30}$$

In order to prove this property, we have to demonstrate that

$$\int_{-\infty}^{\infty} f_P(p) \sin[2\pi(\ell np - a)] dp = 0 \tag{9.31}$$

To do this we make a substitution $\ell np = u$. Thus, we should calculate the integral

$$I = \int_{-\infty}^{\infty} \frac{1}{\sigma\sqrt{2\pi}} \exp\left[-\frac{(t-a)^2}{2\sigma^2}\right] \sin[2\pi(t-a)] dt \tag{9.32}$$

Further substitution $t - a = u$ leads to

$$I = \int_{-\infty}^{\infty} \frac{1}{\sigma\sqrt{2\pi}} \exp\left(-\frac{u^2}{2\sigma^2}\right) \sin(2\pi u) du \tag{9.33}$$

which vanishes since the integral is an odd function. Thus, the function defined in eq. (7.29) represents a probability density of some random variable, denoted $P^{(\varepsilon)}$. Obviously, if $\varepsilon = 0$, $f^{(\varepsilon)}$ and $P^{(\varepsilon)}$ are, respectively, f and P defined in the beginning of this section. Stoyanov (1987) demonstrates that for any $k = 1, 2 \ldots$, we have

$$E\left(P^{(\varepsilon)}\right)^k = E\left(P^k\right) \tag{9.34}$$

i.e., the "perturbed" random variable $P^{(t)}$ has the same moments as those of a "unperturbed" random variable P. Let us calculate now the true reliability associated with the random variable $P^{(\varepsilon)}$:

$$R = Prob(P \le \sigma_y b) = \int_0^{\sigma_y b} f_P^{(\varepsilon)}(p)\, dp$$

$$= \int_0^{\sigma_y b} \frac{1}{p\sigma\sqrt{2\pi}} \exp\left[-\frac{(\ell np - a)^2}{2\sigma^2}\right] \times$$ (9.35)

$$\times \{1 + \varepsilon \sin[2\pi(\ell np - a)]\}\, dp$$

or

$$R = \int_0^{\sigma_y b} \frac{1}{p\sigma\sqrt{2\pi}} \exp\left[-\frac{(\ell np - a)^2}{2\sigma^2}\right] \times$$ (9.36)

$$\times \{1 + \varepsilon \sin[2\pi(\ell np - a)]\}\, dp$$

Introducing again the new variable of integration $\ell np = t$, we reduce reliability to

$$R = \int_{-\infty}^{\ell n\sigma_y b} \frac{1}{\sigma\sqrt{2\pi}} \exp\left[-\frac{(t - a)^2}{2\sigma^2}\right] \{1 + \varepsilon \sin[2\pi(t - a)]\}\, dt$$ (9.37)

with $t - a = u$, we get

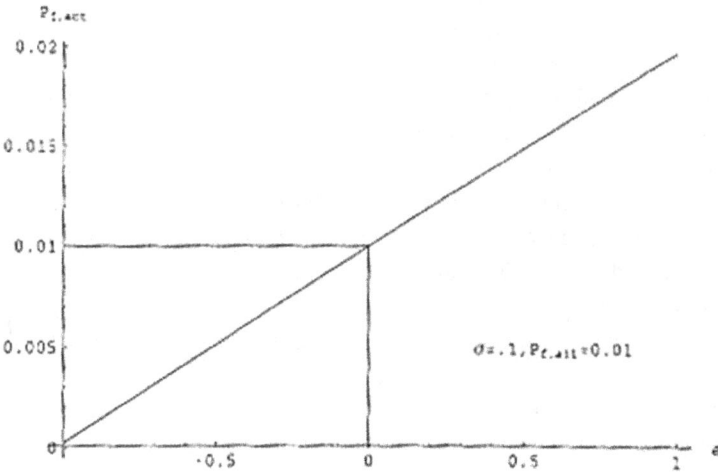

Fig. 9.2: Actual probability of failure as a function of the control parameter ε; the allowed probability of failure is set at 0.01.

$$R = \int_{-\infty}^{\ell n \sigma_y b - a} \frac{1}{\sigma\sqrt{2\pi}} exp\left(-\frac{u^2}{2\sigma^2}\right)\{1 + \varepsilon \sin[2\pi(t-a)]\}dt \tag{9.38}$$

$$= \frac{1}{2} + erf\left(\frac{\ell n \sigma_y b - a}{\sigma}\right) + \frac{\varepsilon}{\sigma\sqrt{2\pi}} \int_{-\infty}^{\ell n \sigma_y b - a} exp\left(-\frac{u^2}{2\sigma^2}\right) \sin(2\pi u)du$$

Evaluation of this integral yields

$$R = \frac{1}{2} + erf\left(\frac{\ell n \sigma_y b - a}{\sigma}\right) + \frac{\varepsilon D}{\sigma\sqrt{2\pi}}\left\{-e^{2\pi^2\sigma^2}\sqrt{\frac{\pi}{2}}\right. \tag{9.39}$$

where

$$D\left(= erfi\left[\frac{2\pi\sigma^2 - i(a+\ell n[\sigma_y b])}{\sqrt{2}\sigma}\right] + erfi\left[\frac{2\pi\sigma^2 + i(-a+\ell n[\sigma_y b])}{\sqrt{2}\sigma}\right]\right) \tag{9.40}$$

where $erfi(z)$ is an imaginary error function, defined as (Wolfram, 1996)

$$erfi(z) = \frac{erf(iz)}{i} \tag{9.41}$$

where $i = \sqrt{-1}$. Let us visualize that the reliability calculations have been performed so that the probability density of the load was postulated to equal $f_P^{(0)}(p)$. Let the required reliability be r, so that the design was performed using expression (9.28). Figures 9.2–9.4 are associated with $r = 0.99$, $r = 0.999$ and $r = 0.9999$, respectively. The figures depict the actual situation in which the parameter ε describing perturbation from reality does not vanish identically. Thus, actual probability of failure $P_{f,act}$ varies with the control parameter ε. Naturally, when $\varepsilon = 0$, the actual probability of failure takes a value $1-r$, or 0.01, 0.001 and 0.0001, respectively, in Figs. 9.2–9.4. An integrand in the integral (7.36) is an oscillatory function. Therefore, depending on the value of σ, the value of the integral is either positive or negative. Hence, different behaviors are exhibited in Figs. 9.2–9.4. Figures 9.2–9.4 demonstrate that one can make "good" errors, i.e., deviation of the model from the reality may turn to be on the safe side. Indeed, if in Fig. 9.2, the control parameter ε tends to −1 from above, the actual probability of failure tends to zero. Likewise, in Fig. 9.3, associated with $\sigma = 0.2$ and $P_{f,all} = 10^{-3}$, if ε reaches unity from below, the actual reliability too will reach unity. Analogously, in Fig. 9.4, with $\sigma = 0.2$ and $P_{f,all} = 10^{-4}$, with ε approaching unity, from below, the actual probability of failure will approach zero.

It is prudent, however, to look at what can go wrong, when the theoretical model differs from the reality. As Fig. 9.2 demonstrates, when ε tends to unity, the actual probability of failure becomes twice as much as the allowable one. This implies, though the use of the frequency interpretation of the probability notion, that the num-

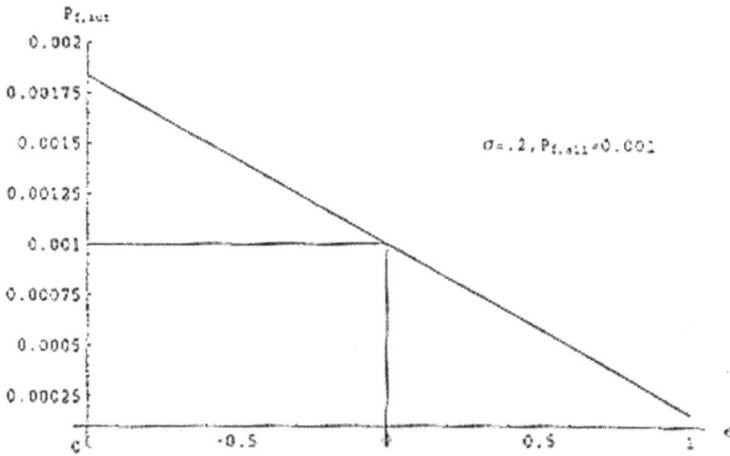

Fig. 9.3: Actual probability of failure as a function of the control parameter ε; the allowed probability of failure is fixed at 0.001.

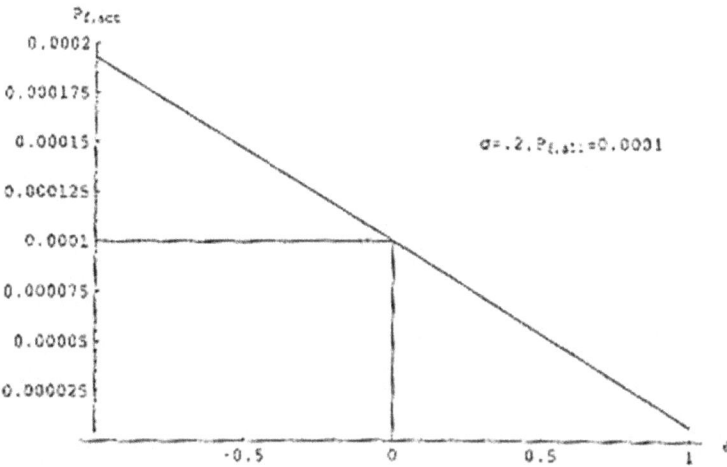

Fig. 9.4: Actual probability of failure as a function of the control parameter ε; the allowed probability of failure is fixed at 0.0001.

ber of unsuccessful performances may be double of the allowed quantity. Nearly analogous situations are portrayed in Figs. 9.3 and 9.4 when ε approaches −1.

It must be stressed that bad things happen to good probabilistic analyses for different parameters of ε: sometimes the unfavorable situation occurs when ε approaches unity (Fig. 9.2), whereas in other circumstances the detrimental effect takes place in the vicinity of $\varepsilon = -1$.

9.4 Conclusion

The leitmotif of this chapter is as follows: Beware of uncritical use of the probabilistic methods when extremely reliable performance of the structure is called for! A colleague of the present writer, in his lecture on the safety of the nuclear power plants conjectured that the probability of failure had to be of order 10^{-42}. As this study shows, the accuracy of the prediction in the absurdly small probabilities of failure may turn out to be illusory!

Our conclusions agree with the statement made by Ekeland (1993): "Even in circumstances in which there are no material interests at stake the probabilities calculated by competent people in good faith may turn out to be systematically underrated." This chapter quantifies Ekeland's qualitative statement in terms of simple examples. We still fall short of the far-going recommendation made by Kalman (1994): "Define randomness without probability." Perhaps, a less antagonistic statement would read, "Define uncertainty without probability." Still, Kalman's idea must become a focus of inquiring thoughts of both probabilistic and non-probabilistic analysts of uncertainty. In this study we have demonstrated the existence of possibly dangerous "underground stones" for those who want to "swim" in the sea of randomness and probability.

Chapter 10
From A. M. Freudenthal's criticisms to optimization and anti-optimization or Wald's Min-Max

New methods for treating uncertainty will become important in virtually all branches of mechanics. (Oden *et al.*, 2003)

It's only the man who can look at the same problem from many different aspects that will make a true leader (Takao Fujisawa; see Sato, 2006)

In this chapter, stochastic approach and non-stochastic and convex modeling of uncertainty are critically contrasted. First, some drawbacks of the probabilistic methods are discussed, attributable to the lack of sufficiently accurate data, then the effect of human error in constructing a probabilistic model for input quantities is elucidated. Extensive quotations from pertinent works of Freudenthal, who is rightfully considered as an architect of modern reliability theory, are utilized to explain some doubts he himself experienced about probabilistic methods. His hints are realized in modern convex modeling of uncertainty, which the writer is advocating and advancing. A set-theoretical, convex description of uncertainty is discussed in detail. Uncertainty is described as a set of constraints unlike the classical probabilistic approach. Moreover, instead of conventional optimization studies, where the minimum possible responses are sought, here an uncertainty modeling is developed as an "anti-optimization" problem of finding the least favorable response under the constraints within the set-theoretical description. The question of how the output quantities of such an anti-optimization process vary when the global knowledge on the uncertainties increases is considered in detail. Response variability of viscoelastic structures is evaluated. Combined probabilistic-convex modeling is proposed for situations where the input quantities should be modeled as stochastic ones, but some of their probabilistic characteristics are unknown but bounded.

10.1 Introduction

Modern deterministic methods have become quite elaborate and include sophisticated mathematical modeling and analysis, highly refined computer evaluation methods, and optimization techniques. However, there was a growing realization among engineers during the recent 25 years or so that the unavoidable uncertainties should be taken into account to produce meaningful designs. This observation led to the development of stochastic methods to analyze the structures. The rapid flourishing of stochastic methods has considerably enhanced the power of the designer, and the concept of reliability has become an important tool which enables designers to deal

https://doi.org/10.1515/9783111354231-010

probabilistically with uncertainty when sufficient knowledge on the random variables and functions is available.

A structure can be modeled as a system subjected to influences, disturbances, excitations, and so forth, designated here as inputs. We are usually interested in the response of this system – its dynamic displacements, stresses, or strains being designated as outputs. In formulating a probabilistic model, we first identify the random variables and/or functions, which can be the characteristics of the inputs and/or of the system itself. To perform the analysis, we need to know the joint probability density functions of these variables.

Within the probabilistic context, the main objective of engineering analysis is the determination of the reliability of the system. Reliability is defined as the probability that the system will adequately perform its intended mission for a specified interval of time when operating under specified environmental conditions.

In this connection, the basic question arises: Are the probabilistic approaches able to deliver the final product – reliability (or its complement, the probability or failure of unreliability) with sufficient precision when there is a lack of sufficient experimental data to validate the assumptions made regarding the joint probability densities of random variables or functions involved?

To be able to answer this question we must consider an example that illustrates the dependence of the probability of failure on the behavior of the probability densities involved at their "tails."

10.2 Influence of imprecise data

Consider how imperfect information on the probability density will affect the structural design. This will model the lack of precise data. Consider a bar under central tensile force N, which is an uncertain variable. We will treat it as a random variable with the probability distribution function $F_N(n)$. For simplicity, assume that N is a Rayleigh distributed random variable, with

$$F_N(n) = Prob(N \leq n) = 1 - \exp\left(-\frac{n^2}{2b^2}\right), n \geq 0 \tag{10.1}$$

where b is a positive parameter, related to the mean load $E(N)$

$$E(N) = b\sqrt{\pi/2}. \tag{10.2}$$

We are interested in the reliability of the bar, defined as the probability of its successful performance or, in probability that the following random event holds as

$$\sigma \leq \sigma_y \tag{10.3}$$

where σ is a random stress and σ_y is a yield stress, which is treated here as a deterministic variable. Now

$$\sigma = \frac{N}{a} \tag{10.4}$$

where a is an area of the cross-section of the bar. Reliability of the bar equals

$$R = Prob(\sigma \le \sigma_y) = Prob(N \le \sigma_y a)$$

$$= F_N(\sigma_y a) = 1 - \exp\left\{ -\frac{\pi}{4}\left[\frac{\sigma_y a}{E(N)}\right]^2 \right\}. \tag{10.5}$$

Let us address the design problem. The cross-sectional area should be determined so
that the reliability is not less that some specified, codified value

$$R \ge r. \tag{10.6}$$

The minimum required cross-section area, a^*, is found from the requirement

$$R = r \tag{10.7}$$

or

$$1 - \exp\left\{ -\frac{\pi}{4}\left[\frac{\sigma_y a}{E(N)}\right]^2 \right\} = r. \tag{10.8}$$

This requirement yields the minimum required cross-sectional area

$$a^* = \frac{E(N)}{\sigma_y}\sqrt{\frac{4}{\pi}\ln\frac{1}{1-r}}. \tag{10.9}$$

In actuality, the data characterizing the distribution function should be derived from
experiments unless some solid theoretical ground exists for their non-experimental
justification; hence, $E(N)$ and σ_y may have some errors, say $\alpha\%$ of error is associated
with $E(N)$ and β percent with σ_y. The cross-sectional area that will be used in actual
design is therefore

$$a_{act} = \frac{E(N)}{\sigma_y}\frac{1+\alpha/100}{1+\beta/100}\sqrt{\frac{4}{\pi}\ln\frac{1}{1-r}}. \tag{10.10}$$

We may ask the question: What is the actual reliability the system is operating with?
To obtain the answer, we substitute eq. (10.10) into eq. (10.5). The result is an actual
reliability

$$R_{act} = 1 - exp\left\{-\frac{\pi}{4}\left[\frac{\sigma_y a_{act}}{E(N)}\right]^2\right\} = 1 - exp\left[\ln(1-r)\left(\frac{1+\alpha/100}{1+\beta/100}\right)^2\right]. \tag{10.11}$$

When $\alpha = \beta = 0$, the reliability R equals r, as it should be. The expression for the actual probability of failure, defined as

$$P_{f,act} = 1 - R_{act} \tag{10.12}$$

reads

$$P_{f,act} = (P_{f,all})^{\left[\frac{1+\frac{\alpha}{100}}{1+\frac{\beta}{100}}\right]^2} \tag{10.13}$$

where $P_{f,all}$ is an "allowed" probability of failure

$$P_{f,all} = 1 - r. \tag{10.14}$$

One can deduce from eq. (10.13) that asymptotically, for small α and β

$$\ln P_{f,act} = \left[1 + 2\left(\frac{\alpha}{100} - \frac{\beta}{100}\right)\right](\ln P_{f,all}) \tag{10.15}$$

so that for negative α and positive β the actual probability of failure may reach values that are much larger than the allowed probabilities of failure.

This simple example strikingly illustrates that one may think that the structure is acceptable, based on the requirement that the probability of failure will not be larger than its allowed level, but in actuality, due to errors in data, the system will be operating in the adverse state.

It should be remarked here that this sensitivity is appearing irrespective of the allowable probability of failure chosen. In nuclear engineering for example, the probability of failure often quoted is $P_t^* = 10^{-7}$. In his paper, Amer (1989) notes that "as an indication of the advancing states of the art in helicopter structural technology, the draft RFP for the U.S. Army Light Experimental Helicopter (LHX) requires an unprecedented reliability against catastrophic failure of 0.999999 at a life of 10,000h (recently revised to 4500h)." So, in this case $P_f^* = 10^{-6}$. In the probabilistic analysis, the decision about the P_f^* should be made. However, for virtually any sufficiently small value of P_f^*, the above strong sensitivity could be illustrated.

It appears instructive to quote here from the book by Feynman (1988), Nobel laureate in physics, "It appears that there are enormous differences in opinion as to the probability of failure with loss of vehicle and of human life. The estimates range from roughly 1 in 100 to 1 in 100,000. The higher figures come from working engineers, and the very low figures come from management. What are the causes and consequences of this lack of agreement? Since 1 part in 100,000 would imply that one could launch a shuttle each day for 300 years expecting to lose only one, we would properly ask,

"What is the cause of management's fantastic faith in the machinery?" "It is true that if the probability of failure was as low as 1 in 100,000, it would take an inordinate number of tests to determine it: You would get nothing but a string of perfect flights in the string so far. But if the real probability is not so small, flights would show troubles, near failures, and possibly actual failures with a reasonable number of trips, and standard statistical methods could give a reasonable estimate . . . There is nothing so wrong with the analysis as believing the answer! Uncertainties appear everywhere in the model. When using a mathematical model, careful attention must be given to the uncertainties in the model."

10.3 Effects of human error

The first harbinger of the new discipline referred to as "Probabilistic mechanics" appeared in 1926 in the book by Mayer (1926). Presently, as Cornell (1981) notes, after more than half a century of intensive development, this subject has passed the age of adolescence and has become a widely accepted discipline with many monographs written and numerous papers appearing each year, and with extensive financial support by the governmental and other organizations throughout the world (Chamis, 1987).

Yet the human errors associated with probabilistic calculations have not been evaluated thoroughly. As a known proverb maintains, "to err is human," however, the consequences of the errors may be of different magnitudes. There is conflicting information in the literature on the percentage of structural failures and malfunctions which are attributable to human errors. Stuart (1990) mentions that a review of statistical surveys of structural failures reveals that up to 90% of these are due to human error. Also, it has been suggested that up to one-half of all structural failures might have been averted had design checking or control been adequate (Matsouseh and Schneider, 1976; Walker, 1980).

Pugsley (1966) gives illuminating discussions on human errors, with attendant wise recommendations. He advises: "Do not 'make do' with inadequate data; . . . insist upon the acquisition, by experiment and the like, of more data." The possible errors associated with lack of adequate data in reliability calculations have been discussed by Ben-Haim and Elishakoff (1990). Yet this important topic has not become a subject of analysis by many researchers contributing to stochastic mechanics. In this chapter, first a simple example illustrating human error is described, and then its ramifications on probabilistic design of structures are explained.

Consider the fundamental problem of structural reliability, namely an element subjected to stress which is a random variable σ and where the yield stress σ_y is a random variable too. The solution of the problem of determining the reliability

$$R = Prob(\sigma < \sigma_y) \tag{10.16}$$

is well-known when σ and σ_y form an independent normal vector; namely if σ is normal with mean a and variance c_1^2, and σ_y is normal with mean b and variance c_2^2

$$f_{\sigma\sigma_y}(\sigma, \sigma_y) = \frac{1}{2\pi c_1 c_2} exp\left[-\frac{(\sigma-a)^2}{2c_1^2} - \frac{(\sigma_y-b)^2}{2c_2^2} \right]. \tag{10.17}$$

The reliability can be expressed as

$$R = \Phi(\beta) \tag{10.18}$$

where β is the so-called safety index

$$\beta = \frac{E(M)}{\sigma_M} \tag{10.19}$$

and M is the safety margin

$$M = \sigma_y - \sigma \tag{10.20}$$

which is normally distributed with

$$E(M) = b - a, \quad \sigma_M = \sqrt{c_1^2 + c_2^2}. \tag{10.21}$$

Therefore, the safety index is

$$\beta = \frac{b-a}{\sqrt{c_1^2 + c_2^2}}. \tag{10.22}$$

Now suppose that the actual joint distribution of σ and σ_y is given by

$$f_{\sigma\sigma_y}(\sigma, \sigma_y) = \frac{1}{2\pi\sqrt{3c_1 c_2}} \left\langle exp\left\{ -\frac{2}{3}\left[\left(\frac{\sigma-a}{c_1}\right)^2 + \left(\frac{\sigma-a}{c_1}\right)^2 \left(\frac{\sigma_y-b}{c_2}\right)^2 + \left(\frac{\sigma_y-b}{c_2}\right)^2 \right] \right\} \right.$$

$$\left. + \left\{ exp -\frac{2}{3}\left[\left(\frac{\sigma-a}{c_1}\right)^2 - \left(\frac{\sigma-b}{c_1}\right) \times \left(\frac{\sigma-b}{c_2}\right) + \left(\frac{\sigma_y-b}{c_2}\right)^2 \right] \right\} \right\rangle. \tag{10.23}$$

Obviously, σ and σ_y do not form a bivariate normally distributed random vector. Yet, one can prove that the stress σ itself is a normally distributed variable $N(a, c_1^2)$; moreover, the yield stress σ_y too is a normally distributed variable $N(b, c_2^2)$. So, despite the fact that σ and σ_y are not jointly normal, they are individually normal. Moreover, σ and σ_y have zero correlation.

The reliability of an element with the joint density of σ and σ_y are not jointly normal; they are individually normal. Moreover, σ and σ_y have zero correlation.

The reliability of an element with the joint density of σ and σ_y given by eq.(10.23) is

$$R = \frac{1}{2}\left[\Phi\left(\frac{b-a}{c_2}\frac{1}{\sqrt{1+a+a^2}}\right) + \Phi\left(\frac{b-a}{c_2}\frac{1}{\sqrt{1-a+a^2}}\right)\right]$$

(10.24)

where

$$a = \frac{c_1}{c_2}$$

(10.25)

and $\Phi\left(.\right)$ is the cumulative distribution function. Equation (10.24) is derived later on. The probability density of the standardized variables

$$Z_1 = \frac{\sigma - a}{c_1}, \quad Z_2 = \frac{\sigma_y - b}{c_2}$$

(10.26)

appears to be extremely close to the density of jointly normal independent variables. Therefore, when one "looks" at this density, one can make a seemingly "safe" assumption that Z_1 and Z_2 are jointly normally distributed. To the best of our knowledge, engineers do not perform a test to check the joint normality of random variables. If we make a "slight" error, by treating Z_1 and Z_2 as independent normal variables, we obtain the estimate of reliability given in eq. (10.18) instead of the expression (10.24). Calculation shows that for a certain combination of parameters, the approximate estimate based on joint normality assumption is higher than the exact reliability, implying that in a real situation the reliability will be lower; this illustrates a detrimental effect of this human error. However, in some circumstances the human error may be somewhat "beneficial": The assumption of independence of σ and σ_y may yield lower estimate than the exact reliability. (This may suggest that if one makes errors, it is "better" to avoid the ones which are on the unsafe side.) The computer codes that are used presently by probabilistic analysts do not incorporate analysis of errors and thus preclude the possibility of knowing what happens in an actual situation.

In order to derive eq. (10.24), we note that the stress and the yield stress σ_y, having a joint probability density given in eq. (10.23), can be conveniently represented as follows:

$$\begin{pmatrix}\sigma \\ \sigma_y\end{pmatrix} = \begin{pmatrix}\sigma^{(1)} \\ \sigma_y^{(1)}\end{pmatrix}A + \begin{pmatrix}\sigma^{(2)} \\ \sigma_y^{(2)}\end{pmatrix}(1-A)$$

(10.27)

where A is a Bernoulli variable,

$$P(A=1) = \frac{1}{2}, \quad P(A=0) = \frac{1}{2}.$$

(10.28)

In addition,

$$E\left(\sigma^{(1)}\right) = E\left(\sigma^{(2)}\right) = a$$

(10.29)

$$E\left(\sigma_y^{(1)}\right) = E\left(\sigma_y^{(2)}\right) = b \tag{10.30}$$

$$Var\left(\sigma^{(1)}\right) = Var\left(\sigma^{(2)}\right) = c_1^2 \tag{10.31}$$

$$Var\left(\sigma_y^{(1)}\right) = Var\left(\sigma_y^{(2)}\right) = c_2 \tag{10.32}$$

$$Cov\left(\sigma^{(1)}, \sigma_y^{(1)}\right) = \frac{c_1 c_2}{2} \tag{10.33}$$

$$Cov\left(\sigma^{(2)}, \sigma_y^{(2)}\right) = -\frac{c_1 c_2}{2} \tag{10.34}$$

and vectors

$$\begin{pmatrix} \sigma^{(1)} \\ \sigma_y^{(1)} \end{pmatrix}, \begin{pmatrix} \sigma^{(2)} \\ \sigma_y^{(2)} \end{pmatrix} \tag{10.35}$$

are independently jointly normal. Then, given $A = 1$, the mathematical expectation and variance of the safety margin equal

$$E(M) = b - a \tag{10.36}$$

$$Var(M) = c_1^2 + c_2^2 - c_1 c_2. \tag{10.37}$$

Also, given $A = 1$, M is normal. Therefore, the conditional reliability is

$$R(A = 1) = \Phi\left(\frac{b - a}{\sqrt{c_1^2 + c_2^2 + c_1 c_2}}\right). \tag{10.38}$$

In complete analogy, given $A = 0$,

$$E(M) = b - a \tag{10.39}$$

$$Var(M) = c_1^2 + c_2^2 - c_1 c_2. \tag{10.40}$$

Thus,

$$R(A = 0) = \Phi\left(\frac{b - a}{\sqrt{c_1^2 + c_2^2 - c_1 c_2}}\right). \tag{10.41}$$

Finally, by the theorem of total probability

$$R = P(M > 0) = P(M > 0|A = 1)P(A = 1) + P(M > 0|A = 0)P(A = 0)$$

$$= \frac{1}{2}\left[\Phi\left(\frac{b-a}{c_2} \frac{1}{\sqrt{1+a+a^2}} \right) + \Phi\left(\frac{b-a}{c_2} \frac{1}{\sqrt{1-a+a^2}} \right) \right] \qquad (10.42)$$

coinciding with eq. (8.24). It appears that the following lessons can be learned from the example considered.

In modern probabilistic codes and in most, if not all, studies the necessary probabilistic information on uncertain quantities is assumed rather than appropriately substantiated through statistical analysis of extensive experimental data. After numerous assumptions are made, some new numerical approaches, often sophisticated ones, are tested on simple examples. On the other hand, the accuracy of the experimental data (if at all present) is not discussed.

The probabilistic analyst in actuality claims: "Give me the joint probability densities of random variables involved and I will calculate the reliability of the structure through sophisticated numerical methods." This reminds us of the well-known statement by Archimedes (circa 287 BCE–circa 212 BCE): "Give me a firm spot on which to stand, and I will move the earth." Likewise, René Decrates (1596–1650) supposedly said: "Give me matter and motion and I will construct a universe" (Bell, 1951). It is clear that this analogy is not perfect. However, the resemblance of the two statements is clear. In this connection the following quotation of Blekhman et al. (1983) appears to be in order: "Significantly, the weakness of numerous works on stochastic models – sometimes ruling out any application – lies in the choice of statistical hypothesis, especially of assumptions regarding the probabilistic features of the given accidental quantities and functions. These features are often regarded as fully known (like an assumption of a normal distribution with known parameters), or as capable of determination. In real situations, it mostly turns out that the needed information is lacking. Moreover, as we have seen above, even small errors in probabilistic data may lead to large errors in estimating probabilities of failure." In this respect, the reader is highly recommended to consult with the article by Neal *et al.* (1992).

It should be borne in mind that, as Wentzel (1980) notes, probabilistic methods are "frequently regarded as a kind of 'magic wand' which produces information out of a void. This is a fallacy; the theory of probability only enables information to be *transformed*, and conclusions on inaccessible phenomena to be drawn from data on observable one."

The following conclusions appear to be relevant:
(1) the determination of the probability densities involved should be an integral part of any valid probabilistic project;
(2) the effect of human error should be introduced;

(3) the probabilistic approach may not always be an answer to the need for incorporating uncertainty; uncertainty is not always equivalent to randomness; in many situations the information about uncertainty may be of a non-stochastic nature; in these circumstances the approach based on fuzzy sets, or a new, convex modeling might be employed.

10.4 Convex models of uncertainty

As we have seen, probabilistic modeling requires extensive knowledge of the distribution functions involved. Is probabilistic modeling the only way one could describe uncertainty? Does uncertainty equal randomness? It turns out that the answer is no. Indeed, the indeterminacy about the uncertain variables involved could be stated in terms of these variables belonging to some sets, as elucidated by Schweppe (1973), Blekhman *et al.* (1983);

(1) The uncertain parameter x is bounded

$$|x| \leq a; \tag{10.43}$$

(2) The uncertain function has envelope bounds,

$$x_{lower}(t) \leq x(t) \leq x_{upper}(t) \tag{10.44}$$

where $x_{lower}(t)$ and $x_{upper}(t)$ are deterministic functions which delimit the range of variation of $x(t)$;

(3) The uncertain function has an integral square bound

$$\int_{-\infty}^{\infty} x^2(t)dt \leq a. \tag{10.45}$$

Instead of precise information on the probability content of random events, we possess imperfect, scarce knowledge on uncertain quantities. This description of uncertainty is a set-theoretic, *non-probabilistic one*. We will limit ourselves to the representation of uncertain phenomena by convex sets and will refer to this approach as convex modeling as was done by Ben-Haim and Elishakoff (1990).

In order to illustrate the fundamental issues in convex modeling, we consider dynamics and failure of structures with initial imperfections. In a monograph by Ben-Haim and Elishakoff (1990), particular examples of a bar and a cylindrical shell with axisymmetric initial imperfections were studied. In contrast to earlier studies, the uncertainty involved in the initial imperfections will be described by convex modeling.

Dynamics and failure of isotropic, unstiffened cylindrical shells are dealt with in a number of studies. These papers could be classified according to various criteria. One of them is whether the model of the initial imperfections and/or elastic properties is deterministic or probabilistic. Deterministic models of imperfection were adopted by Coppa and Nash (1964), Bieneck *et al.* (1966), Kornev (1969), and Zincik and Tenny-son (1980). Probabilistic modeling of initial imperfections for impact buckling problems was developed by Lindberg (1965) and Elishakoff (1979) in different contexts.

A "practical" criterion for the safe operation of the structure under dynamic loading was suggested by Hoff (1965): A safe operation region of the structure is one where "admissible finite disturbances of its initial state of static or dynamic equilibrium are followed by displacements whose magnitude remains within allowable bounds during the required lifetime of the structure." Applying Hoff's definition, one may assume that a shell with initial imperfections fails under axial forces when its dynamic response (deflection, strain or stress) first reaches an upper-bound level Q^+ or a lower-bound level, $-Q^-$, where Q^+ and Q^- are prescribed positive numbers that represent borderlines between failure and safe operation.

We will study impact loading of structures with initial imperfections. To simulate an often-encountered practical situation, we will assume that only limited information is available for the initial imperfections. In particular, we will assume that the initial imperfection Fourier coefficients fall within a given ellipsoidal set, which could be determined experimentally. With this limited information we determine the bounds of the total response. The ratio of the failure boundary to the worst-case response will define the minimum factor of safety, depending upon the definition of failure.

The differential equation governing the motion of a structure reads

$$L(w) + N\frac{\partial^2 w}{\partial x_1^2} + m\frac{\partial^2 w}{\partial t^2} = -N\frac{\partial^2 w_0}{\partial x_1^2} \qquad (10.46)$$

where $w_0(x)$ is the initial imperfection function, $w(x,t)$ is the additional displacement of the structure, N is the axial loading, m is the material density, and L is the elastic operator of the beam, plate or the shell. The eq. (10.46) is supplemented by boundary conditions

$$A(w) = 0 \ on \ \Gamma, \qquad (10.47)$$

where Γ is the boundary, as well as by initial conditions

$$w(x,t) = \frac{\partial w}{\partial t} = 0 \ at \ t = 0. \qquad (10.48)$$

The set of natural frequencies is found from the equation

$$L(w) + m\frac{\partial^2 w}{\partial t^2} = 0 \qquad (10.49)$$

where

$$w(\rho, t) = \psi_n(\rho)\exp(i\omega_n t) \qquad (10.50)$$

where $\psi_n(\rho)$ are the mode shapes, ω_n are the natural frequencies, ρ is the scalar for a one-dimensional structure and a vector of space coordinates for the two- or three-dimensional structures.

The initial imperfection function is expanded in the following Fourier series:

$$w_0(\rho) = \sum_{n=1}^{\infty} A_n \psi_n(\rho).$$
(10.51)

Similarly, we expand the additional displacement of the structure in a series as

$$w(\rho, t) = \sum_{n=1}^{\infty} G_n(\tau) \psi_n(\rho)$$
(10.52)

which leads to the following differential equation for $G_n(\tau)$:

$$\frac{d^2 G_n}{dt^2} + f_{1n}(\alpha) G_n(\tau) = f_{2n} A_n$$
(10.53)

where $f_{1n}(\alpha)$ is a function of the system parameters, as well as the non-dimensional load,

$$\alpha = \frac{N}{N_{cl}}$$
(10.54)

and f_{2n} depends in the system parameters, N_{cl} is the classical buckling load. The eq. (10.53) is supplemented by the initial conditions

$$G_n(0) = 0, \quad \frac{dG_n(0)}{dt} = 0$$
(10.55)

so that $G_n(t)$ can be represented in most general cases as

$$G_n(t) = A_n \Phi_n(t)$$
(10.56)

where generally $\Phi_n(t)$ has the following form

$$\Phi_n(t) = \begin{cases} \Phi_n^{(1)}(t), if\, \alpha > 1 \\ \Phi_n^{(2)}(t), if\, \alpha < 1 \\ \Phi_n^{(3)}(t), if\, \alpha = 1 \end{cases}$$
(10.57)

For beams, plates, and the shells $\Phi_n^{(1)}(t)$ contains hyperbolic cosine functions, and $\Phi_n^{(2)}(t)$ contains trigonometric cosine functions, whereas $\Phi_n^{(3)}(t)$ is a polynomial in t. The total displacement of the structure reads

$$w_\tau(\rho, t) = w_0(\rho) + w(\rho, t)$$

$$= \sum_{n=1}^{\infty} A_n [1 + \Phi_n(t)] \psi_n(\rho).$$
(10.58)

Let us assume that we have only limited deterministic information on the initial imperfections. In particular, the only available information is that the dominant N initial imperfection Fourier coefficients in eq. (10.58) fall within an ellipsoidal domain:

$$Z(\Omega,\theta) = \{A^T = (A_{m1}, A_{m2}, \ldots, A_{mN}): A^T\Omega A \le \theta^2\} \qquad (10.59)$$

where Ω is a positive definite symmetric matrix, θ^2 is a positive constant and $m_1, m_2, \ldots, m_N$ are the indices of the dominant imperfection amplitudes. We will assume that the rest of the Fourier coefficients vanish identically. Thus, eq. (10.58) becomes:

$$w(\rho, t) = \chi(\rho)^T A. \qquad (10.60)$$

Here χ is an $N \times 1$ vector whose nth element is

$$\chi_n(\rho, t) = [1 + \Phi_n(\tau)]\psi_n(\rho). \qquad (10.61)$$

The problem is formulated as one of finding the initial imperfection vector that maximizes the total displacement, given an imperfection ellipsoid of the initial imperfections.

We will denote this maximizing vector of initial imperfections A_{worst}. The maximum is formally represented as

$$v(\rho, t) = max_{A \varepsilon Z(\Omega,\theta)} w_T(\rho, t). \qquad (10.62)$$

The set of extreme points A of $Z(\Omega,\theta)$ is the ellipsoidal shell

$$C(\Omega,\theta) = \{A: A^T\Omega A = \theta^2\}. \qquad (10.63)$$

The set $C(\Omega,\theta)$ is a convex set, and $w_T(\rho, t)$ is a linear function of A. Hence, the maximum deflection will be reached on the extreme points of $Z(\Omega,\theta)$, i.e., on the ellipsoidal shell $C(\Omega,\theta)$. In other words:

$$v(\xi, \tau) = max_{A \varepsilon C(\Omega,\theta)} w_T(\rho, t). \qquad (10.64)$$

The closed form solution of $v(\rho, t)$ is obtained by the method of Lagrange multipliers, as shown by Elishakoff and Ben-Haim (1990) and Lindberg (1992a, 1992b). Define the Lagrangean as:

$$H(A) = \chi^T A + \lambda(A^T\Omega A - \theta^2), \quad A\Theta \equiv A. \qquad (10.65)$$

For the extremum, we require

$$\frac{\partial H}{\partial A} = \chi + 2\lambda\Omega A = 0 \qquad (10.66)$$

or

$$A = -\frac{1}{2\lambda}\Omega^{-1}\phi. \tag{10.67}$$

Combining this expression with the constraint on $A, A^T\Omega A = \theta^2$, results in

$$A^T\Omega A = \frac{1}{4\lambda^2}\chi = \theta^2 \tag{10.68}$$

or

$$\lambda^2 = \frac{1}{4\theta^2}\chi^T\Omega^{-1}\chi. \tag{10.69}$$

This results in the vector causing the worst (maximum) deflection:

$$A_{worst} = \frac{\theta}{\sqrt{\chi^T\Omega^{-1}\chi}}\Omega^{-1}\chi. \tag{10.70}$$

This yields a closed form solution for the maximum displacement, least favorable response (L.F.R.) in view of eq. (10.68)

$$L.F.R. = \phi^T A_{worst} = \theta\sqrt{\chi(\xi,\tau)^T\Omega^{-1}\chi(\xi,\tau)}. \tag{10.71}$$

It is remarkable that A_{worst} is a function of the vector $\chi(\rho,\tau)$, which means that A_{worst} depends on time t and on the space coordinates ρ. The dependence of A_{max} on δ and t should be interpreted as follows: the initial imperfection vector which maximizes the total displacement at time instance t at the point ρ depends on t and ρ. In other words, different initial imperfections will maximize the total displacement at different ρ and t.

The analysis presented in this section is a further generalization of the approach developed by Elishakoff and Ben-Haim (1990). Additional models of convex analysis of initial imperfections were given by Lindberg (1992a, 1992b) and Ben-Haim (1993). Contrasting of probabilistic and convex modeling in buckling context was provided in recent investigations by Elishakoff et al. (1991a, 1991b). The other models of convex excitations are considered in a monograph by Ben-Haim and Elishakoff (1990). A particular case of convex modeling, namely, ellipsoidal modeling, received wide coverage in studies by Schweppe (1973, 1968), Chernousko (1982, 1994), Kurzhanskii (1980), and other investigators, in the control problems contexts.

10.5 Davies-Hammond bound

In this section we describe a method developed recently at the Institute of Sound and Vibration in Southhampton by Davies and Hammond (1988,1986). Hereinafter, we will follow Davies and Hammond (1986). The main objective is to predict upper bounds of the response of linear and nonlinear systems. The concept of an analytic signal is of importance in this context. A complex-values function

$$f(z) = u(z) + iv(z), z = x + iy \tag{10.72}$$

is analytic in a specified domain D if it is both single valued and differentiable at all points in D. Analytic functions satisfy (Cauchy-Riemann equations)

$$\frac{\partial u}{\partial x} = \frac{\partial v}{\partial y}, \frac{\partial u}{\partial y} = -\frac{\partial v}{\partial x}. \tag{10.73}$$

These equations imply that if the real part of an analytic function is known, then the imaginary part can be determined within the additive constant. We consider a function of a single real variable, which can be considered to be a complex function analytic everywhere in the upper half plane and evaluated along the real axis:

$$f(x + i0) = u(x + i0) + iv(x + i0). \tag{10.74}$$

The relationship between $u(x)$ and $v(x)$

$$v(x) = \frac{1}{\pi} \int_{-\infty}^{\infty} \frac{u(p)}{x - p} dp \tag{10.75}$$

is known as the Hilbert transform of $u(x)$.

The signals of interest are time histories and are therefore functions of t. These signals will be assumed to be continuous and bounded for all t. Therefore, the Hilbert transform of the signal will give the imaginary part of the analytic signal. Given a signal, then the analytic signal, of which $x(t)$ is the real part, will be denoted by

$$s_x = x(t) + i\dot{x}(t) \tag{10.76}$$

where $\dot{x}(t)$ is the Hilbert transform of .

If $x(t)$ is the excitation and $y(t)$ the response of a linear system then

$$y(t) = \int_{-\infty}^{\infty} x(\tau)h(t - \tau)d\tau \tag{10.77}$$

where $h(t)$ is the system's impulse response. We will consider $y(t)$ to be the real part of a complex analytic function $s_y(t)$ whose imaginary part $\dot{y}(t)$ is the Hilbert transform of $y(t)$

$$s_y(t) = y(t) + i\ddot{y}(t) \tag{10.78}$$

Furthermore, the $s_y(t)$ impulse response function can also be written as

$$s_h(t) = h(t) + i\dot{h}(t). \tag{10.79}$$

The relationship between $s_x(t)$, $s_y(t)$ and $s_h(t)$ is Davies and Hammond (1988), Davies (1985), Davies and Hammond (1986), and Dyne *et al.* (1988)

$$s_y(t) = \frac{1}{2} \int_{-\infty}^{\infty} s_x(t)s_h(t-\tau)d\tau. \tag{10.80}$$

The upper bound for the response of the system is derived from taking the modulus of each side, which yields

$$A_y(t) \leq \frac{1}{2} \int_{-\infty}^{\infty} A_x(\tau)A_h(t-\tau)d\tau \tag{10.81}$$

where $A_x(t)$ and $A_h(t)$ are the envelopes

$$s_x(t) = A_x(t)\exp[i\Phi_x(t)] \tag{10.82}$$

$$s_y(t) = A_y(t)\exp[i\Phi_y(t)] \tag{10.83}$$

$$s_h(t) = A_h(t)\exp[i\Phi_h(t)] \tag{10.84}$$

and $\Phi_x(t)$, $\Phi_y(t)$, and $\Phi_h(t)$ are the phases.

Doctoral thesis by Davies (1985) describes the effectiveness of the inequality eq. (10.81) for different types of systems. Envelopes of the response of nonlinear oscillators are discussed by Davies and Hammond (1988) where a structure subjected to a shock excitation was studied. Computation has shown that the factor by which the upper bound exceeds the response envelope peak is affected by the physical properties of the system. The overshooting factor of two is typical, but for the system with fast convection speed, this factor is reduced to about 1.2.

10.6 Bounds by Drenick and Shinozuka

Let us briefly review the bounds for the response of a single-degree-of-freedom system, which has been developed by Drenick and his associates (1968, 1970, 1977, 1984, 1979) and Shinozuka (1970) in the context of earthquake excitation. Although the hints on the necessity to bound the systems' response appeared earlier, Drenick and Shinozuka were the pioneers of such an approach in structural engineering.

Consider again a single-degree-of-freedom structure. The response of the system is written as in eq. (10.77). In order to find the maximum response, we employ the Cauchy-Schwarz inequality

$$\left(\int_{-\infty}^{\infty} h(t-\tau)x(\tau)d\tau \right)^2 \leq \int_{-\infty}^{\infty} h^2(t-\tau)d\tau \int_{-\infty}^{\infty} x^2(\tau)d\tau. \tag{10.85}$$

Equality in eq. (10.85) is obtained when $x(\tau)$ is proportional to $h(1-\tau)$. Therefore, the maximum in eq. (10.85) is achieved if

$$x(\tau) = ah(t-\tau) \tag{10.86}$$

where a is the proportionally constant. The excitation satisfies the inequality

$$\int_{-\infty}^{\infty} x^2(\tau)d\tau \leq M^2. \tag{10.87}$$

The maximizing excitation satisfies the equality

$$\int_{-\infty}^{\infty} x^2(\tau)d\tau = M^2. \tag{10.88}$$

Hence,

$$a = M \left[\int_{-\infty}^{\infty} h^2(t-\tau)d\tau \right]^{-1/2}. \tag{10.89}$$

The least favorable response equals

$$L.F.R. = Mh(t-\tau)\left[\int_{-\infty}^{\infty} h^2(t-\tau)d\tau \right]^{-1/2}. \tag{10.90}$$

It is remarkable that the excitation $x(\tau)$ during the interval $[0,t]$, which maximizes the response at time t, is proportional to $h(t-\tau)$. We denote

$$\int_{-\infty}^{\infty} h^2(t-\tau)d\tau = N^2. \tag{10.91}$$

Equation (8.89) becomes

$$a = \frac{M}{N}. \tag{10.92}$$

The least favorable response takes the form

$$L.F.R. = MN. \tag{10.93}$$

The bound given in eq. (10.93) belongs to Drenick. Equation (10.87) stipulates that we know the bound on excitation energy. However, we may be able to obtain more information than just a bound on the excitation energy. Shinozuka (1987) has suggested characterizing the excitation uncertainty by specifying an envelope of the Fourier amplitude spectrum. In particular we presume knowledge of an envelope on the magnitude of the Fourier spectrum of the excitation:

$$|X(\omega)| \leq X_e(\omega) \tag{10.94}$$

where

$$X(\omega) = \int_{-\infty}^{\infty} x(t)e^{-i\omega t}d\omega. \tag{10.95}$$

The bound by Shinozuka reads

$$|y(t)| \leq \frac{1}{2\pi} \int_{-\infty}^{\infty} |H(\omega)||X_e(\omega)d\omega. \tag{10.96}$$

For envelopes $X_e(\omega)$ Shinozuka (1970) suggested three different analytical forms. Ideas of bounding the response which are spectral-distribution free were further developed by Shinozuka and Deodatis (1988a, 1988b, 1991). In the following section the use of linear programming will be illustrated for convex analysis of structures subjected to base excitation.

10.7 Convex modeling of base excitation

Tremendous advances have been made in recent years in probabilistic modeling of earthquake excitation. These studies are based on the presumed knowledge of probabilistic information on the excitation. This knowledge in most situations is not highly reliable. An alternative approach was pioneered by Drenick (1968, 1970, 1977), Drenick *et al.* (1984), Bedrosian *et al.* (1980), Drenick and Yun (1979), and Shinozuka (1970). Within these analyses, the description of base excitation is not probabilistic, but rather several bounds on the excitation are assumed to be known. Drenick (1968, 1970) used a constraint on the total energy which the earthquake was likely to develop at the location of the structure.

In this section we utilize linear programming techniques in an analogous context, for the model structure with the base excitation. We use the knowledge of the base acceleration bound, as well as of its velocity or displacement. Our main concern is the following question: how the maximum possible response is modified, if at all, as the amount of global information on the excitation increases?

Consider a single degree of freedom system, subjected to base excitation:

$$m\ddot{x} + c\dot{x} + kx = -m\ddot{x}_b \qquad (10.97)$$

where $x(t)$ is the response of the structure, m is its mass, c and k are the damping and stiffness coefficients, respectively, and $\ddot{x}_b$ is the base acceleration. This acceleration function is treated here as an uncertain function. We will assume that $\ddot{x}_b$ is a monotonously decreasing function, so that the right–hand side of eq. (10.1), namely $m\ddot{x}_b$ is a monotonously increasing function. The base force

$$f_b(t) = -m\ddot{x}_b \qquad (10.98)$$

can be represented as follows:

$$f_b(t) = \sum_{i=1}^{L} A_i \sin\Omega_i t \qquad (10.99)$$

where Ω_i are representative excitation frequencies, A_i are appropriate amplitudes, and L are the number of terms in series eq. (10.99). The response $x(t)$ is written in terms of the convolution integral

$$x(t) = \int_0^t f(\tau)h(t-\tau)d\tau \qquad (10.100)$$

with $h(t)$ the impulse response function

$$h(t) = \frac{\exp(-\zeta\omega_n t)}{m\omega_d}\sin\omega_d t$$

$$\omega_d = \omega_n\sqrt{1-\zeta^2} \qquad (10.101)$$

where $\omega_n^2 = k/m$ is the natural frequency squared and $\zeta = \left(\frac{c}{2}\right)km$ is the damping ratio. Bearing in mind eq. (10.99), the response is written as

$$x(t) = \sum_{i=1}^{L} A_i \Phi_i(t) \qquad (10.102)$$

where

$$\Phi_i(t) = \frac{1}{m\omega_d}\int_0^t \sin(\Omega_i\tau)\exp[-\zeta\omega_n(t-\tau)] \times \sin[\omega_d(t-\tau)]d\tau. \qquad (10.103)$$

If the coefficients A_i are fully specified, the response to the excitation in eq. (10.99) is calculated in a straightforward fashion. Usually, however, the excitation is an uncertain function. Realistic situations may be represented by the knowledge of a few bounds on the excitation function, such as a bound on base acceleration

$$\ddot{x}_b \le a_0. \tag{10.104}$$

Sometimes we may possess more information on the excitation function. For instance, we may know as well that the base velocity is also a bounded quantity

$$\dot{x}_b \le a_1 \tag{10.105}$$

or that the base displacement is bounded

$$x_b \le a_2. \tag{10.106}$$

The combined sets $Z_2 = \{\ddot{x}_b \le a_0\} \cap (x_b \le a_0)\}$ or $Z_3 = \{(\ddot{x}_b \le a_0) \cap (x_b \le a_0) \cap (x_b \le a_0)\}$ are convex, since they represent intersections of convex sets, as shown by Elishakoff and Pletner (1991). Here we will apply the linear programming method to tackle the problem. The central goal of the present investigation is to learn how the global information on the response is modified as the knowledge of the excitation is increased. Equations (10.104–10.106) read in terms of coefficients A_i

$$\sum_{i=1}^{L} A_i sin\Omega_i t \le a_0 \tag{10.107}$$

$$\sum_{i=1}^{L} A_i \frac{1 - cos\Omega_i t}{\Omega} \le a_1 \tag{10.108}$$

$$\sum_{i=1}^{L} A_i \left[\frac{t}{\Omega_i} - \frac{sin\Omega_i t}{\Omega_i^2} \right] \le a_2. \tag{10.109}$$

The objective function is given by eq. (10.100). We are interested in finding such a combination of A_i, which will cause the response as defined in eq. (10.100) to attain a maximum. The problem can be stated in matrix notation as follows:

$$maximize\ x(t) = A^T \Phi(t) \tag{10.110}$$

$$subject\ to\ C_a A \le b. \tag{10.111}$$

Where in the case of three constraints the matrix C_a reads

$$C_a = \begin{bmatrix} sin\Omega_1 t & \cdots & sin\Omega_L t \\ \frac{1}{\Omega_1}(1 - cos\Omega_1 t) & \cdots & \frac{1}{\Omega_L}(1 - cos\Omega_L t) \\ \frac{1}{\Omega_1}(\Omega_1 t - sin\Omega_1 t) & \cdots & \frac{1}{\Omega_L^2}(\omega_L t - sin\Omega_L t) \end{bmatrix} \tag{10.112}$$

and

$$A^T = [A_1\}A_2 \cdots A_L] \tag{10.113}$$

$$b^T = [a_0 \ a_1 \ a_2].$$
(10.114)

As is seen from eqs. (10.110)–(10.111), we are dealing with the maximization of the linear function $x(t)$ of the parameters $A_i(i=1, \ldots, L)$ subjected to linear constraints. The question arises whether or not this can be posed as a mathematical programming problem.

We have assumed that the base acceleration is a monotonously decreasing function. In addition, we visualize that we are interested in the response of the structure during the time interval $(0,T)$. We then use the odd periodic continuation of the excitation, so that only the sine terms, as given in eq. (10.99) will appear in the Fourier series with the coefficients

$$A_i = \frac{2}{T} \int_0^T (-m\ddot{x}_b) \sin\frac{i\pi t}{T} dt$$
(10.115)

where the serial index, i is also a number of half sine waves of the period T. Let us visualize the function $-m\ddot{x}_b$, which is an increasing function, and the function $\sin(i\pi t/T)$, for the particular case in which $i = 3$. The integral can be divided into three parts. The integral I_1 between zero and $T/3$ is negative; the integral I_2 between $T/3$ and $2T/3$ is positive, but smaller than the absolute value of I_1, since $-m\ddot{x}_b$ is an increasing but negative function. The integral I_3 over the interval $2T/3$ to T is negative. Therefore, the sum $I_1 + I_2 + I_3$ is negative. This implies that A_3 is negative. By analogous reasoning, all A_i are negative.

In order to achieve standard linear programming formulation in which the unknown coefficients must be non-negative, we introduce new positive coefficients

$$\bar{A}_i = -A_i.$$
(10.116)

The objective function becomes

$$x(t) = \tilde{A}^T \Psi(t)$$
(10.117)

where

$$\Psi(t) = -\Phi(t)$$
(10.118)

subject to constraints

$$-C_a \tilde{A} \le b.$$
(10.119)

In eqs. (10.54)–(10.56) we should substitute

$$\Omega_i = \frac{i\pi}{T}.$$
(10.120)

In case of two mode approximation, $L = 2$, the linear programming procedure is capable of simple representation. We use the system from Clough and Penzien (1975) with values

$$m = 2 \left[\frac{kips \cdot s^2}{in} \right], \quad k = 60 \left[\frac{kips}{in} \right], \quad c = 0.438 \left[\frac{kips \cdot s}{in} \right]. \tag{10.121}$$

Limiting ourselves to the time interval (0,12), and optimizing at $t = 5s$ while using the representative frequencies

$$\Omega_1 = \frac{\pi}{12} = 0.262 \left[\frac{rad}{s} \right], \quad \Omega_2 = \frac{3\pi}{12} = 0.785 \left[\frac{rad}{s} \right] \tag{10.122}$$

we get the following expression for the response in terms of the uncertain amplitudes $\tilde{A}_i$

$$x = -0.0157\tilde{A}_1 + 0.131\tilde{A}_2. \tag{10.123}$$

The constraint coefficients matrix C_a becomes

$$C_a = \begin{bmatrix} 0.966 & -0.707 \\ 2.831 & 2.174 \\ 5.005 & 7.513 \end{bmatrix} \tag{10.124}$$

Letting the upper bounds vector b equal

$$b^T = [1.05.010.0]. \tag{10.125}$$

The graphical solutions for this problem given by Elishakoff and Pletner (1991) are useful for gaining insight into linear programming solutions, but in the general case the simplex algorithm must be used. This requires the use of slack variables, in this case S_1, S_2, and S_3, in order to arrive at the equality constraints needed by the simplex algorithm. Solution by simplex for this example yields the following (Elishakoff and Pletner, 1991):

    most favorable response (in absolute value), 0.01625 in;
    least favorable response (in absolute value), 0.01744 in.

Let us now consider a proportionally damped coupled dynamic system with N degrees of freedom, the motion of which is governed by the following set of equations:

$$M\ddot{x} + C\dot{x} + Kx = F_{base} \tag{10.126}$$

We can express F_{base} as a function of base acceleration

$$F_{base} = -Mj\ddot{x}_b(t) \tag{10.127}$$

where j denotes vector $j^T = [1, 0, \ldots, 1]$. In order to uncouple the system, we first solve the eigenvalue problem

$$\det\left[\boldsymbol{M}^{-1}\boldsymbol{K} - \lambda I\right] = 0 \qquad (10.128)$$

and obtain the natural frequencies ω_i, and respective natural modes u_i, which constitute the columns of the modal matrix $\boldsymbol{P}$. Defining generalized coordinates

$$y = \boldsymbol{P}x \qquad (10.129)$$

and pre-multiplying eq. (10.126) by $\boldsymbol{P}^T$, we obtain the decoupled system for the proportionally damped system

$$\boldsymbol{M}_0\ddot{y} + \boldsymbol{C}_0\dot{y} + \boldsymbol{K}_0 y = -\boldsymbol{P}^T \boldsymbol{M}j\ddot{x}_b \qquad (10.130)$$

where $\boldsymbol{M}_0, \boldsymbol{C}_0$ and $\boldsymbol{K}_0$ are, respectively, the diagonal mass, damping and stiffness matrices of the system in generalized coordinates. Pre-multiplying system (10.77) by $\boldsymbol{M}_0^{-1}$, we get N equations in canonic form

$$\ddot{y}_i + 2\zeta_i\omega_i\dot{y}_i + \omega_i y_i = \phi_i\ddot{x}_g \quad i=1, \ldots, N \qquad (10.131)$$

where ϕ_i are the components of the vector $\boldsymbol{M}^{-1}\boldsymbol{P}^T \boldsymbol{M}j$.

Each of these equations can be solved by the convolution integral technique. We represent the base acceleration again as in

$$\ddot{x}_g = \sum_{i=1}^{L} A_i sin\Omega_j t. \qquad (10.132)$$

The Fourier coefficients are written in the familiar form to eq. (10.62) as

$$A_i = \frac{2}{T}\int_0^T (-\ddot{x}_b)sin\frac{i\pi t}{T} dt. \qquad (10.133)$$

The response of the jth DOF (degree of freedom) in generalized coordinates can now be found as a linear combination of A_i as follows:

$$y_j(t) = \sum_{i=1}^{L} A_i \bar{\Phi}_i^{(j)}(t) \qquad (10.134)$$

$\bar{\Phi}_i^{J}$ can be derived directly by substituting m_j, ζ_i and ω_j for m, ζ and Ω_n, respectively, and multiplying by, $\boldsymbol{M}_j$ are the diagonal terms of the matrix $\boldsymbol{M}$. The vector y must be pre-multiplied by the inverse of the modal matrix $\boldsymbol{P}$ to obtain an expression for the response of each DOF in the primitive coordinates. This results in

$$\Phi_i^{(j)} = \sum_{r=1}^{N} \hat{P}_{jr}\Phi_i^{(ir)} (i=1, \ldots, L; j=1\ldots, N) \qquad (10.135)$$

where P_{jr} is the rth term of the jth row of the matrix P^{-1}. This again is a mathematical programming problem. It is now possible to estimate the extremal response of the jth DOF by solving the linear programming problem. Maximize (or minimize)

$$x_j(t) = \tilde{A}^T \Psi^{(j)}(t), (j=1, \ldots, N) \tag{10.136}$$

subject to

$$C_d \tilde{A} \le b \tag{10.137}$$

where $\Psi^j = - -\Phi^j$, and Φ^j is a vector of components of which are Φ^j_i, $i=1, \ldots, L$. Other symbols are defined in previous sections.

In the textbook by Clough and Penzien (1975, p. 560), a three-story building subjected to base type excitation is approximated by a lumped three DOF structure. The system has the following characteristics:

$$
\begin{aligned}
m_1 &= 1.0 \left[\frac{kips}{in} s^2\right], \quad k_1 = 60 \left[\frac{kips}{in}\right] \\
m_2 &= 1.5 \left[\frac{kips}{in} s^2\right], \quad k_2 = 120 \left[\frac{kips}{in}\right] \\
m_3 &= 2.0 \left[\frac{kips}{in} s^2\right], \quad k_3 = 180 \left[\frac{kips}{in}\right].
\end{aligned} \tag{10.138}
$$

The mass and stiffness matrices become

$$
M = \begin{bmatrix} 2.0 & 0 & 0 \\ 0 & 1.5 & 0 \\ 0 & 0 & 1.0 \end{bmatrix} \left[\frac{kips}{m} s^2\right] \tag{10.139}
$$

$$
K = \begin{bmatrix} 60 & -60 & 0 \\ -60 & 180 & -120 \\ 0 & -120 & 300 \end{bmatrix} \left[\frac{kips}{in}\right]. \tag{10.140}
$$

Solution of eigenvalue problem yields

$$\omega_1 = 4.59 \left[\frac{rad}{s}\right], \quad \omega_2 = 9.82 \left[\frac{rad}{s}\right], \quad \omega_3 = 14.58 \left[\frac{rad}{s}\right] \tag{10.141}$$

$$
u_1 = \left\{ \begin{array}{c} 1.0 \\ 0.649 \\ 0.302 \end{array} \right\}, \quad u_2 = \left\{ \begin{array}{c} 1.0 \\ -0.607 \\ -0.679 \end{array} \right\}, \quad u_3 = \left\{ \begin{array}{c} -0.393 \\ 1.0 \\ -0.960 \end{array} \right\}. \tag{10.142}
$$

Bearing in mind that the damping ratio ζ is given as 5% for each mode, the decoupled equations of motion in generalized coordinates can now be explicitly written

$$\ddot{y}_1 + 0.459\dot{y}_1 + 21.07y_1 = -3.6682\ddot{x}_g$$

$$\ddot{y}_2 + 0.982\dot{y}_2 + 96.43y_1 = 2.0605\ddot{x}_g \qquad (10.143)$$

$$\ddot{y}_1 + 1.458\dot{y}_1 + 212.58y_1 = -1.8341\ddot{x}_g$$

Assuming that the base acceleration is approximated by only two modes over the time interval (0,15 s), and that the representative excitation frequencies are

$$\Omega_1 = \frac{\pi}{15} = 0.209 \left[\frac{rad}{s} \right],$$

$$\qquad (10.144)$$

$$\Omega_2 = \frac{3\pi}{15} = 0.628 \left[\frac{rad}{s} \right]$$

the response vectors $\Psi^{(j)}$ become, at the time instant $t = 5s$

$$\Psi^{(1)} = \left\{ \begin{array}{c} -0.0365 \\ -0.0756 \end{array} \right\} \quad \Psi^{(2)} = \left\{ \begin{array}{c} -0.0963 \\ -0.1189 \end{array} \right\} \quad \Psi^{(3)} = \left\{ \begin{array}{c} -0.1141 \\ -0.1088 \end{array} \right\} \qquad (10.145)$$

The constraint coefficient matrix C_a assumes the form

$$C_a = \begin{bmatrix} 0.886 & 0 \\ 2.387 & 3.183 \\ 4.130 & 7.958 \end{bmatrix}. \qquad (10.146)$$

Supposing that the upper bounds vector **b** is

$$b^t = [1.3 \ 5.0 \ 10.0]. \qquad (10.147)$$

Solution by linear programming Leunberger (1984) yields following numerical results:
    least favorable response of the first mass, 0.095 in;
    least favorable response of the second mass, 0.1975 in;
    least favorable response of the third mass, 0.2197 in.

One should stress that in order to use classic linear programming solution we have assumed that the base acceleration was a monotonously decreasing function. This led to the conclusion that the coefficients A_j are all positive. However, positiveness of the coefficients is not a sufficient condition the function $-m\ddot{x}_b$ to be a monotonically increasing function. Indeed, this is clearly seen when one retains only one term in expansion eq. (8.99). Positiveness of A_1 does not guarantee that the function $-m\ddot{x}_b = m\Omega_1^2 A_1 \sin \Omega_1 t$ is a monotonically increasing function. The topic for multi-term approximation should be further investigated. Anyway, the positveness property of Fourier coefficients A_i can be abandoned by representing them as $A_j = B_j - C_j$ where both B_j and C_j are positive. The extended problem can again be solved by the classic linear programming methods.

10.8 Robust control under bounded excitation

There is a vast literature dedicated to robust control of systems subjected to bounded uncertainties. For authoritative reviews readers may consult with recent review by the dean of uncertainty modeling in the science of control, Professor George Leitmann of University of California, Berkeley (1993) and Corless (1993). Selected studies conducted by Leitmann *et al.* (1979, 1993). Other representative studies are by Schmitendorf (1986), Schmitendorf and Barmish (1986), Subbotin and Chentsov (1981), Ben-Haim (1990a, 1990b). In their paper Kelly *et al.* (1987), studied the robust control in conjunction with base isolation in order to assume arbitrarily small motion of a seismically excited structure. The proposed method required control force application only at the base (first) floor. The efficacy of the scheme was illustrated by extensive simulations for a prototype six-story building. Ellipsoidal modeling in the context of control was pioneered by Fred Schweppe of Massachusetts Institute of Technology (1968, 1973) and developed extremely thoroughly by Felix Chernousko of the Russian Academy of Sciences (1982, 1983, 1981, 1982, 1983, 1994) and Alexandr Kurzhanskii (1980).

10.9 Response variability of viscoelastic structures

Vibration and stability of viscoelastic structures have been dealt with in a number of monographs by Bland (1975) and Flügge (1975). In these studies, material properties of the structure have been fixed at some deterministic parameters. However, it is well established that the viscoelastic properties of structures exhibit a large scatter (Koltunov, 1976). This scatter is usually accounted for by considering the material properties as random variables. Huang and Cozarelli (1972, 1973) were apparently the first investigators to include material uncertainty in their analyses. Recently, Hilton *et al.* (1991) extended the elastic-viscoelastic analogies to the stochastic case due to random linear viscoelastic material properties. Both normal and beta distributions were considered for random variables modeling the uncertainty in the data.

In probabilistic analyses, the needed probabilistic information for analysis was postulated as given. For example, Huang and Cozarelli (1973) utilize a log normal density or a truncated log-normal density. However, extensive experimental data is needed to substantiate the probability densities with regards to the data.

More often, the necessary data is simply lacking, or only fragmentary information is available about the parameters. In such circumstances the usefulness of the results of probabilistic modeling, when it is based on limited data, may be questionable, especially when the scatter is large, and when extremely small probabilities of failure (in the range between $P_f = 10^{-10}$ and $P_f = 10^{-7}$) are sought. Indeed, scatter in material properties was treated within the realm of the stochastic finite element method (Ste fanoe, 2009), which requires knowledge of the auto-correlation function of random material properties. However, as Shinozuka (1987) mentions, "it is recognized that it is rather difficult to estimate experimentally the autocorrelation function, or in the case of weak homogeneity, the spectral density function of the stochastic variation of material properties. In view of this, the upper bound results are particularly important, since the bounds derived . . . do not require knowledge of the autocorrelation function."

In the stochastic context, Shinozuka (1987), and Shinozuka and Deodatis (1988), and Deodatis and Shinozuka (1989) derived upper or lower bounds on the probabilistic characteristics of the response in terms of probabilistic characteristics of the material variability.

Convex modeling of the material properties of viscoelastic structures was proposed by Elishakoff *et al.* (1994). The transverse vibrations of a viscoelastic bar due to harmonic excitation were studied. Viscoelastic parameters were treated as varying in a solid "ball" in the four-dimensional space, thus modeling the scatter in material properties. The maximum possible response needed for design of such beams was analytically determined.

In the paper by Elishakoff *et al.* (1994) the authors developed a non-probabilistic method to predict the variability in buckling loads of composite plates and shells resulting from the scatter in elastic moduli. The available measurements of moduli were fitted by the uncertainty ellipsoid. The lower bound of the buckling load was derived. This lower bound, rather than the nominal buckling load should be used for design of structures susceptible to buckling.

10.10 Nagging question

The natural, nagging question arises, "Who is right? Those who disagree with the probabilistic methods or those who revere the probabilistic methods?" Paradoxically, it appears that both probabilists and non-probabilists may be correct! In this connection the following ancient story appears to be extremely instructive. The judge was operating the court from his house since the courthouses were absent (Fig. 10.1). Once two men accused each other. The judge decided to listen to the complainants separately. He listened carefully to the first suitor, and in the end the judge told him, "You are right!" Then the judge listened to the second claimant at length and informed him. "You are right!" The wife of the judge, who was at home, happened to overhear the suit. She could not refrain herself from asking the judge, "You told both suitors, that they were right. But this is impossible. They could not be equally correct!" The judge then repeated after some thinking, now to this wife, "You are right."

Thus, the judge acted more like a meditator in the situation, which he deemed to be bridgeable. Likewise, in the situation under discussion too, the probabilistic ideas and the set-theoretic ideas can be bridged.

When sufficient information is available on uncertain quantities one can successfully utilize the probabilistic models. When only limited information is available, one must *not* resort to the probabilistic model, by *arbitrarily* choosing the distribution function of random variables or functions. In this circumstance the convex modeling of uncertainty appears to be a viable alternative. In some *intermediate* situations one may possess sufficient information to justify a probabilistic approach, which may turn out to be the only feasible approach, but for the parameters of the distribution we may have extremely limited data. In these circumstances the combined probabilistic-convex modeling will be most logical, as will be illustrated in Section 12.

10.11 Critical comparison of probabilistic and convex approaches

A natural question arises: can one perform, on the same engineering problem, a direct comparison between probabilistic and convex approaches? Such a question was kindly communicated to the present writer by Professor Stephen Harry Crandall of the Massachusetts Institute of Technology. In this section this question will be addressed using an example of the column with uncertain initial imperfections, resting in nonlinear elastic foundation.

It is known that an initial geometric imperfection in a structure, inevitably present due to the very nature of the manufacturing process, has a significant effect on the buckling load of the structure. The load carrying capacity of many shell structures is imperfection–sensitive in the sense that the load they carry can be significantly reduced due to the presence of small imperfections. This conclusion, which constituted the major breakthrough in the research in buckling of structures (about 200 years after Euler first uncovered the expression for the buckling load of a uniform column) is mainly due to the work of Koiter (1970, 1963) and Budiansky and Hutchinson (1964, 1979). However, despite the acceptable theoretical explanation of the buckling behavior of shells, the use of the concept of imperfection sensitivity in engineering practice is still in the *ad hoc* stage and engineers prefer to rely on the *knockdown factor* (NASA, 1968), chosen so that its product with the classical buckling load yields a lower bound to available experimental data. The apparent reluctance to take advantage of theoretical findings stems from the fact that most imperfection studies are conditional on advance deterministic knowledge of the geometric imperfections of the particular structure, which is rarely possible. In the ideal case the imperfections can be measured experimentally and incorporated in the theoretical analysis to predict the buckling loads. This approach, while justified for single prototype-like structures, is impracticable as a general method of behavior prediction. Information of a particular structure would be too specific and not strictly valid for other realizations of the same structure, even those obtained by the same manufacturing process.

In the light of these considerations and bearing in mind the large scatter of the experimental results, it appears obvious that practical applications of the imperfection-sensitivity theories are conditional on their being combined with taking into account the uncertainty of initial imperfections and the loads the structure can support.

Fig. 10.1: Discussion of the nagging question: who is right? (from the paper Elishakoff I., Essay on Uncertainties in Elastic and Viscoelastic Structures: From A. M. Freudenthal's Criticisms to Modern Convex Modeling, *Computers and Structures* Vol. 56(6), 871–895, 1995a, by permission).

The notion of stochasticity of the initial imperfections was introduced by Bolotin (1962). For selective reviews on random initial imperfections the reader may consult with studies of Roorda (1972), and Amazigo (1976), Elishakoff (1978, 1979, 1980, 1986, 1983, 1988a, 1988b) and Elishakoff and co-workers (1982, 1985, 1987, 1989).

Some difficulties arising in probabilistic modeling of imperfections in buckling of structures were studied by Elishakoff and Nordstrand (1991). Convex modeling of uncertain initial imperfections was initiated by Ben-Haim and Elishakoff (1989, 1990, 1989). The critical contrast between the probabilistic and convex modeling of uncertainty for buckling of structures was accomplished by Elishakoff *et al.* (1994). Detailed comparison of exact solutions and perturbation solutions within the convex modeling was performed in the paper by Elishakoff *et al.* (1991).

We will briefly review the pertinent conclusions of the above paper by Elishakoff *et al.* (1994). The model problem of the beam on elastic foundation was considered. The governing equation and boundary conditions read

$$EI\frac{d^4w}{dx^4} + P\frac{d^2w}{dx^2} + k_1w - k_3w^3 = -P\frac{d^2w}{dx^2}$$

$$w = \frac{d^2w}{dx^2} = 0 \ at \ x = 0 \ and \ x = L \qquad\qquad (10.148)$$

where $\bar{w}$ is the initial imperfection, w is the additional deflection due to axial load P, EI is the bending rigidity, both k_1 and k_3 are positive constants, representing, respectively, the linear and non-linear spring constants of the elastic foundation, $L = $ length.

The initial imperfection is expanded as

$$\bar{w}(x) = \sum_{m=1}^{\infty} \bar{X}_m \sin\left(\frac{m\pi x}{L}\right). \tag{10.149}$$

The solution of eq. (10.148) is sought also in the form of Fourier series

$$w(x) = \sum_{m=1}^{\infty} X_m \sin\left(\frac{m\pi x}{L}\right). \tag{10.150}$$

The series (10.149) and (10.150) are truncated and only M terms are retained in each of them.

Substitution of eqs. (10.149) and (10.150) into eq. (10.148) and application of Galerkin's method yields the following set of non-linear algebraic equations:

$$a_m X_m - a(X_m + \bar{X}_m) - \frac{s}{8} \frac{m_*^2}{m^2} I_m = 0 \tag{10.151}$$

where

$$S = \frac{2k_3}{k_1}, \quad a_m = \frac{\pi^2 m^2 + k_1 \left(\pi^2 m^2\right)}{\pi^2 m_*^2 + k_1 / \left(\pi^2 m^2\right)}$$

$$I_m = \sum_{p=1}^{\infty} \sum_{q=1}^{\infty} \sum_{r=1}^{\infty} X_p X_q X_r [\delta_{p+qs+m} - \delta_{|p-q|s+m|} - \delta_{p+q|r-m|}$$

$$+ \delta_{|p-q||r-m|} \delta_{p,q} \delta_{r,m}] \tag{10.152}$$

where $\delta_{p,q}$ is the Kronecker's delta, m is the number of modes corresponding to the buckling of the corresponding linear column without initial imperfections.

The limit load a^* is defined as

$$\left[\frac{da}{dF}\right]_{a=a*} = 0 \tag{10.153}$$

where

$$F = F(w, \bar{w}) = \int_0^L \left[\frac{1}{2}\left(\frac{dw}{dx}\right)^2 + \frac{dw}{dx}\frac{d\bar{w}}{dx}\right] dx \tag{10.154}$$

representing the end shortening of the column. According to this definition, the buckling load a^* represents a maximum load the structure can sustain.

Numerical method to find a^* was developed by Elishakoff *et al.* (1994). The stochastic analysis was based on the assumption that the initial imperfection's Fourier coefficients $\bar{X}_m$ have truncated normal distribution with the following density:

$$f_{\bar{X}_m}(\bar{x}_m) = \begin{cases} c_m exp\left(-\frac{x_m^2}{b_m^2} \right), & |x_m| \leq a_m \\ 0 & , |x_m| \geq a_m \end{cases} \tag{10.155}$$

where each a_m is a maximum possible value for the random variable $\bar{X}_m$, b_m and c_m are positive parameters. Within stochastic modeling the problem was solved by the means of the Monte Carlo method. The outcome of the stochastic analysis is the reliability function. The reliability for the column with a random imperfection to a prescribed axial load a is defined as the probability that the structure does not fail prior to a. In other words, reliability equals the probability that the limit load a^*, which turns out to be a random variable due to stochasticity assumption for, exceeds a:

$$R(a) = Prob[a^* > a]. \tag{10.156}$$

If we design a column based on the stochastic approach, the value of the load corresponding to reliability R equal to a codified, required reliability r is the maximum admissible axial load. The latter load is referred to as a *design load* within the stochastic modeling. The design loads associated with $r = 0.9$, $r = 0.99$, $r = 0.999$, or $r = 0.9999$ have been calculated by Elishakoff *et al.* (1994).

The same problem has also been analyzed by convex modeling, in order to compare the designs obtained by alternative processing of the same information. If we still expand the initial deflection as a Fourier series (8.149), then a simple non-probabilistic model for the initial imperfection is that its Fourier coefficients vary in a hyper-cuboid set (a solid "box),

$$\{Z(\bar{X}):|\bar{X}_m| \leq a_m, (a_m \geq 0, m=1, 2, \ldots, M)\}. \tag{10.157}$$

The objective in these new circumstances is to find the minimum, the least-favorable limit load for all possible initial imperfections $\bar{X} = \{\bar{X}_1, \bar{X}_2, \ldots, X_M\}$ belongs to set $Z(\bar{X})$. If we design the column based on the non-stochastic approach, the above minimum limit load is the maximum admissible value of the axial load. In this way we arrive at an alternative way of determining the admissible axial load, which can be applied to an ensemble of columns with bounded Fourier coefficients.

It is remarkable that in some circumstances both approaches, although being of fundamentally different nature, may yield close values for the design axial loads. If probabilistic information is unavailable, one should not propose a probabilistic model, based on an arbitrary assumption of the distribution on the Fourier coefficients. Rather, in such circumstances one should use the non-stochastic approach to model uncertainty. On the other hand, when the full probabilistic information is available and the initial imperfection's Fourier coefficients have a relatively small deviation, use of the non-stochastic approach will be inadvisable and purely stochastic analysis should be conducted (implying, in terms of the story in Fig. 10.1 that the judge will be wrong to always conclude "You are right!"; but he can do so in some very special circumstances).

Remarkably, even when probabilistic information is available to substantiate the probabilistic analysis, if the density of the initial imperfections is rather "flat" (i.e., if the distribution is close to the uniform one), one may prefer a simpler non-stochastic, convex analysis, since it yields admissible axial loads comparable with the results of stochastic approach. For further details one may consult with Elishakoff *et al.* (1994).

10.12 Combined probabilistic–convex method

In the previous section we have critically contrasted the two competing methods of uncertainty analysis: probabilistic method and the set-theoretical method. As it was pointed out each method can be used in different circumstances; in some cases both methods yield comparable results; in other cases application of the convex modeling is preferable over the probabilistic one. The natural question arises, can these competing methods be combined?

Fig. 10.2: How much reliability should we require? (from the paper Elishakoff I., Essay on Uncertainties in Elastic and Viscoelastic Structures: From A. M. Freudenthal's Criticisms to Modern Convex Modeling, *Computers and Structures*, Vol. 56(6), 871–895, 1995a, by permission).

The answer to this question is affirmative. These methods can be combined when sufficient information is available to justify the probabilistic method, but some parameters are not known precisely. The uncertainty in these parameters can then be dealt with by the convex modeling.

Such a case was studied by Elishakoff and Colombi (1993a, 1993c). Some pertinent details of this investigation will be given here to elucidate the possible hybridization of these two methods.

Motion of the system is governed by the equation

$$m\ddot{X} + c\dot{X} + kx = F(t) \tag{10.158}$$

Where m is the mass, $k =$ spring constant, $c =$ damping coefficient. The excitation $F(t)$ is a weakly stationary random process with zero mean and an autocorrelation function:

$$R_F(\tau) = E\left(F^2\right)e^{-\alpha|\tau|}\left[\cos(\beta\tau) + \left(\frac{\alpha}{\beta}\right)\sin\left(\frac{\beta}{|\tau|}\right)\right] \tag{10.159}$$

where $E(F^2)$ is mean-square excitation, parameters α and β are parameters of excitation, and τ denotes the time delay. Two parameters, α and β model the presence of number of parameters in the fitted or hypothesized probabilistic characteristics of the random acoustic excitation. The spectral density of the excitation

$$\begin{aligned}\Phi_{FF}(\omega) &= \frac{1}{2\pi}\int_{-\infty}^{\infty} R_{FF}(\tau)e^{i\omega t}d\tau \\ &= \frac{E(F^2)}{\pi}\frac{2\alpha(\alpha^2 + \beta^2)}{\left(\alpha^2 + \beta^2\right)^2 + 2\left(\alpha^2 - \beta^2\right)\omega^2 + \omega^4}.\end{aligned} \tag{10.160}$$

The mean square response of the system is written as

$$E(X^2) = \int_{-\infty}^{\infty} \Phi_{FF}(\omega)|H(\omega)|^2 d\omega \tag{10.161}$$

where $H(\omega)$ is the frequency response function

$$\begin{aligned}H(\omega) &= \frac{1}{m(i\omega)^2 + c(i\omega) + k} \\ &= \frac{1}{m\left[(i\omega)^2 + 2\zeta_1(i\omega)\omega_n + \omega_n^2\right]}.\end{aligned} \tag{10.162}$$

In eq. (10.162) ζ_1 is the viscous damping factor, $\zeta_1 = \dfrac{c}{2m\omega_n}$, $\omega_n = \sqrt{k/m}$ is the natural frequency. Equation (8.151) becomes

$$E(X^2) = \frac{2\alpha(\alpha^2 + \beta^2)E(F^2)}{m^2\pi} \times \int_{-\infty}^{\infty} \frac{1}{\left(\alpha^2 + \beta^2\right)^2 + 2\left(\alpha^2 - \beta^2\right)\omega^2 + \omega^4} \times \frac{d\omega}{\left|(i\omega)^2 + 2\zeta_1 i\omega\omega_n + \omega_n^2\right|^2}. \tag{10.163}$$

The result obtained through the use of the residue theorem reads:

$$E(X^2) = f(\alpha, \beta)E(F^2) \tag{10.164}$$

$$f(a,\beta) = \frac{2a}{m} \frac{\left(a^2 + \beta^2\right)\left(a_3 - a_1 a_2\right)}{a_4 \left(a_3^2 + a_1^2 a_4 - a_1 a_2 a_3\right)} \tag{10.165}$$

with notations

$$a_1 = 2\left(a + \zeta_1 \omega_n\right)$$

$$a_2 = a^2 + \beta^2 + 4\zeta_1 \omega_n a + \zeta_1^2 \omega_n^2 + \omega_d^2$$

$$a_3 = 2\left(\zeta_1 \omega_n a^2 + \zeta_1 \omega_n \beta^2 + a\zeta_1^2 \omega_n^2 + a\omega_d^2\right)$$

$$a_4 = \left(a^2 + \beta^2\right)\left(\omega_*^2 + \zeta_1^2 \omega_n^2\right), \omega_d = \omega_n\left(1 - \zeta_1^2\right)^{1/2}. \tag{10.166}$$

Note that when the maxima of the frequency response function and the spectral density of excitation are well separated, and the damping coefficient ζ_1 is much smaller than unity, the method of approximate integration Elishakoff (1983) can be employed to yield the following approximation

$$E\left(X^2\right) \approx \Phi_{FF}(\omega_n) \int_{-\infty}^{\infty} |H(\omega)|^2 d\omega$$

$$+ |H(K_1)|^2 \int_{-\infty}^{\infty} \Phi_{FF}(\omega) d\omega \tag{10.167}$$

Here ω_n is the frequency at which the frequency response function has its maximum, and K_1 is the frequency which maximizes $\Phi_{FF}(\omega)$. In the vicinity of ω_n, the spectral density of excitation is assumed to be a slowly varying function, whereas in the vicinity of β, the frequency response function varies slowly. This version of approximate integration generalizes the one discussed by Elishakoff (1983). Indeed, for moderate values of a only the first term will suffice, whereas for small values of a inclusion of the second term is necessary. Evaluation of eq. (10.167) yields

$$E\left(X^2\right) \approx \Phi_{FF}(\omega_n) \int_{-\infty}^{\infty} |H(\omega)|^2 d\omega + E\left(F^2\right)|H(\gamma)|^2 = E\left(F^2\right)f(a,\beta) \tag{10.168}$$

$$f(a,\beta) = \frac{2a\left(a^2 + \beta^2\right)}{\left(a^2 + \beta^2\right)^2 + 2\left(a^2 - \beta^2\right)\omega_n^2 + \omega_n^4} \times \frac{1}{2\zeta_1\omega_n^3 m} + \frac{1}{\left(\omega^2 - K_1\right)^2 + \left(2\zeta_1\omega_n K_1\right)^2}. \tag{10.169}$$

The error due to approximation attains its maximum in the vicinity $K_1 \approx \omega_n \approx \beta$. In this region the exact expression given in eqs. (10.165) and (10.166), although cumbersome, should be used. When ω_n and β are far apart, the approximation given by eq. (10.169) holds.

Assume now that a and β are uncertain parameters. We basically have three ways of describing the situation:

(a) the parameters α and β are possible values taken by the respective random variables A and B. This situation occurs when we have sufficient experimental data on the random variables A and B so that the hypotheses about their joint density function $P_{AB}(\alpha, \beta)$ can be checked.

Consider the case when A and B are random variables. Their joint probability density $p_{AB}(\alpha, \beta)$ is assumed to be given. Then, according to the total probability formula

$$E(X^2) = \int\limits_{-\infty}^{\infty} \int\limits_{-\infty}^{\infty} E(X^2 | A = \alpha, B = \beta) \times p_{AB}(\alpha, \beta) d\alpha d\beta \qquad (10.170)$$

and the integration extends over the region where α and β are varying. In eq. (10.170) the value $E(X^2 | A = \alpha, B = \beta)$ is a conditional mean-square displacement calculated on a condition that the random variable A takes on value α, and random variable B takes on value β.

(b) There is no information on the loading, which is treated deterministically. Instead, a fragmentary knowledge is present; namely, that some governing parameters belong to some bounded set. In a sufficiently large class of situations this set is convex. In these circumstances an alternative analysis, namely, convex modeling, developed by Ben-Haim and Elishakoff 1990), should be applied Elishakoff (1990a), Elishakoff (1990b), Elishakoff and Ben-Haim (1990).

(c) There is probabilistic information on the loading process, and the form of the probabilistic characteristics can be estimated. However, there is scant knowledge on some governing parameters. In these circumstances probabilistic methods can be successfully combined with convex modeling. This is done in the following section.

Consider now a more realistic situation when α are uncertain but not "enough" information is available to obtain their probabilistic characteristics. Under these circumstances, the new, convex modeling of uncertainty should be applied (Ben-Haim and Elishakoff, 1990). We assume that we possess only scarce information on α and β; namely, that their parameters belong to some bounded convex sets.

$$Z(\Omega, \theta) = \{(\alpha, \beta) : G(H) \leq \theta^2\} \qquad (10.171)$$

where vector H has its elements α and β, and G is a quadratic form, θ^2 is a positive constant.

The proper choice of the form $G(H)$ and of a constant 0 is performed based on experimental information on loading, to determine, for example, the smallest ellipse enclosing all available, although scant data.

We assume that H_0 is a nominal vector, with nominal values of α_0 and β_0; for example, H_0 may correspond to average values of α and β. Structures which operate under similar conditions will experience common patterns of forces, belonging to the ellipsoidal set.

The mean square response for the force parameters $a_0 + \delta_1, \beta_0 + \delta_2$ to the first order in δ_1 and δ_2 is:

$$E\left[X^2(a_0 + \delta_1, \beta_0 + \delta_2)\right] = E\left(F^2\right)f(a_0 + \delta_1, \beta_0 + \delta_2) \qquad (10.172)$$

Where

$$f(a_0 + \delta_1, \beta_0 + \delta_2) = f(a_0, \beta_0) + \frac{\partial f}{\partial a}\bigg|_{H=H_0}\delta_1 + \frac{\partial f}{\partial \beta}\bigg|_{H=H_0}\delta_2 + \ldots \qquad (10.173)$$

where $H_0^T = (a_0, \beta_0)$.

We want to evaluate lower and upper limits of the mean square values as δ_1 and δ_2 vary within an ellipse. For convenience we define the gradient vector as

$$(gradf)_0^T = \left(\frac{\partial f}{\partial a}, \frac{\partial f}{\partial \beta}\right)_0 \qquad (10.174)$$

where T denotes transposition and subscript means that the derivatives are evaluated at the nominal values, a_0 and β_0. Consider a particular case of eq. (10.171), when the deviations δ_1 and δ_2 from the nominal parameters forming a vector δ vary in the ellipse

$$Z(\Omega, \theta^2) = \left\{\delta: \delta^T \Omega \delta \leq \theta^2\right\} \qquad (10.175)$$

where Ω is a positive-definite matrix

$$\Omega = \begin{bmatrix} \omega_{11} & \omega_{12} \\ \omega_{21} & \omega_{22} \end{bmatrix}. \qquad (10.176)$$

We are interested in the maximum possible mean-square response the system may possess:

$$E\left(X^2; \Omega, \theta\right) = max_{\delta \in Z(\Omega, \theta^2)} E\left(F^2\right)[f(a_0, \beta_0) + (gradf_0, \delta)] \qquad (10.177)$$

where (f_1, f_2) is an inner product so that two last terms in eq. (10.177) represent $(gradf)_0^T \delta$.

Equation (10.177) demands finding the maximum of the linear functional $(gradf)_0^T \delta$ on the convex set $Z(\Omega, \theta^2)$. According to the Kelly-Weiss theorem (Ben-Haim and Elishakoff, 1990) the extreme value will occur on the set of extreme points of Z which is a collection of vectors $\eta = (\eta_1, \eta_2)$ in the set

$$C(\Omega, \theta^2) = \left\{\eta: \eta^T \Omega \eta = \theta^2\right\}. \qquad (10.178)$$

Equation (10.177) becomes

$$E\left(X^2; \Omega, \theta\right) = max_{\eta \in C(\Omega, \theta^2)} E\left(F^2\right)\left[f(a_0, \beta_0) + (gradf)_0^T \eta\right]. \qquad (10.179)$$

We wish to maximize

$$\Psi \equiv f(a_0, \beta_0) + (gradf)_0^T \eta \tag{10.180}$$

subject to the requirement that vector η satisfies the equality $\eta^T \Omega \eta = \theta^2$. We employ the Lagrange multiplier method. We construct Lagrangean as

$$H = f(a_0, \beta_0) + (gradf)_0^T \eta + \lambda (\eta^T \Omega \eta - \theta^2). \tag{10.181}$$

A necessary condition for extremum reads:

$$\frac{\partial H}{\partial \eta} = (gradf)_0 + 2\lambda\Omega \eta = 0. \tag{10.182}$$

We premultiply this equation by Ω^{-1} from the left to arrive at

$$\eta = -\frac{1}{2\lambda}\Omega^{-1}(gradf)_0. \tag{10.183}$$

Substituting equation (10.183) into the constraint itself results in

$$\frac{(gradf)_0^T (\Omega^{-1})^T \Omega \Omega^{-1}(gradf)_0}{4\lambda^2} = \theta^2 \tag{10.184}$$

or, since Ω is a symmetric matrix, we obtain the following expression for the Lagrange multiplier squared

$$\lambda^2 = \frac{1}{4\theta^2}(gradf)_0^T \Omega^{-1}(gradf)_0. \tag{10.185}$$

In view of eq. (10.183) we arrive at the vector η which makes $E(X^2; \Omega, \theta)$ extremal:

$$\eta = \pm \frac{\theta}{\sqrt{(gradf)_0^T \Omega^{-1}(gradf)_0}}\Omega^{-1}(gradf)_0. \tag{10.186}$$

Maximum mean-square displacement becomes

$$max_{a,\beta}E[X^2(a,\beta)] = E[X^2(a_0,\beta_0)] \times \left[\left(1 + \theta\sqrt{\left(gradf\right)_0^T \Omega^{-1}\left(gradf\right)_0}\right)\right], \tilde{f} = f/f_0 \tag{10.187}$$

whereas the minimum mean-square displacement reads

$$min_{a,\beta}E[X^2(a,\beta)] = E[X^2(a_0,\beta_0)] \times \left[\left(1 - \theta\sqrt{\left(grad\tilde{f}\right)_0^T \Omega^{-1}\left(grad\tilde{f}\right)_0}\right)\right]. \tag{10.188}$$

Equation (10.177) yields the maximum possible response the system may possess. It can be directly incorporated into the design procedures of the structure.

The simplest convex modeling of uncertainty is interval analysis (Elishakoff, 1995), exposed in numerous books (Moore,1966, 1979; Alefeld and Herzberger, 1983; Neumaier, 1990; Jaulin *et al.*, 2001; Hansen and Walster, 2003; Huynh *et al.*, 2008; Moore *et al.*, 2009; Gustafson, 2017; Ceberio and Kreinovich, 2021).

Interval analysis is extremely attractive due to its philosophy of simplicity. However, it may lead to large overestimation. As an example, consider the interval variables X and Y. Lower bounds of X and Y are $\underline{x}$ and $\underline{y}$, respectively. Upper bounds of X and Y are $\bar{x}$ and $\bar{y}$, respectively. According to the definition, the difference between two intervals $X - Y$ is a third interval $Z = [\underline{z}, \bar{z}]$ with $\underline{z} = \underline{x} - \bar{y}$, $\bar{z} = \bar{x} - \underline{y}$. Now in the particular case that these two intervals are equal to each other, i.e., $\underline{x} = \underline{y}$, $\bar{x} = \bar{y}$, *then* $X - Y = X - X = [\underline{x} - \bar{x}, \underline{x} - \bar{x}]$. Depending on the value of $\underline{x}$ $\underline{x}$ and $\bar{x}$ $\bar{x}$ the overestimation of the response that must equal 0 as the difference of equal quantities may be extremely large. Several "cures" have been introduced to deal with above "illnesses." For example, in the case at hand, one can try the same quantity to appear only once in an expression. Hence, straightforward evaluation of $X - X^2$, where $X = [0, 1]$ yields $X - X^2 = [-1, 1]$, whereas if we rewrite the expression $X - X^2$, following Alefeld and Claudio (1998) as $1/4 - (X - 1/2)(X + 1/2)$ we get $X - X^2 = [0, 1/2]$ which constitutes a much sharper interval than $[-1, 1]$.

To reduce the 'catastrophic' effects (as these were characterized by Muhanna and Mullen (2001)) of the dependency phenomenon, the so-called *generalized interval analysis* (Hansen, 1975) and the *affine arithmetic* (Stolfi J. and De Figueiredo, 2003; Comba and Stolfi, 1993) have been introduced in the literature. More recently, within the framework of static structural analysis, the *parameterized interval analysis* (Elishakoff and Miglis, 2012a, 2012b; Elishakoff, 2013) and the *improved interval analysis by Extra Unitary Interval* (EUI) (Muscolino and Sofi, 2012a, 2012b, 2013) and have been proposed. These two procedures were contrasted and combined in the study by *Santoro et al.* (2015).

There are further studies on the dependency phenomenon in the interval analysis. The interested reader can also consult a definite book by Sainz *et al.* (2014) on modal interval analysis, as well as papers by Ioakimidis (2019, 2020, 2021) utilizing quantifier elimination technique to derive sharp intervals of response quantities. Alternative techniques were reported in the papers by Elishakoff, Gabriele and Wang (2016) and Popova and Elishakoff (2020). Additionally, Elishakoff and Bekel (2013) proposed to present uncertain data as bounded by Lamé's super ellipsoids that tend to either a rectangle or the ellipsoid depending on the governing parameter. Jiang, Zhang *et al.* (2015) utilized the multidimensional parallelepiped, instead. Qiu, Wu *et al.* (2021) utilized data-based polyhedron model.

10.13 Concluding remarks

We would like to start the concluding remarks by quoting (yet again!) Alfred Freudenthal (1968), widely recognized as a founder of modern stochastic mechanics, criticizing the usual deterministic approach, "it seems absurd to strive for more and more refinement of methods of stress-analysis, if, in order to determine the dimensions of the structural elements, its results are subsequently compared with so-called working stress, derived in a rather crude manner by dividing the values of somewhat dubious material parameters obtained in conventional materials tests by still more dubious empirical numbers called safety factors."

According to Grandori (1991), Freudenthal's conclusion is that the problem of structural safety "can be put on a rotational basis commensurate with the development of modern methods of stress analysis only through the consideration of the statistical dispersion of the operation loads as well as structural resistance." This is one way "to remove the concept of structural safety from the realm of metaphysics to that of physical reality." The key idea of this rational approach according to Freudenthal, is the selection of an "acceptable risk of failure."

One could paraphrase the above remark of Freudenthal (1968) by mentioning that "it seems inconsistent to strive for more and more refinement in stochastic calculations, if, in order to determine the dimensions of the structural elements via the probabilistic analysis, its final result – reliability – is subsequently compared with so-called required reliability while there are no codes or guidelines to specify the latter and, moreover, there is a strong sensitivity to small changes in the required reliability."

Actually, Freudenthal himself realized the limitations of the probabilistic methods. He stressed (Freudenthal, 1961), that "when dealing with probabilities a clear distinction should be made between conditions arising in design of inexpensive mass products in which the probability figures are derived by statistical interpretation of actual observations or measurements (since a sufficiently large number of observations are actually obtainable), and conditions arising in design of structures of complex systems. In the latter, probability figures are used simply as a scale or measure of reliability that permits the comparison of alternative designs. The figures can never be checked by observations or measurements since they are obtained by extrapolations so far beyond any possible range of observation that such extrapolation can no longer be based on statistical arguments but could only be justified by relevant physical reasoning. Under these conditions the absolute probability figures have no real significance." The greatness of A. M. Freudenthal (see also Bažant and Le, 2017) can clearly be felt from this passage, since the pioneer of the probabilistic methods criticizes his own creation, both honestly and spiritually.

Here a particular personal encounter appears to be in order. Elishakoff's research in difficulties of probabilistic modeling was inspired in the following way. In 1980, the mastermind of modern buckling theory of structures, Professor Warner Tjrdus Koiter of Delft University of Technology had organized a European Mechanics Symposium on

Buckling of Structures. There the present writer had presented a lecture entitled "Random Buckling of Shells," coauthored with Professor Johann Arbocz of Delft University of Technology. After the lecture, Professor Koiter posed a question, which the present writer can recall almost verbatim: "In order for society to adopt the probabilistic methods, the resulting reliability of the structure must be extremely high or, alternatively, the probability of failure must be extremely small. Yet, in order to calculate the above probabilities, one must utilize deterministic theories which are of an approximate nature. Can one be sure that such small probabilities of failure can be accurately predicted?" To reply to this question, at least partially, the present writer undertook an investigation (Elishakoff, 1983) based on simple models where exact solution was available. It turned out that in some circumstances, due to use of the approximate deterministic theory the error in evaluating the probability of failure may be quite substantial. Professor Koiter's cardinal question, and its ramifications, was always kept in the present writer's mind; indeed, this question has had a great impact on the present writer's research activities since then. Several investigations have been directly associated with Professor Koiter's inquiry. The present writer had an opportunity to investigate the substantial errors which can be introduced by an ergodicity assumption Scheurkogel et.al., (1981), Scheurkogel and Elishakoff (1985) (an interest in this subject was triggered by the kind suggestion of Professor Bernard Budiansky of Harvard University) or neglect of cross-correlations Elishakoff (1977, 1983b, 1986, 1988a, 1988b, 2020a) (the subject pioneered by Professor Vladimir V. Bolotin of Russian Academy of Sciences) while utilizing the probabilistic approach.

Probabilistic and convex modeling of uncertainty may refer to different aspects of a phenomenon and yield answers to specific questions in each case. We feel that the pragmatic approach for accounting for uncertainty should be flexible too. Different models can live in peace with each other, as long as they serve different purposes. Therefore, the answer to the question as to possible contradiction between probabilistic and convex modeling appears to be negative, as was also stressed by Ben-Haim and Elishakoff (1990). Convex modeling may prove preferable over the probabilistic one when there is incomplete probabilistic information available to justify the utilization of probabilistic ideas.

We can summarize that there are different methods to deal with uncertainty:

(a) theory of probability and random processes;

(b) theory of "fuzzy" sets (applicable when the partial knowledge has a "fuzzy" character (Dubois and Prade, 1980; Blockley, 1980; *Chiang et al.,* 1987; Wood *et al.,* 1990; Bernardini, 1992; Bignoli, 1993; Zadeh, 1975a; 1975b; 1975c; 1975d; Yager, 1982) (this topic was not touched upon in this essay);

(c) convex modeling of uncertainty (set-theoretic approach), which more generally could also be referred to as an *anti-optimization* under uncertainty (the term was coined by Elishakoff (1990), although its predecessor name of worst-case design would be fully appropriate, as discovered by Wald (1950). The ideas of convex modeling now are referred to as info-gap modeling by Ben-Haim (2001; 2006). These ideas were further developed by Hlavacek, Chleboun and Babuška (2004); Kanno (2011), Takewaki (2013), Takewaki *et al.* (2013), and many others who followed the works by Schweppe (1973), Chernousko and Ben-Haim and Elishakoff (1990).

This discussion is again in perfect agreement with the statement by Freudenthal (1972) "that ignorance of the cause of variation does not make such variation random." In other words, Freudenthal realized that not every uncertain phenomenon must be analyzed as a random one. In our opinion, this realization was one of the major thoughts in Freudenthal's work. He however did not live long enough to develop the alternative to the probabilistic methods of uncertainty, when the uncertainty cannot be identified with randomness. The convex modeling of uncertainty in theoretical and applied mechanics was pioneered in the monograph by Ben-Haim and Elishakoff (1990). The combined probabilistic-convex modeling of uncertain excitations and associated response is developed in the recent monograph by Elishakoff *et al.* (1994).

Yet, when one reads and re-reads the above statement of Freudenthal, one must realize the vision of this great scientist. Although he pioneered the probabilistic approach, he did not have a fetishistic believe in it. The present writer hopes that the researchers who are following the footsteps of Freudenthal in developing the probabilistic methods in structures, will carefully read Freudenthal's criticisms of probabilistic methods.

In this respect quotation of Professor Giuseppe Grandori (1991) of Milan's Polytechnic appears to be most instructive, "the probabilistic approach to structural safety is today a well-established paradigm. As to the current state of this paradigm, however, one can notice an asymmetry similar to that observed by Freudenthal, in the traditional approach. An overwhelming part of the research effort, in fact, has been and still is devoted to estimating failure probabilities. By contrast, only sporadic research deals with the problem of choosing an acceptable risk of failure.

It is true that the adaptation of a probabilistic approach is in any case progress, even in the case when acceptable risk levels are conventionally defined, because it allows us to treat different structures with homogeneous criteria. However, the concept of structural safety will not leave the "realm of metaphysics" unless we devise a method for justifying the choice of risk acceptability levels" (see also Grandori *et al.,* 1998).

As Freudenthal remarks (1972), "an approach based on the direct specification of a very low failure probability alone suffers from a major shortcoming: there is no intrinsic significance to a particular failure probability since no *a priori* rationalization can be given for the adaptation of a specific quantitative probability level in preference to any other, so that the selection of this level remains an arbitrary decision." Bolotin (1961) too, in the first, Russian, edition of his monograph advocates that probabilistic methods should be used to compare various designs. He stresses that "small probabilities of failure, if they are correctly found, still retain their importance as some objective characteristics of possibility of random event taking place. They become sensible when comparing them with each other, allowing contrast of the risk of failure of different structures or of the same structure in different working conditions." Bolotin then remarks to the effect that "there is a wide range of the problems of structural mechanics in which the use of statistical methods is the most adequate means of investigation – yet there are problems – where statistical methods may only play the role of auxiliary methods of analysis. Here the statistical and deterministic methods could successfully coexist, complementing each other. In these problems the overestimation of the role of statistical methods can only be harmful." Maybe these views would appear as too pessimistic for modern *energetic* computational stochastic mechanicians who try to develop extremely efficient and sophisticated methods to obtain the estimates of the probability of failure. Yet, the natural question, what value must be adopted for required reliability, r (Fig. 10.2) must be seriously addressed in parallel, and with no less enthusiasm.

The answer appears to be a matter of value judgment, with allowance made for the large number of design factors such as functional purpose, responsibility, maintenance, weight, human safety, legal consideration, economic aspects, and social significance. The designed quantities should depend *inter alia* on the cost consequences of failure, including damage or total loss of material goods, and especially human injury or fatality (for the challenging project of setting a value on human life, see an interesting and provocative article by Lind (1978)). It appears that the nagging question "how much reliability?" cannot be answered by individual scientists, but must be referred to competent national and international committees (Fig. 10.2) and undoubtedly to the society at large; the taxpayers provide resources for constructing the engineering facilities; in addition, taxpayers may be affected by the unreliability of these facilities. Therefore, public both should be educated on safety issues, and input and decisions should be sought from the society at large.

In one of his papers Koiter (1969) mentions the totally different subject, "flexible bodies like thin shells require a flexible approach." Likewise, structural engineers dealing with uncertainties in flexible structures are well advised to exercise flexibility in choosing the methods of analyses: *probabilistic, convex* (Wald's min-max) or *fuzzy subsets based modeling*. It appears that presently there is a Babel Tower situation when proponents of different methods do not talk each other's language; specialist dealing with uncertainty must be both flexible and tolerant; engineers as pragmatic creatures may have whole spectrum of methods which either compete with each other, or complement each other, depending on circumstances.

This will allow for more flexibility in choosing methods to analyze the structures under uncertainty. Three possible approaches to handling uncertainty could be "right" in different circumstances; each may have preference over the others, depending on the degree of information we possess. This gives rise to the idea of the uncertainty triangle (Fig. 10.1).

Before deciding which "corner" of this triangle to apply, engineers and researchers should critically evaluate the amount and character of the information they possess to make "right" choices in analyzing structures with uncertain parameters and/or
loads. To sum up, the central statement of this review is the message that *uncertainty
and randomness are not reciprocal.* Engineers and pragmatic professionals should not
overlook the convex modelling – a successfully function alternative to the probabilistic paradigm.

We would like to repeat the concluding remarks by quoting Alfred Freudenthal
(1968), widely recognized as a founder of modern stochastic mechanics, criticizing the
usual deterministic approach, "it seems absurd to strive for more and more refinement
of methods of stress-analysis, if, in order to determine the dimensions of the structural
elements, its results are subsequently compared with so-called working stress, derived
in a rather crude manner by dividing the values of somewhat dubious material parameters obtained in conventional materials tests by still more dubious empirical numbers
called safety factors."

According to Grandori (1991), Freudenthal's conclusion is that the problem of
structural safety "can be put on a rotational basis commensurate with the development of modern methods of stress analysis only through the consideration of the statistical dispersion of the operation loads as well as structural resistance." This is one
way "to remove the concept of structural safety from the realm of metaphysics to that
of physical reality." The key idea of this rational approach according to Freudenthal,
is the selection of an "acceptable risk of failure."

One could paraphrase the above remark of Freudenthal by mentioning that "it
seems inconsistent to strive for more and more refinement in stochastic calculations,
if, in order to determine the dimensions of the structural elements via the probabilistic analysis, its final result – reliability – is subsequently compared with so-called required reliability while there are no codes or guidelines to specify the latter and,
moreover, there is a strong sensitivity to small changes in the required reliability."

Actually, Freudenthal himself realized the limitations of the probabilistic methods. He stressed (1961), that "when dealing with probabilities a clear distinction
should be made between conditions arising in design of inexpensive mass products in
which the probability figures are derived by statistical interpretation of actual observations or measurements (since a sufficiently large number of observations are actually obtainable), and conditions arising in design of structures of complex systems. In
the latter, probability figures are used simply as a scale or measure of reliability that
permits the comparison of alternative designs. The figures can never be checked by
observations or measurements since they are obtained by extrapolations so far beyond any possible range of observation that such extrapolation can no longer be
based on statistical arguments but could only be justified by relevant physical reasoning. Under these conditions the absolute probability figures have no real significance."
The greatness of A. M. Freudenthal can clearly be felt from this passage, since the pioneer of the probabilistic methods criticizes his creation, both honestly and spiritually.

Here a particular personal encounter appears to be in order. Elishakoff's research in difficulties of probabilistic modeling was inspired in the following way. In 1980, the mastermind of modern buckling theory of structures, Professor Warner Tjrdus Koiter of Delft University of Technology had organized a European Mechanics Symposium on Buckling of Structures. There the present writer had presented a lecture entitled "Random Buckling of Shells," coauthored with Professor Johann Arbocz of Delft University of Technology. After the lecture, Professor Koiter posed a question, which the present writer can recall almost verbatim: "In order for society to adopt the probabilistic methods, the resulting reliability of the structure must be extremely high or, alternatively, the probability of failure must be extremely small. Yet, in order to calculate the above probabilities, one must utilize deterministic theories which are of an approximate nature. Can one be sure that such small probabilities of failure can be accurately predicted?" To reply to this question, at least partially, the present writer undertook an investigation (1983) based on simple models where exact solution was available. It turned out that in some circumstances, due to use of the approximate deterministic theory the error in evaluating the probability of failure may be quite substantial. Professor Koiter's cardinal question, and its ramifications, was always kept in the present writer's mind; indeed, this question has had a great impact on the present writer's research activities since then. Several investigations have been directly associated with Professor Koiter's inquiry. The present writer had an opportunity to investigate the substantial errors which can be introduced by an ergodicity assumption Scheurkogel et.al., (1981) Scheurkogel and Elishakoff (1985) (an interest in this subject was triggered by the kind suggestion of Professor Bernard Budiansky of Harvard University) or neglect of cross-correlations Elishakoff (1977, 1979, 1983, 1986, 1988, 1988) (the subject pioneered by Professor Vladimir V. Bolotin of Russian Academy of Sciences) while utilizing the probabilistic approach.

Probabilistic and convex modeling of uncertainty may refer to different aspects of a phenomenon and yield answers to specific questions in each case. We feel that the pragmatic approach for accounting for uncertainty should be flexible too. Different models can live in peace with each other, as long as they serve different purposes. Therefore, the answer to the question as to possible contradiction between probabilistic and convex modeling appears to be negative, as was also stressed in Ref. (1990). Convex modeling may prove preferable over the probabilistic one when there is incomplete probabilistic information available to justify the utilization of probabilistic ideas.

We can summarize that there are different methods to deal with uncertainty:
(a) theory of probability and random processes;
(b) theory of "fuzzy" sets (applicable when the partial knowledge has a "fuzzy" character (this topic was touched upon in chapter 7);

(c) convex modeling of uncertainty (set-theoretic approach), which more generally could also be referred to as an *anti-optimization* under uncertainty (the term was coined by Elishakoff (1990), although its predecessor name of worst-case design would be fully appropriate, as discovered by Wald (1950). The ideas of convex modeling now are referred to as info-gap modeling by Ben-Haim (). These ideas were further developed by Hlavacek et al. (2004); Kanno (2011), Takewaki (2013), Takewaki *et al.* (2013), and many others who followed the works by Schweppe (1973), Chernousko (1994) and Ben-Haim and Elishakoff (1990).

This discussion is again in perfect agreement with the statement by Freudenthal (1972) "that ignorance of the cause of variation does not make such variation random." In other words, Freudenthal realized that not every uncertain phenomenon must be analyzed as a random one. In our opinion, this realization was one of the major thoughts in Freudenthal's work. He however did not live long enough to develop the alternative to the probabilistic methods of uncertainty, when the uncertainty cannot be identified with randomness. The convex modeling of uncertainty in theoretical and applied mechanics was pioneered in the monograph by Ben-Haim and Elishakoff (1990). The combined probabilistic-convex modeling of uncertain excitations and associated response is developed in the recent monograph by Elishakoff *et al.* (1994).

Yet, when one reads and re-reads the above statement of Freudenthal, one must realize the vision of this great scientist. Although he pioneered the probabilistic approach, he did not have a fetishistic believe in it. The present writer hopes that the researchers who are following the footsteps of Freudenthal in developing the probabilistic methods in structures, will carefully read Freudenthal's criticisms of probabilistic methods.

In this respect quotation of Professor Giuseppe Grandori (1991) of Milan's Polytechnic appears to be most instructive, "the probabilistic approach to structural safety is today a well-established paradigm. As to the current state of this paradigm, however, one can notice an asymmetry similar to that observed by Freudenthal, in the traditional approach. An overwhelming part of the research effort, in fact, has been and still is devoted to estimating failure probabilities. By contrast, only sporadic research deals with the problem of choosing an acceptable risk of failure.

It is true that the adaptation of a probabilistic approach is in any case progress, even in the case when acceptable risk levels are conventionally defined, because it allows us to treat different structures with homogeneous criteria. However, the concept of structural safety will not leave the "realm of metaphysics" unless we devise a method for justifying the choice of risk acceptability levels" (see also Grandori *et al.*, 1998).

As Freudenthal remarks (1972), "an approach based on the direct specification of a very low failure probability alone suffers from a major shortcoming: there is no intrinsic significance to a particular failure probability since no *a priori* rationalization can be given for the adaptation of a specific quantitative probability level in preference to any other, so that the selection of this level remains an arbitrary decision." Bolotin (1961) too, in the first, Russian, edition of his monograph advocates that probabilistic methods should be used to compare various designs. He stresses that "small probabilities of failure, if they are correctly found, still retain their importance as some objective characteristics of possibility of random event taking place. They become sensible when comparing them with each other, allowing contrast of the risk of failure of different structures or of the same structure in different working conditions." Bolotin then remarks to the effect that "there is a wide range of the problems of structural mechanics in which the use of statistical methods is the most adequate means of investigation – yet there are problems – where statistical methods may only play the role of auxiliary methods of analysis. Here the statistical and deterministic methods could successfully coexist, complementing each other. In these problems the overestimation of the role of statistical methods can only be harmful." Maybe these views would appear as too pessimistic for modern *energetic* computational stochastic mechanicians who try to develop extremely efficient and sophisticated methods to obtain the estimates of the probability of failure. Yet, the natural question, what value must be adopted for required reliability, r (Fig. 10.2) must be seriously addressed in parallel, and with no less enthusiasm.

The answer appears to be a matter of value judgment, with allowance made for the large number of design factors such as functional purpose, responsibility, maintenance, weight, human safety, legal consideration, economic aspects, and social significance. The designed quantities should depend *inter alia* on the cost consequences of failure, including damage or total loss of material goods, and especially human injury or fatality (for the challenging project of setting a value on human life, see an interesting and provocative article by Lind (1978)). It appears that the nagging question "how much reliability?" cannot be answered by individual scientists, but must be referred to competent national and international committees (Fig. 10.2) and undoubtedly to the society at large; the taxpayers provide resources for constructing the engineering facilities; in addition, taxpayers may be affected by the unreliability of these facilities. Therefore, public both should be educated on safety issues, and input and decisions should be sought from the society at large.

In one of his papers Koiter (1969) mentions the totally different subject, "flexible bodies like thin shells require a flexible approach." Likewise, structural engineers dealing with uncertainties in flexible structures are well advised to exercise flexibility in choosing the methods of analyses: *probabilistic, convex or fuzzy subsets based modeling*. It appears that presently there is a Babel Tower situation when proponents of different methods do not talk each other's language; specialist dealing with uncertainty must be both flexible and tolerant; engineers as pragmatic creatures may have whole spectrum of methods which either compete with each other, or complement each other, depending on circumstances.

This will allow for more flexibility in choosing methods to analyze the structures under uncertainty. Three possible approaches to handling uncertainty could be "right" in different circumstances; each may have preference over the others, depending on the degree of information we possess. This gives rise to the idea of the uncertainty triangle (Fig. 10.1).

Before deciding which "corner" of this triangle to apply, engineers and researchers should critically evaluate the amount and character of the information they possess to make "right" choices in analyzing structures with uncertain parameters and/or loads. To sum up, the central statement of this review is the message that *uncertainty and randomness are not reciprocal.* Engineers and pragmatic professionals should not overlook the convex modelling – a successfully function alternative to the probabilistic paradigm.

10.14 Needs for further research

The need for further research is tremendous. This is both for analytical and numerical aspects of the convex methodology as well as for its experimental verification.

It appears that the funding agencies presently prefer more traditional, probabilistic studies. This is mainly due to three reasons:

(a) Lack of deep understanding of difficulties associated with stochastic modelling.
(b) Fear that L.F.R.s may be too conservative. This fear is justifiable in cases when only a very scant knowledge is present on uncertainties. However, in these extreme circumstances the stochastic approach would yield unmeaningful details. On the other hand, one can demonstrate that when the global knowledge increases the bounds of response becomes less conservative; sometimes stochastic approach may be more conservative.
(c) Due to the complete discard of introducing uncertainty in the past and unacceptance of probabilistic methods, the stochastic community is doing a good job of "educating" funding agencies that uncertainty should be accounted for. Unfortunately, most funding agencies do not know yet that probabilistic or the fuzzy-subsets-based treatments are not the only ways to take into account uncertainty.

It appears that the researchers dealing with set-theoretic models of uncertainty should dedicate efforts to "educate" *both* the "deterministic" and "stochastic" communities. Convex modeling of uncertainty, and in general set-theoretic modeling of it (the constraints should not necessarily be treated as convex) do not supplant probabilistic ideas. Convex modeling rather complements both the probabilistic approach and the fuzzy subsets-based treatment (see Fig. 1.1).

Note that Gangadharan *et al.* (1993,1999) utilized the anti-optimization methodology to compare alternative structural models. Since the publication of the monograph by Ben-Haim and Elishakoff (1990) numerous papers were dedicated to convex models of uncertainty. Some representative ones were composed by Lindberg (1992a,1992b), Pantelides (1995,1996), Ganzerli and Pantelides (1999, 2000), Pantelides and Ganzerli (2001),

Kang *et al.* (2011), Qiu *et al.* (1995, 2004, 2008a, 2008b), Wang (2003), Wang *et al.* (2008, 2009a, 2009b, 2011a, 2011b), Jiang *et al.* (2011, 2012a, 2012b, 2013a, 2013b, 2018) and others. Several books appeared on this topic. These include books by Elishakoff *et al.*, 1994; Ben-Haim, 1996, 2001, 2006; Elishakoff, 1999; Elishakoff *et al.*, 2001; Chernousko (1994); Svetlitskii (2003); Hlavacek *et al.*, 2004; Qiu and Wang, 2008; Ben-Tal *et al.*, 2009; Elishakoff and Ohsaki, 2010; Takewaki (2013); Takewaki *et al.* (2013); Skalna, 2018; Jiang *et al.*, 2020; Chakraverty and Rout, 2020; Chakraverty, 2021.

Chapter 11
Probability and convexity concepts are perfectly compatible

We sail within a vast sphere, ever drifting in uncertainty, driven from end to end. Blaise Pascal,
Pensées (Thoughts), 1660

This chapter is devoted to two objectives: to illustrate that the probability and convexity concepts are not antagonistic and to introduce a new non-probabilistic convex model for structural reliability analysis. It is shown that the new measure of safety is easier to evaluate than the corresponding measure utilizing interval analysis. Moreover, the interrelation between the classical probabilistic method and the convex modeling method is demonstrated. The purpose of this book is not to replace the probabilistic approach with the convex modeling method, but to illustrate that the probability and convexity concepts are perfectly compatible. Some numerical examples are presented to illustrate the feasibility of the proposed methodology.

11.1 Introduction

In practical engineering, the problems arising from uncertainties are increasingly prominent (Wang, 2002). With the growing complexity of engineering structures, the development of better ways of handling uncertainty has become the focus of both analysis and design (Oberkampf *et al.*, 2004). Traditional methods for dealing with uncertainty engineering problems, namely, probabilistic analysis and fuzzy sets-based theory need to have sufficient information to determine the probability distributions or the membership functions (Elishakoff, 1998), respectively. However, experimental data is often limited so that one cannot meet the requirements for the available data to justify either the probabilistic reliability model or the fuzzy reliability model appropriately (Ben-Haim and Elishakoff, 1990; Qiu and Wang, 2008; Elishakoff and Ohsaki, 2010). In addition, the probabilistic model may turn out to be hyper-sensitive to perturbations of parameters so that the small errors in them may lead to a serious error in calculating the structural reliability. This sensitivity shows that the absence of adequate statistical information in describing the probability model and the needed subjective assumptions may yield unreliable results.

While the applicability conditions of the probabilistic model or fuzzy sets-based model often may not be sufficiently substantiated, the approximation to the bounds on the uncertain information can be obtained more easily than the probability distributions in the probabilistic analysis or the membership functions in fuzzy sets-based treatment. Ben-Haim (1997) (see also Guo *et al.*, 2001) first proposed the concept of the non-probabilistic safety of structures, which was based on the convex model. He ar-

https://doi.org/10.1515/9783111354231-011

gued that the system is reliable if it allowed the accuracy of uncertainty to fluctuate within a certain range, without providing a specific non-probabilistic safety measure. Based on interval analysis, Elishakoff (1995) was the first to propose a quantitative measure of the non-probabilistic safety.

The notion of "non-probabilistic reliability" was embraced by several investigators. Guo *et al.* (2001) quantified the uncertain structural parameters as interval variables and proposed another measure of the "non-probabilistic reliability," which was taken as the shortest distance from the origin to the failure surface (Qiu, Müller *et al.*, 2004; Qiu, Chen *et al.*, 2004; Wang, 2003) suggested a non-probabilistic model of convex reliability by using the partial order relation of the superscribed hyper-rectangle or interval vectors. However, the above non-probabilistic reliability interval variables appear to be inconvenient for optimization, which is often the main problem of concern.

In this section, a new non-probabilistic safety measure for structural reliability is proposed. Based on the non-probabilistic convex model, the ratio of the volume of the safe region to the total volume of the region of the variation of the uncertain variables is suggested as the measure of structural non-probabilistic safety (Wang *et al.*, 2008). Compared to interval analysis, a convex model method provides the ability to have uncertain variables more explicitly expressed by continuously differentiable mathematical equations.

In Section 6 of this chapter, the comparison between the non-probabilistic convex model and the probabilistic model is performed, generalizing Elishakoff's (1999a) earlier work on this subject. We are looking for the interrelation between seemingly totally different analyses of uncertainty: the classical probabilistic method and the convex modeling method. The goal of this chapter is not to stress excessively the advantage of convexity over probability or fuzziness, or to substitute the probabilistic method, but to provide a complement to probabilistic analysis.

11.2 Compatibility of convexity and probability

Much has been written of the scientific method in engineering.
    The question is, is there a single scientific method in engineering or anywhere else? There are many methods of arriving at the truth, though often truth itself is uncertain because criteria are needed to determine what constitutes truth in special fields. Hardy Cross, *Firm Foundations for Towers*

In practical engineering, due to physical imperfections, model inaccuracies, and system complexities, almost all structures can be characterized with physical and geometrical uncertainties to various degrees. Therefore, in structural problems, the objective function and the constraint conditions will be influenced by the uncertainties in the structural parameters. The uncertainty usually is dealt with and solved by most scientists and engineers using the probabilistic approach or non-probabilistic convex modeling methods.

Over the past decades, the probabilistic approach became the most popular for analysis of systems with uncertainties in system parameters and inputs, and has gained a large popularity in the area of finite element analysis. The uncertainty is quantified by using the theory of stochastic functions (processes/fields). This method can provide numerical characteristics, including mean value (expectation) and standard deviation of system response, and even probability density function for the response of a simple system.

Recently, the development of non-probabilistic convex modeling methods for non-deterministic analysis mainly originated from criticism of the credibility of probabilistic analysis when it is based on limited information. Indeed, the probabilistic reliability analysis relies on a precise description of the random inputs at a certain confidence level. The data, however, are frequently extremely difficult to obtain, especially when only a limited number of samples are available or very few repeatable events are encountered. As revealed by Elishakoff (1995a, 1995c), unjustified assumptions in constructing a probabilistic model for input quantities may be dangerous, and even small errors between the assumed distribution data and the real ones may yield misleading predictions in the probabilistic reliability analysis.

Therefore, convex modeling methods have been considered as a beneficial supplement to the traditional probabilistic model. In particular, when sufficient samples are not available and thus the probabilistic distribution data of the inputs cannot be readily extracted from the existing results of physical tests or measurements, the convex model which bounds all possible values of the uncertainties within a convex set without assuming a probability distribution, might be attractive for the analysis and design of structures.

Although the expressions and the scope of application are different, the above two methods may turn out to coincide with each other. Moreover, in Section 6 of this chapter, we will further illustrate the compatibility by evaluation a specific example. Naturally, the purpose of this volume is not to advocate for the replacement of the probabilistic approach by the non-probabilistic convex modeling, but to suggest that the latter may be a complement to probabilistic analysis when the information of the structural problem is missing or insufficient.

11.3 Problem statement

Based on the description of interval uncertainty, we resort to structural non-probabilistic reliability issues with the non-probabilistic set-theoretic stress-strength interference model for the two-dimensional problems. They defined the measure of structural safety, which was defined as the ratio of the area of the safe region to the total area of the region associated with the variation of the basic interval variables as shown in Fig. 11.1:

$$R_{set} = \eta(M(\delta_R, \delta_S) > 0) = \frac{S_{safe}}{S_{total}} \qquad (11.1)$$

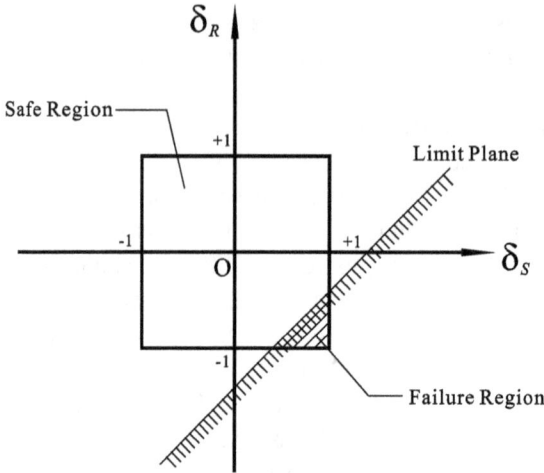

Fig. 11.1: Scheme for the model of the structural non-probabilistic reliability based on interval variables.

In this chapter, a new measure of structural reliability is proposed for two- or multi-dimensional problems; the key point consists in the convention of the range of the uncertain variables into an equivalent spherical region by using standard substitution. The geometric relationship between the spherical region and the failure plane of the structure then allows one to derive the reliability of the system.

The convex model represents the structural uncertainty parameters, belonging to the hyper-ellipsoid as follows

$$\left(\frac{x_1 - X_1^c}{X_1^r}, \ldots, \frac{x_n - X_n^c}{X_n^r}\right)^T \left(\frac{x_1 - X_1^c}{X_1^r}, \ldots, \frac{x_n - X_n^c}{X_n^r}\right) \leq 1 \qquad (11.2)$$

where X_i^c represents the median value of the interval variable; X_i, X_i^r denotes the radius of X_i, and T meaning transposition operation.

By standardizing the variables (Thoft-Christensen and Baker, 1982), the inequality can be rewritten as

$$\sum y_k^2 \leq 1, \ y_k = \frac{x_k - X_k^c}{X_k^r}, \ k = 1, 2, \ldots, n \qquad (11.3)$$

The margin of safety M is expressed as the function of structural stress S, and the structural strength R as follows

$$M(R,S) = R - S \qquad (11.4)$$

Given the specified values of S and R, it is possible to judge if the structural state belongs to the safety region or to the failure region. The basic variable space is therefore divided into two parts, namely the safe region and the failure region, by the failure plane or the limit state plane, i.e.,

$$M(R,S) = R - S = 0 \qquad (11.5)$$

The positive value of M identifies the safe region of basic variables, while the negative value of M represents the failure region, i.e.,

$$M(R,S) > 0, \ R, S \in safe\,region \qquad (11.6)$$

and

$$M(R,S) < 0, \ R, S \in failure\,region \qquad (11.7)$$

11.4 Measure of structural non-probabilistic safety based on convex model

In realistic problems, the uncertain parameters describing structural function are bounded variables, rather than specific, crisp values. Therefore, stress and strength must be defined as stress interval and strength interval, respectively. The measure of non-probabilistic structural safety must therefore be the function of the stress interval as well as the strength interval. When the overlapping between the stress interval and the strength interval occurs – even when the median value S_c of the stress is smaller than the median value R_c of the strength – it cannot be ensured that the stress will not take on values in excess of the strength. Thus, the set associated with stress being larger than the strength will be not empty. This fact can be stated as follows

$$\eta(M(R,S) < 0) > 0 \qquad (11.8)$$

where $\eta(X)$ represents the possibility of the event X.

In this chapter, the possibility that the stress is larger than the strength will be referred to by us the non-probabilistic unreliability measure, which can be defined for the two-dimensional case as the ratio of the area of failure region to the total area of basic variables' region, i.e.,

$$F_{set} = \eta(M(R,S) < 0) = \frac{S_{failure}}{S_{total}} \qquad (11.9)$$

Accordingly, the system is reliable when the stress is smaller than the strength, and the ratio of the area of safe region to the total area region of basic variables is defined as the non-probabilistic safety measure. This definition can be expressed mathematically as

$$R_{set} = \eta(M(R,S) > 0) = \frac{S_{safe}}{S_{total}}$$ (11.10)

In the following, some specific cases of the stress and the strength based on the convex model will be considered for illustrating the non-probabilistic measure of the structural safety. Based on the convex model (Ben-Haim and Elishakoff, 1990), the range of the uncertain parameters is expressed as a hyper-ellipsoid, which is reduced to an ellipse

$$\frac{(R - R_c)^2}{R_r^2} + \frac{(S - S_c)^2}{S_r^2} \leq 1$$ (11.11)

With normalized variables, the above elliptical domain of uncertain variables is converted to the two-dimensional circle of radius equal unity

$$x^2 + y^2 \leq 1, \ x = \frac{R - R_c}{R_r}, \ y = \frac{S - S_c}{S_r}$$ (11.12)

The limit state becomes

$$M = R - S = R_r x - S_r y + R_c - S_c = 0$$ (11.13)

Consider some specific cases. When the mean values of the stress and the strength are equal, i.e., $R_c = S_c$, the failure plane passes through the origin of the coordinate system. The circular domain of basic variables is divided into two parts evenly, where $r = 1$ means the radius equals unity. Thus, we obtain the numerical value of non-probabilistic safety measure as

$$R_{ellipsoidal} = \eta(M(x,y) > 0) = \frac{S_{safe}}{S_{total}} = 0.5$$ (11.14)

In practical problems, $R_c \geq S_c$. It is instructive to reproduce here the formula of the distance from a point to a straight line, the distance d from the origin to the failure plane

$$d = \frac{|R_r \cdot X_0 - S_r \cdot Y_0 + R_c - S_c|}{\sqrt{R_r^2 + S_r^2}} = \frac{R_c - S_c}{\sqrt{R_r^2 + S_r^2}}$$ (11.15)

where $(X_0, Y_0) = (0,0)$ is the coordinate of origin.

The chord $\overline{AB}$ and the corresponding central angle θ are obtained as

$$\overline{AB} = 2\sqrt{1 - d^2}, \ \theta = \angle AOB = 2\cos^{-1} d$$ (11.16)

Obviously, the total area of the circular domain S_{total} of radius unity is π, while the shaded area indicates the failure region $S_{failure}$, which is denoted by the difference with the fan-shaped area $S^{\bullet}_{OAB}$, and the triangular area $S_{\Delta OAB}$. The failure area $S_{failure}$ equals (Fig. 11.2, 11.3, 11.4):

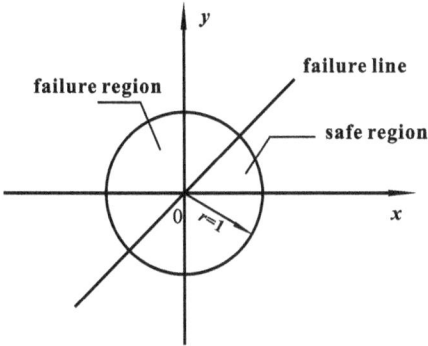

Fig. 11.2: Safe and failure regions for the same mean values of strength and stress.

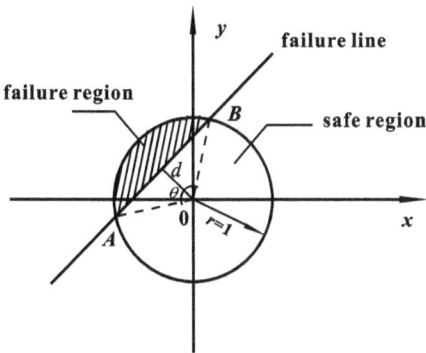

Fig. 11.3: Safe and failure regions for case $R_c \geq S_c$.

$$S_{failure} = S^{\bullet}_{OAB} - S_{\Delta OAB} \tag{11.17}$$

$$S^{\bullet}_{OAB} = \frac{1}{2}\theta \cdot R^2 = \frac{1}{2} \cdot 2\cos^{-1}d \cdot 1^2 = \cos^{-1}d \tag{11.18}$$

and

$$S_{\Delta OAB} = \frac{1}{2}\overline{AB} \cdot d = \frac{1}{2} \cdot 2\sqrt{1-d^2} \cdot d = d\sqrt{1-d^2} \tag{11.19}$$

Thus, we get

$$F_{ellipsoidal} = \frac{S_{failure}}{S_{total}}$$

$$= \frac{1}{\pi}\left[\cos^{-1}\frac{R_c - S_c}{\sqrt{R_r^2 + S_r^2}} - \frac{R_c - S_c}{\sqrt{R_r^2 + S_r^2}}\sqrt{1 - \left(\frac{R_c - S_c}{\sqrt{R_r^2 + S_r^2}}\right)^2}\right] = \frac{\cos^{-1}d - d\sqrt{1 - d^2}}{\pi}$$

$$(11.20)$$

and

$$R_{ellipsoidal} = \frac{S_{safe}}{S_{total}}$$

$$= 1 - \frac{1}{\pi}\left[\cos^{-1}\frac{R_c - S_c}{\sqrt{R_r^2 + S_r^2}} - \frac{R_c - S_c}{\sqrt{R_r^2 + S_r^2}}\sqrt{1 - \left(\frac{R_c - S_c}{\sqrt{R_r^2 + S_r^2}}\right)^2}\right]$$

$$= 1 - \frac{\cos^{-1}d - d\sqrt{1 - d^2}}{\pi} \qquad (11.21)$$

When the interference between the stress and the strength as shown in Fig. 11.3 does not take place, the event that the stress is bigger than the strength constitutes an impossible event. In other words, the possibility that the stress is bigger than the strength is zero and the system turns out to be absolutely reliable. In other words,

$$F_{ellipsoidal} = \eta(M(x,y) < 0) = \frac{S_{failure}}{S_{total}} = 0 \qquad (11.22)$$

and

$$R_{ellipsoidal} = \eta(M(x,y) > 0) = \frac{S_{safe}}{S_{total}} = 1 \qquad (11.23)$$

When the system is in the critical state, the failure plane and the value domain are just tangent as shown in Fig. 11.5, so the distance from the origin to the failure plane equals unity, i.e.,

$$d = \frac{R_c - S_c}{\sqrt{R_r^2 + S_r^2}} = r = 1 \qquad (11.24)$$

Thus, $F_{ellipsoidal} = 0$ and $R_{ellipsoidal} = 1$, we also obtain

$$S_r^2 + R_r^2 = \left(R_c - S_c\right)^2$$

$$(R_c^2 - R_r^2) + (S_c^2 - S_r^2) = 2R_cS_c$$

$$\overline{R}\cdot\underline{R} + \overline{S}\cdot\underline{S} = \frac{1}{2}(\overline{R}\cdot\overline{S} + \overline{R}\cdot\underline{S} + \underline{R}\cdot\overline{S} + \underline{R}\cdot\underline{S}) \qquad (11.25)$$

For the general nonlinear limit state function as shown in Fig. 11.6, the above concept of the non-probabilistic safety measure can still be applied, as a ratio of the appropriate areas.

When the limit state function is multidimensional, the hyper-ellipsoid enclosed by uncertain variables will be subdivided into the safe region and the failure region by a hyper-surface. In this case, the failure measure of structural non-probabilistic reliability based on the convex model can be defined by the ratio of the hyper-volume of failure region over the hyper-volume of total region; the measure of structural non-probabilistic safety can be defined by the ratio of the hyper-volume of safe region to the hyper-volume of total region.

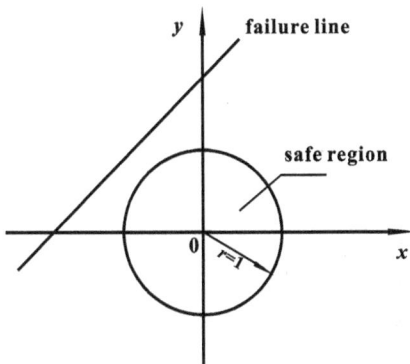

Fig. 11.4: Scheme for the totally reliable state.

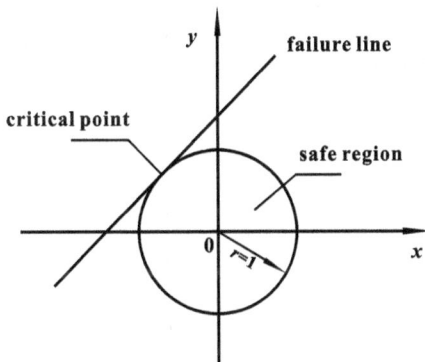

Fig. 11.5: Scheme for the critical state.

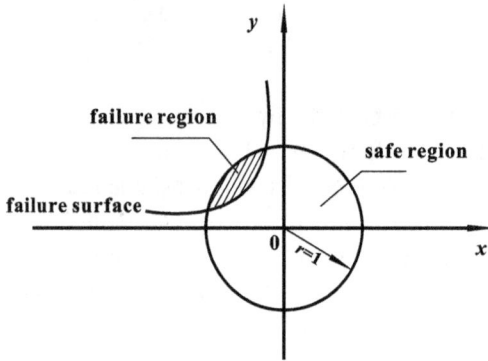

Fig. 11.6: Nonlinear limit state function.

11.5 Comparison between interval analysis and convex model method

In this section, we will compare the accuracy of the convex ellipsoidal model method and interval analysis method in solving the problem of non-probabilistic reliability. The uncertain region based on the above two methods is shown in the paper by Wang *et al.* (2008b).

From eqs. (11.20) and (11.21), we are able to compute the non-probabilistic failure measure and the non-probabilistic safety measure by the convex model method with uncertain parameters R and S, respectively. Meanwhile, the non-probabilistic failure measure $F_{inteval}$ and the non-probabilistic safety measure $R_{inteval}$ have been proposed in the literature by interval analysis as follows

$$F_{interval} = \begin{cases} 0, & \underline{R} > \overline{S} \\ \dfrac{(\overline{S} - \underline{R})^2}{8S_r R_r}, & \overline{S} > \underline{R} \\ 1, & \underline{R} < \overline{S} \end{cases} \tag{11.26}$$

and

$$R_{interval} = 1 - F_{interval} = \begin{cases} 1, & \underline{R} > \overline{S} \\ 1 - \dfrac{(\overline{S} - \underline{R})^2}{8S_r R_r}, & \overline{S} > \underline{R} \\ 0, & \underline{R} < \overline{S} \end{cases} \tag{11.27}$$

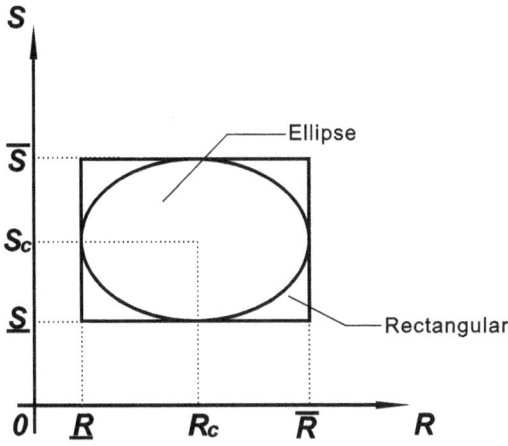

Fig. 11.7: Rectangle containing ellipse.

Non-probabilistic measure $R_{inteval}$ is the functions of four variables, i.e., (R_c, R_r, S_c, S_r). In order to simplify the process so that result is more concise, the failure line in eq. (11.13) is rewritten as

$$y = kx + b \qquad (11.28)$$

where $k = R_r/S_r$ denotes the slope of the failure line, $b = (R_c - S_c)/S_r$ denotes the intercept of the failure line. In most cases, $b \geq 0$ since $R_c \geq S_c$.

The non-probabilistic failure measure and the non-probabilistic safety measure by the convex model method are the functions of d, which is the distance from the origin to the failure line. Obviously, the intercept b can be expressed as follows

$$b = \frac{R_c - S_c}{S_r} = d\sqrt{1 + k^2} \qquad (11.29)$$

Thus, the non-probabilistic failure measure $F_{interval}$ given in eq. (11.26) and based on interval analysis can be rewritten in another form, i.e.,

$$F_{interval} = \begin{cases} 0, & \underline{R} > \overline{S} \\ \dfrac{\left(1 + k - d\sqrt{1 + k^2}\right)^2}{8k}, & \overline{S} > \underline{R} \\ 1, & \overline{R} < \underline{S} \end{cases} \qquad (11.30)$$

Naturally, the non-probabilistic safety measure $R_{interval}$ in eq. (11.27) becomes

$$R_{interval} = 1 - F_{interval} = \begin{cases} 1, & \underline{R} > \overline{S} \\ 1 - \dfrac{\left(1 + k - d\sqrt{1+k^2}\right)^2}{8k}, & \overline{S} > \underline{R} \\ 0, & \overline{R} < \underline{S} \end{cases} \qquad (11.31)$$

Let us compare eqs. (11.20), (11.21) and (11.30) and (11.31). We observe that the non-probabilistic measure proposed by the convex model is a constant when d is available. However, the measure given by interval analysis is a function of k even if d is constant. In order to conduct the comparison between the two methods, a new variable, namely ΔF is introduced as the difference between the non-probabilistic failure measure based on interval analysis and the non-probabilistic failure measure based on the convex model. Likewise, ΔR is defined as the difference between the non-probabilistic safety measure based on interval analysis and the non-probabilistic safety measure based on the convex model, i.e.,

$$\Delta F = F_{interval} - F_{ellipsoidal} = \frac{\left(1 + k - d\sqrt{1+k^2}\right)^2}{8k} - \frac{\cos^{-1}d - d\sqrt{1-d^2}}{\pi} \qquad (11.32)$$

and

$$\Delta R = R_{interval} - R_{ellipsoidal} = \left(1 - F_{interval}\right) - \left(1 - F_{ellipsoidal}\right) = F_{ellipsoidal} - F_{interval} = -\Delta F \qquad (11.33)$$

Different values are assigned to d, and a set of curves $F_{ellidsoidal}$ and $F_{interval}$ which are varied with the slope k are obtained as shown Fig. 11.8(a) and (b), respectively. In the same way, the curves k vs ΔF are shown in Fig. 11.8(c). From figure, it is easy to determine that ΔF is greater than zero when d is assigned values from 0.5 to 0.9. In other words, the non-probabilistic failure measure based on interval analysis $F_{interval}$ is greater than the non-probabilistic failure measure based on convex model $F_{ellipsoidal}$. Moreover, when k is in the interval [0,1], the extreme value of $F_{interval}$ is obtained. When k changes from 0.4 to 2.7, ΔF increases with d. We conclude that in this case the interval approach is more conservative in solving non-probabilistic problems while the convex model method is more economical (see also Figs. 11.9 – 11.24).

11.6 Interrelation between probability and convex models

Elishakoff (1999) studied the probabilistic method for the specific reliability problem, where random variables obey the uniform distribution within the ellipse. A natural question arises: are probabilistic and anti-optimization (Wald's min-max) methods interrelated? In order to find the interrelation between the probabilistic methods and

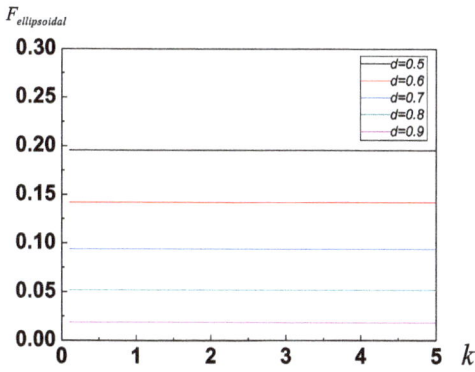

(a) Variation of $F_{ellipsoidal}$ versus k for different values of d

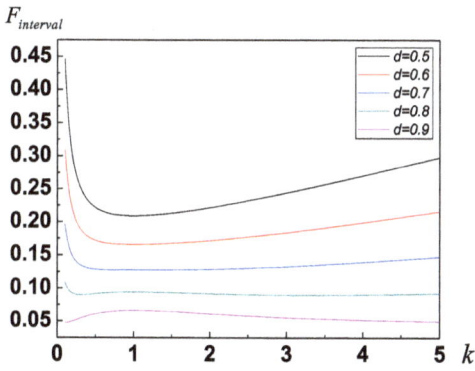

(b) Variation of $F_{interval}$ versus k for different values of d

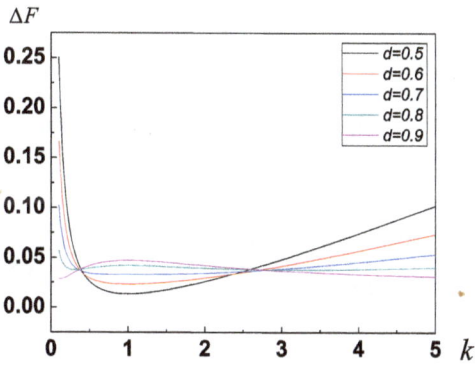

(c) The curves k vs ΔF for different values of d

Fig. 11.8: $F_{ellidsoidal}$, $F_{interval}$ and ΔF vs k for different values of d.

convex modeling of uncertainty consider the function of structural stress S and the structural strength R in eq. (11.4) from the probabilistic point of view. Let a joint probability density be given by

$$f_{RS}(r,s) = \begin{cases} \dfrac{1}{R_r S_r \pi}, & for \ \left(\dfrac{r-R_c}{R_r}\right)^2 + \left(\dfrac{s-S_c}{S_r}\right)^2 \le 1 \\ 0, & otherwise \end{cases} \tag{11.34}$$

In other words, let R and S have an uniform distribution in an ellipse; R_c, R_r, S_c and S_r are constants. Reliability then reads

$$R_{prob} = \eta(M(R,S) > 0) = \iint_{R-S>0} f_{RS}(r,s)drds \tag{11.35}$$

The line $R=S$ intersects the uncertainty region in points A_1 and A_2 as shown in Fig. 11.9.

The abscissas R_1 and R_2 of the points A_1 and A_2 are found by solving the equation

$$\left(\dfrac{R-R_c}{R_r}\right)^2 + \left(\dfrac{S-S_c}{S_r}\right)^2 = 1 \tag{11.36}$$

Thus, solving with respect to R, we get expressions for R_1 and R_2, respectively

$$R_1 = \dfrac{R_c S_r^2 + S_c R_r^2 - R_r S_r \sqrt{R_r^2 + S_r^2 - \left(R_c - S_c\right)^2}}{R_r^2 + S_r^2} \tag{11.37}$$

and

$$R_2 = \dfrac{R_c S_r^2 + S_c R_r^2 + R_r S_r \sqrt{R_r^2 + S_r^2 - \left(R_c - S_c\right)^2}}{R_r^2 + S_r^2} \tag{11.38}$$

Obviously, due to the limit plane $R=S$, the x-coordinates S_1 and S_2 are equal to the y- coordinates R_1 and R_2, respectively (see Fig. 11.9))

$$S_1 = R_1 = \dfrac{R_c S_r^2 + S_c R_r^2 - R_r S_r \sqrt{R_r^2 + S_r^2 - \left(R_c - S_c\right)^2}}{R_r^2 + S_r^2} \tag{11.39}$$

and

$$S_2 = R_2 = \dfrac{R_c S_r^2 + S_c R_r^2 + R_r S_r \sqrt{R_r^2 + S_r^2 - \left(R_c - S_c\right)^2}}{R_r^2 + S_r^2} \tag{11.40}$$

In order to calculate the probabilistic reliability eq. (11.35), the double integration must be performed. However, the different location for the line $R=S$ will produce different ranges of integral variables R and S. Five different cases will emerge as shown in Fig. 11.10, which shows the different locations of the points A_1 and A_2. Hence, the corresponding integral expresses of the probabilistic reliability for the five cases are given as follows

$R_{prob} = \eta(R - S > 0)$

$$
= \begin{cases}
\displaystyle\int_{S_c-S_r}^{r} \sqrt{1-\left(\frac{r-R_c}{R_r}\right)^2} \int_{R_1}^{R_2} \frac{1}{R_r S_r \pi} \, drds, & \text{for } S_c > S_2 > S_1 \\[4ex]
\displaystyle 1-\int_{r}^{S_c+S_r} \sqrt{1-\left(\frac{r-R_c}{R_r}\right)^2} \int_{R_1}^{R_2} \frac{1}{R_r S_r \pi} \, drds, & \text{for } S_2 > S_1 > S_c \\[4ex]
\displaystyle 1-\int_{R_c-R_r}^{s} \sqrt{1-\left(\frac{s-S_c}{S_r}\right)^2} \int_{S_1}^{S_2} \frac{1}{R_r S_r \pi} \, dsdr, & \text{for } S_2 > S_c > S_1 \text{ and } R_c > R_2 > R_1 \\[4ex]
\displaystyle \int_{s}^{R_c+R_r} \sqrt{1-\left(\frac{s-S_c}{S_r}\right)^2} \int_{S_1}^{S_2} \frac{1}{R_r S_r \pi} \, dsdr, & \text{for } S_2 > S_c > S_1 \text{ and } R_2 > R_1 > R_c \\[4ex]
\displaystyle \frac{1}{4} + \int_{S_c-S_r}^{r} \sqrt{1-\left(\frac{r-R_c}{R_r}\right)^2} \int_{R_1}^{R_c} \frac{1}{R_r S_r \pi} \, drds + \\[4ex]
\displaystyle \int_{s}^{R_c+R_r} \sqrt{1-\left(\frac{s-S_c}{S_r}\right)^2} \int_{S_c}^{S_2} \frac{1}{R_r S_r \pi} \, dsdr, & \text{for } S_2 > S_c > S_1 \text{ and } R_2 > R_c > R_1
\end{cases}
$$

(11.41)

The unreliability measure based on probabilistic method reads

$$P_f = \eta(R - S < 0) = 1 - R_{prob} \tag{11.42}$$

By normalized two-dimensional space shown in eq. (11.12), the joint probability density $f_{RS}(r,s)$ in eq. (11.34) IS rewritten as

$$f_{RS}(r,s) \Rightarrow f(x,y) = \begin{cases} \frac{1}{\pi}, & x^2 + y^2 \le 1 \\ 0, & \text{otherwise} \end{cases} \tag{11.43}$$

In this way, it is easier to determine structural reliability from geometric consideration rather than from direct integration. Thus, the structural probability of failure P_f and the reliability R_{prob} are expressed as follows

$$P_f = \frac{1}{\pi}\left[\arccos\frac{R_c - S_c}{\sqrt{R_r^2 + S_r^2}} - \frac{R_c - S_c}{\sqrt{R_r^2 + S_r^2}}\sqrt{1 - \left(\frac{R_c - S_c}{\sqrt{R_r^2 + S_r^2}}\right)^2}\right] \tag{11.44}$$

and

$$R_{prob} = 1 - \frac{1}{\pi}\left[\arccos\frac{R_c - S_c}{\sqrt{R_r^2 + S_r^2}} - \frac{R_c - S_c}{\sqrt{R_r^2 + S_r^2}}\sqrt{1 - \left(\frac{R_c - S_c}{\sqrt{R_r^2 + S_r^2}}\right)^2}\right] \tag{11.45}$$

Hence, comparison of eqs. (11.20), (11.21) and (11.44), (11.45) leads to the conclusion that

$$F_{ellipsoidal} = P_f, \ or \ R_{ellipsoidal} = R_{prob} \tag{11.46}$$

It is remarkable that, under the same circumstances of uncertain information the non-probabilistic safety measure coincides with the probabilistic estimate.

11.7 Transition from probabilistic to convex modeling

In this section, a truncated normal distribution model will be discussed as the expression of the probability density function to describe the random variables. When the value of parameters in the probability density function changes, the variability of random variables either will increase or decrease. With the increase of variability, the random variable tends to one with uniform distribution. If so, the two results of reliability, based on probabilistic methods and convex modeling methods, may show a good agreement. In other words, a nice transition from probabilistic to convex modeling may be implemented. In Sections 7.1 and 7.2, the transition process with reference to the univariate and bivariate analysis will be carried out in detail.

11.8 Univariate analysis based on stochastic model

When dealing with single random variable in practical structures, the truncated normal distribution model is often employed. The probability density reads

$$f(\xi) = \begin{cases} c \cdot \exp\left(-\frac{(\xi - \xi_c)^2}{b^2}\right), & |\xi - \xi_c| \le \xi_r \\ 0 & , |\xi - \xi_c| > \xi_r \end{cases} \tag{11.47}$$

where $f(\xi)$ is the probability density function of ξ, ξ_c means the central value and ξ_r is a bound of variation of the uncertainty for the random variable ξ with respect to ξ_c value; b represents a parameter. The normalization constant c can be derived from

$$c = \left[2b \cdot erf\left(\frac{\xi_r}{b}\right) \right]^{-1}$$ (11.48)

in which $erf(\cdot)$ is the error function and is defined as

$$erf(x) = \int_0^x e^{-t^2} dt$$ (11.49)

Figure 11.11 shows the probability density function of the random variable ξ for different values of b. If ξ_r is given, then the probability density $f(\xi)$ depends exclusively on b, namely, a large b corresponds to a large deviation in ξ. When $b^2 \gg \xi_r^2$, ξ is nearly uniformly distributed, as is indicated by the curve associated with $b = 1.0$ in Fig. 11.11.

Monte Carlo simulation method can be utilized to calculate possible values of the simple random variable ξ with given parameters b and ξ_r in eq. (11.47). Suppose the system is safe if $\xi \leq \xi^*$. Then the structural reliability is expressed as follows:

$$R_{prob} = \eta\left(\xi \leq \xi^*\right)$$ (11.50)

in which ξ^* means the admissible value of random variable ξ.

We turn now to the two-dimensional case.

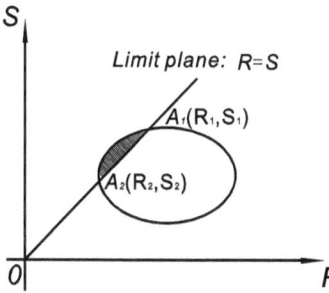

Fig. 11.9: Scheme for the model of the structural reliability based on probabilistic theory.

11.9 Bivariate analysis based on stochastic model

In this case, the truncated normal probability density function of two variables varying within an ellipse will be used to describe the distribution for strength R and stress S. The new expression of above statements is written as

$$f_{RS}(r,s) = \begin{cases} c \cdot \exp\left(-\frac{(r-R_c)^2}{a^2 p^2} - \frac{(s-S_c)^2}{b^2 p^2}\right), & \frac{(r-R_c)^2}{R_r^2} + \frac{(s-S_c)^2}{S_r^2} \leq 1 \\ 0, & \frac{(r-R_c)^2}{R_r^2} + \frac{(s-S_c)^2}{S_r^2} > 1 \end{cases}$$ (11.51)

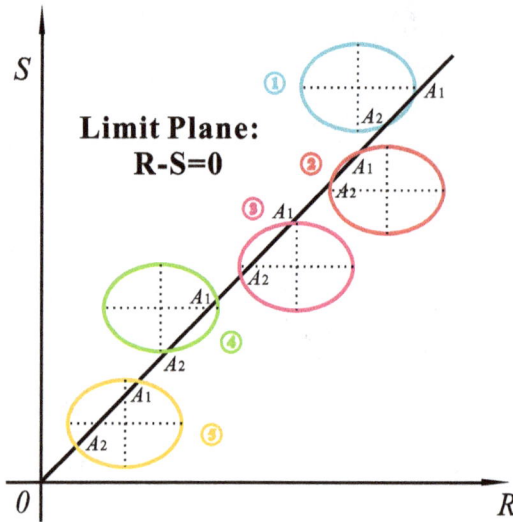

Fig. 11.10: Different cases about the intersections between the limit line and the uncertainty region.

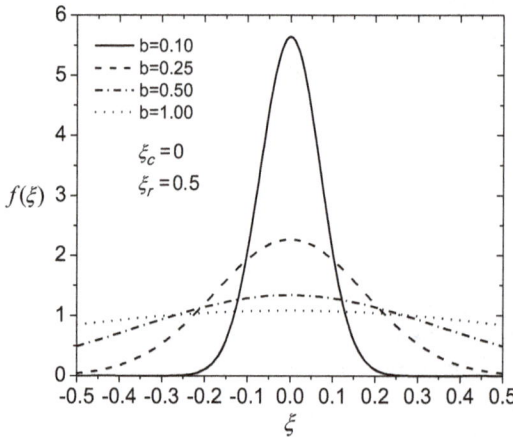

Fig. 11.11: Probability density function for a truncated normally distributed random variable.

where $f_{RS}(r, s)$ is the joint probability density function of R and S, R_c and S_c indicate respective central values; R_r and S_r are the bounds of uncertainty for the random variables R and S, respectively; a and b represent coefficients, the normalization constant c is derived from

$$c = \left[4ab\rho^2 \cdot erf\left(\frac{R_r}{a\cdot\rho}, \frac{\sqrt{S_r^2 - \frac{S_r^2}{R_r^2}r^2}}{b\cdot\rho} \right) \right]^{-1} \tag{11.52}$$

in which two-dimensional error function $erf(x,y)$ is defined as

$$erf(x,y) = \int_0^x\int_0^y e^{-(\xi^2+\eta^2)} d\eta d\xi \tag{11.53}$$

The coefficient ρ introduced in eq. (11.51), in order to describe our possible ignorance on the distribution. Although the probability distribution is specified as truncated normal, we may not precisely know its exact values. It is of utmost interest then to evaluate the reliability as a function of ρ.

As the case of univariate analysis, if the coefficients a and b as well as the values R_r and S_r are known, then the probability density $f_{RS}(r,s)$ depends exclusively on ρ. Thus, the variability of random variables R and S increases with the increase of ρ; when ρ is large enough, the random variables tend to a uniform distribution within a range of ellipse, as shown by Fig. 11.12.

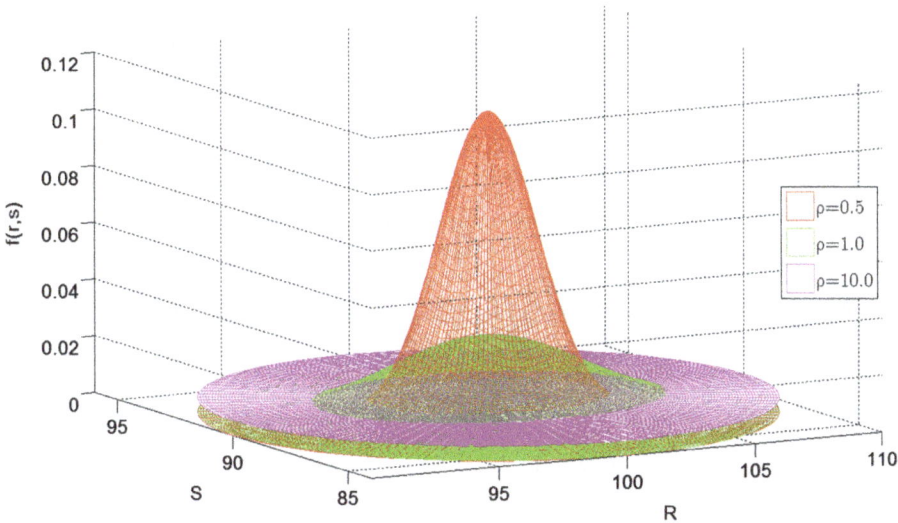

Fig. 11.12: Probability density function of random variables R and S with different parameters ρ.

Monte Carlo simulation can be carried out to obtain possible values of random variables R and S when the parameters a, b, R_r, and S_r are given. Consider the limit state function in eq. (11.5) as the criterion to judge the structural safety measure. It is easy to see that the reliability of structure based on probabilistic methods can be expressed as follows

$$R_{prob} = \eta(R - S > 0) = \frac{k}{n} \tag{11.54}$$

where n is the number of random samples created by Monte Carlo simulations, and k means the number of points, which satisfies the criterion $R - S > 0$.

Consequently, as the increasing deviation of R and S, these two random variables tend to be uniformly distributed within a given ellipse; and hence the results of reliability based on probabilistic analysis and convex models are anticipated to show a good agreement. Thus, a new concept of the transition from probabilistic to convex modeling will be presented.

11.10 Numerical examples

In this section, some examples will be used to illustrate the validity of the present non-probabilistic safety measure based on convex models.

As the first example, we consider a bar with two random variables involved. The stress interval is $S^I = [84, 96]$ MPa, and the strength interval is $R^I = [90, 110]$ MPa. The safety margin is $M = R - S$.

According to the convex modeling method the value domain of R and S is determined as

$$\left(\frac{R - 100}{10}\right)^2 + \left(\frac{S - 90}{6}\right)^2 \leq 1 \tag{11.55}$$

The non-probabilistic failure measure and safety measure are computed by the presented formulations as

$$F_{ellipsoidal} = S_{failure}/S_{total} = 0.099/\pi = 0.0315, \quad R_{ellipsoidal} = S_{safe}/S_{total} = 0.9685 \tag{11.56}$$

The failure measure and safety measure based on non-probabilistic interval model are obtained as

$$F_{interval} = S_{failure}/S_{total} = 0.075, \quad R_{interval} = S_{safe}/S_{total} = 0.925 \tag{11.57}$$

We should prefer the measure that results in a smaller reliability, or a greater failure measure, in order to be on the safe side.

By Monte Carlo simulation, 10,000 random samples whose joint probability density function is given in eq. (11.51) are generated within the above ellipse, for $R_c = 100$, $R_r = 10$, $S_c = 90$, $S_r = 6$. Suppose the coefficients a and b equal to 4 and 3, respectively. For various parameters ρ, different distribution patterns are obtained, as shown in Figs. 11.13–11.18

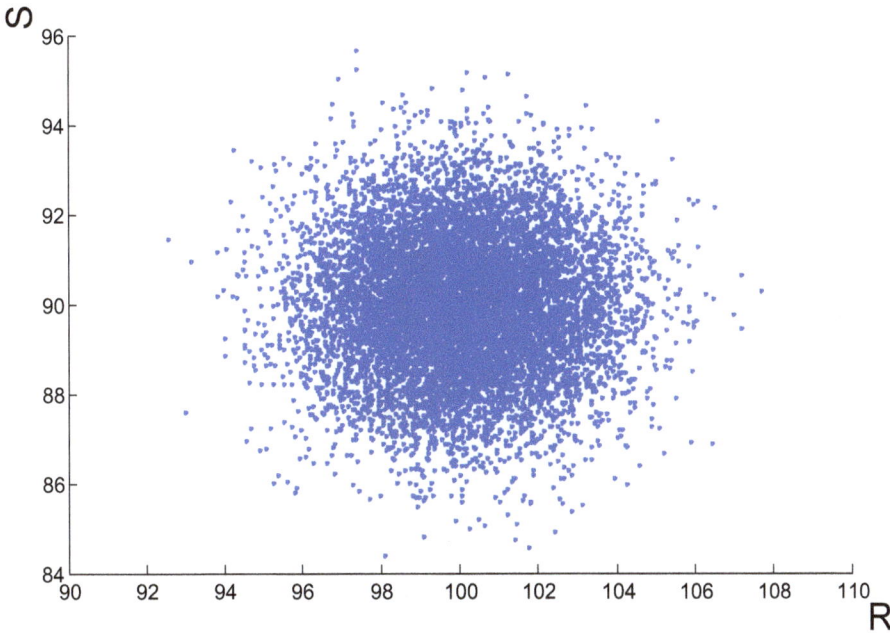

Fig. 11.13: Realization of 10,000 random samples with given parameter $\rho = 0.5$ obtained by Monte Carlo simulation.

They present the increasing variability in R and S. According to eq. (11.54), the structural reliability R_{prob} for the different ρ are calculated and listed in Table 11.1.

By analyzing the numerical results in Table 11.1, we find that larger deviation of R and S, the more conservative safety measures are resulting. When given parameter ρ is large enough, the two random variables are nearly uniformly distributed, and the result of structural safety measure based on probability will agree with the one from convex model, which seems to constitute a nice transition from probabilistic to convex modeling. This is compatible with the principle of maximum entropy (Ziegler, 2010).

Indeed, it appears instructive to quote Claude Shannon: "My greatest concern was what to call it. I thought of calling it 'information', but the word was overly used, so I decided to call it 'uncertainty'. When I discussed it with John von Neumann, he had a better idea. Von Neumann told me, 'You should call it **entropy**'" (Wikipedia, 2020). Thus, the maximum entropy implies maximum uncertainty. Uniform density for the random variable signifies its maximum uncertainty and thus, the worst-case scenario. In this sense, this density is directly associated with the worst-case scenario predicted by interval analysis or convex, ellipsoidal modeling.

As is seen, the ellipsoidal reliability measure is 0.9685. Probabilistic reliability changes with ρ, as expected. Our ignorance of the probability distribution is clearly illustrated by Table 11.1. Indeed, with increased value of ρ, the reliability decreases.

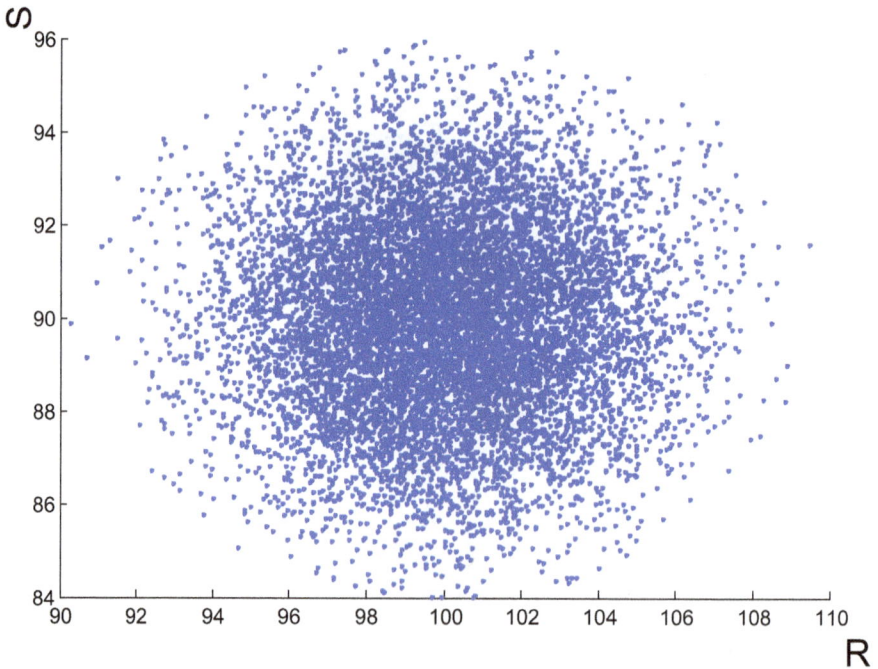

Fig. 11.14: Realization of 10,000 random samples with given parameter $\rho = 1.0$ obtained by Monte Carlo simulation.

To be on the safe side we should choose the minimum reliability value that is attained when $\rho = 200.0$. Remarkably, the resulting value of reliability 0.9686 nearly coincides with 0.9685 provided by the ellipsoidal analysis. Thus, when the knowledge on probability density is imprecise, one is better off to use convex modeling. It brings to the sought result much faster than the reliability calculation. This observation is also consistent with the maximum entropy principle: non-uniform distribution is associated with less entropy than the uniform distribution, just as a fair coin has more entropy than the coin that is not fair. Thus, when there is a lack of knowledge on probability density, or even of its one or more parameters, one can be advised to employ either interval or convex analysis. We are obliged to Professor Franz Ziegler, whose comment led to the above introduction of coefficient ρ as the coefficient of ignorance (Ziegler, 20101).

As the second example, we consider a cantilever beam as shown in Fig. 11.25. The cantilever is subjected to three concentrated forces applied at distances $b_1 = 1.0m$, $b_2 = 2.0m$, and $b_3 = 5.0m$ from the fixed end, respectively. The structure is considered to fail if $|m_{\max}| \geq m_{cr}$, where $m_{\max}$ is the maximum actual moment and m_{cr} is the limit moment. Suppose that the basic interval variables are: $m_{cr}^I = [25, 27]kN \cdot m$,

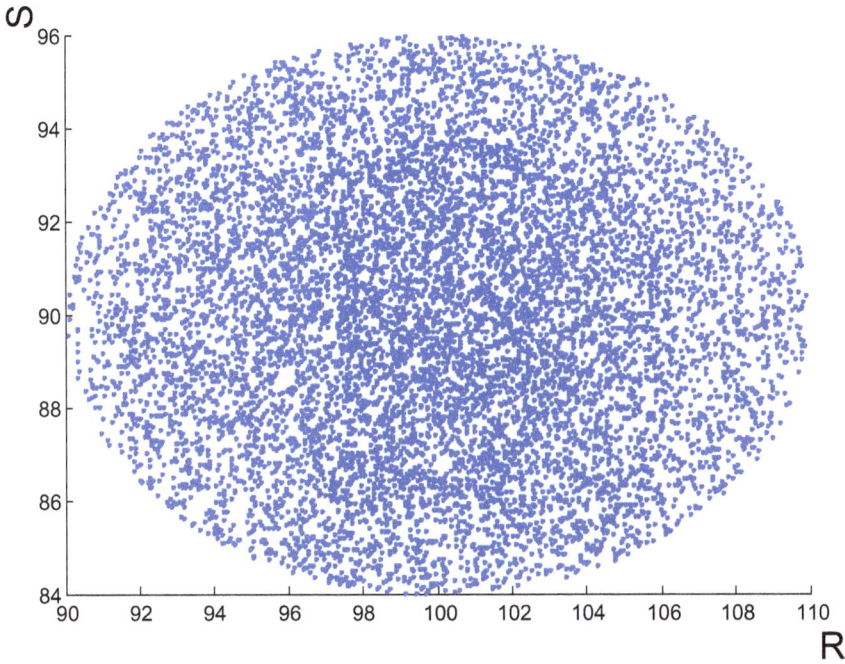

Fig. 11.15: Realization of 10,000 random samples with given parameter $\rho = 5.0$ obtained by Monte Carlo simulation.

$p_1^I = [2.8,\ 3.2]kN$, $p_2^I = [4.4,\ 5.6]kN$, and $p_3^I = [1.7,\ 2.3]kN$. The safety margin of the structure is $M = m_{cr} - b_1 p_1 - b_2 p_2 - b_3 p_3$.

Let p_1 be assigned to a median value $p_1 = 3\ kN$. The safety margin of the structure is rewritten as $M = m'_{cr} - b_2 p_2 - b_3 p_3$, where $m'^I_{cr} = [22,\ 24]\ kN \cdot m$.

Within the ellipsoidal modeling, the uncertain parameters are expressed as belonging to the set

$$\left(\frac{m'_{cr} - m'^c_{cr}}{m'^r_{cr}}\right)^2 + \left(\frac{p_2 - p_2^c}{p_2^r}\right)^2 + \left(\frac{p_3 - p_3^c}{p_3^r}\right)^2 \leq 1 \tag{11.58}$$

where $m'^c_{cr} = 23\ kN \cdot m$, $m'^r_{cr} = 1\ kN \cdot m$, $p_2^c = 5\ kN$, $p_2^r = 0.6\ kN$, $p_3^c = 2\ kN$, and $p_3^r = 0.3\ kN$.

By using the normalized coordinate system, the above inequality becomes

$$x^2 + y^2 + z^2 \leq 1,\ for\ x = \frac{m'_{cr} - m'^c_{cr}}{m'^r_{cr}},\ y = \frac{p_2 - p_2^c}{p_2^r},\ z = \frac{p_3 - p_3^c}{p_3^r} \tag{11.59}$$

The non-probabilistic failure and safety measures turn out to equal, respectively,

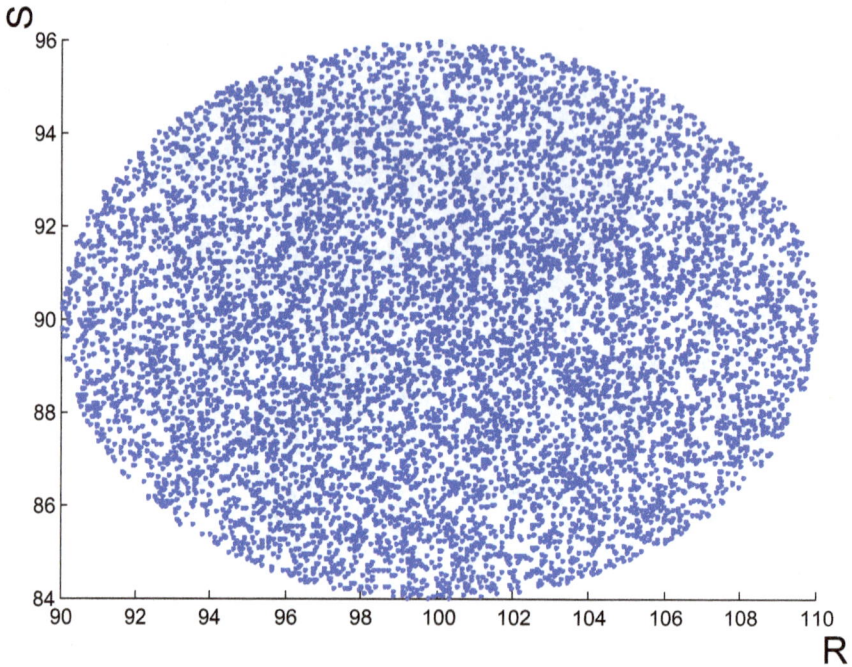

Fig. 11.16: Realization of 10,000 random samples with given parameter $p = 20.0$ obtained by Monte Carlo simulation.

$$F_{ellipsoidal} = V_{failure}/V_{total} = 0.0043, \quad R_{ellipsoidal} = V_{safe}/V_{total} = 0.9957 \quad (11.60)$$

where $d = |0 - 1.2 \times 0 - 1.5 \times 0 - 2| / \sqrt{1 + 1.2^2 + 1.5^2} = 0.9235$ is the distance from the origin to the limit state plane, $V_{total} = (4/3)\pi = 4.1888$ is the total volume of the sphere enclosed by the uncertain parameters, $V_{failure} = \int_{0.9235}^{1} \pi \cdot (1 - x^2)dx = 0.0179$ is the volume of the failure region, $V_{safe} = V_{total} - V_{failure} = 4.1709$ is the volume of the safe region.

Based on non-probabilistic interval model, the failure and the safety measures equal, respectively,

$$F_{interval} = V_{failure}/V_{total} = 0.004, \quad R_{interval} = V_{safe}/V_{total} = 0.996 \quad (11.61)$$

where

$$V_{total} = 2.0 \times 1.2 \times 0.6 = 1.44 \quad (11.62)$$

and

$$V_{failure} = \frac{1}{3} \times \frac{1}{2} \times 0.14 \times 0.35 \times 0.7 = 0.00572 \quad (11.63)$$

and

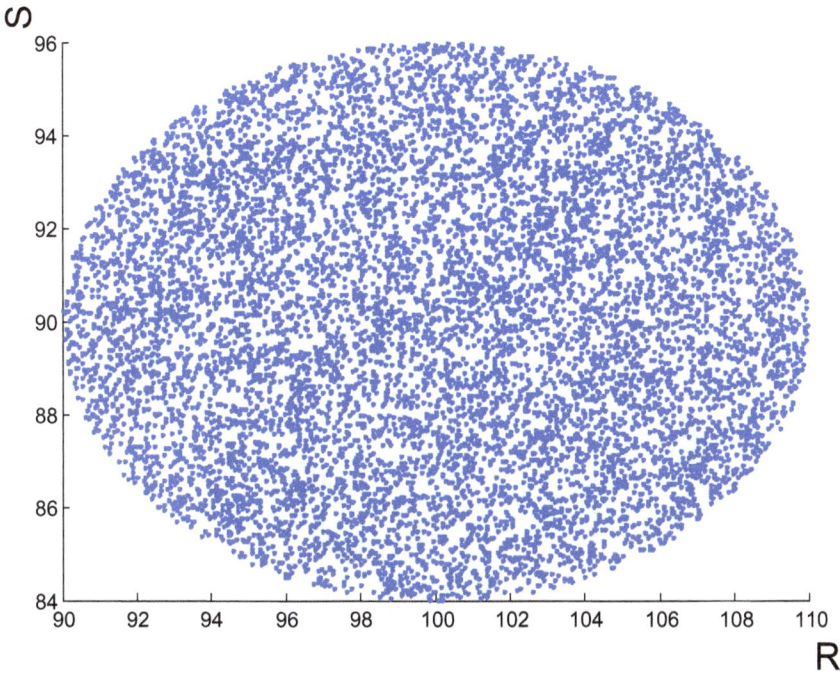

Fig. 11.17: Realization of 10,000 random samples with given parameter $\rho = 100.0$ obtained by Monte Carlo simulation.

$$V_{safe} = V_{total} - V_{failure} = 1.43 \qquad (11.64)$$

Let us consider now a multidimensional case with p_1 is treated as an interval variable, namely $p_1^I = [0,\ 6]\ kN$. With convex modeling method, the uncertain parameters are expressed as

$$\left(\frac{m_{cr} - m_{cr}^c}{m_{cr}^r}\right)^2 + \left(\frac{p_1 - p_1^c}{p_1^r}\right)^2 + \left(\frac{p_2 - p_2^c}{p_2^r}\right)^2 + \left(\frac{p_3 - p_3^c}{p_3^r}\right)^2 \leq 1 \qquad (11.65)$$

Where $m_{cr}^c = 26\ kN \cdot m,\ m_{cr}^r = 1\ kN \cdot m,\ p_1^c = 3\ kN,\ p_1^r = 3\ kN,\ p_2^c = 5\ kN,\ p_2^r = 0.6 kN,\ p_3^c = 2\ kN,\ p_3^r = 0.3\ kN$.

By normalizing the coordinate system, the above inequality becomes

$$x^2 + y^2 + z^2 + u^2 \leq 1$$

$$for\ x = \frac{m_{cr} - m_{cr}^c}{m_{cr}^r}, y = \frac{p_2 - p_2^c}{p_2^r}, z = \frac{p_3 - p_3^c}{p_3^r}, u = \frac{p_1 - p_1^c}{p_1^r} \qquad (11.66)$$

The non-probabilistic failure measure and safety measure equal, respectively,

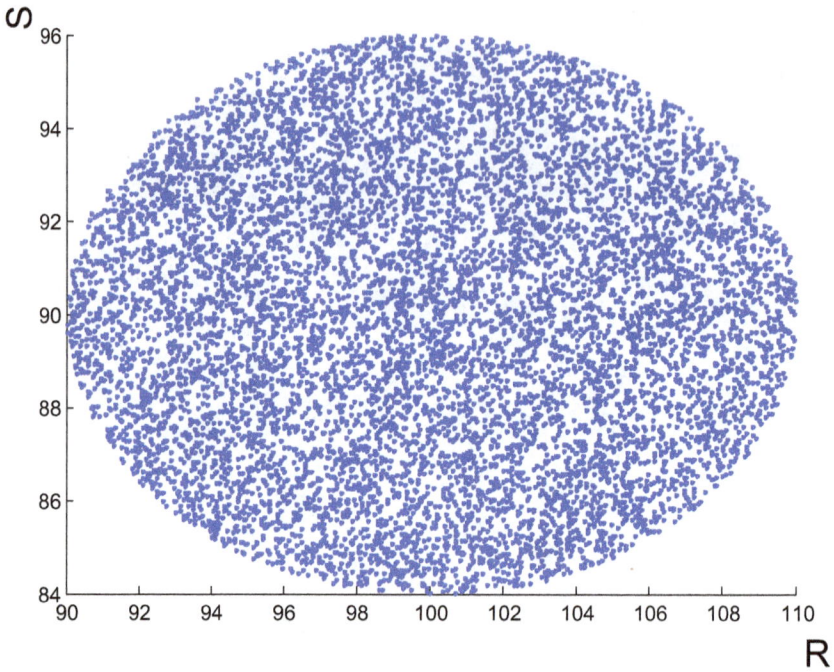

Fig. 11.18: Realization of 10,000 random samples with given parameter $p = 200.0$ obtained by Monte Carlo simulation.

Table 11.1: Comparison between R_{prob} and $F_{ellipsoidal}$ with different parameters p.

	$p = 0.5$	$p = 1.0$	$p = 5.0$	$p = 20.0$	$p = 100.0$	$p = 200.0$
R_{prob}	0.9999	0.9984	0.9808	0.9719	0.9694	0.9686
$F_{ellipsoidal}$	0.9685	0.9685	0.9685	0.9685	0.9685	0.9685

$$F_{ellipsoidal} = V_{failure}/V_{total} = 0.0451, \quad R_{ellipsoidal} = V_{safe}/V_{total} = 0.9549 \tag{11.67}$$

where

$$d = 3 - 3u/\sqrt{4.69}, \quad V_{total} = \int_{-1}^{1} \frac{4}{3}\pi(1 - u^2)^{\frac{3}{2}} du = \pi/2 \tag{11.68}$$

is the total volume of the hyper-sphere

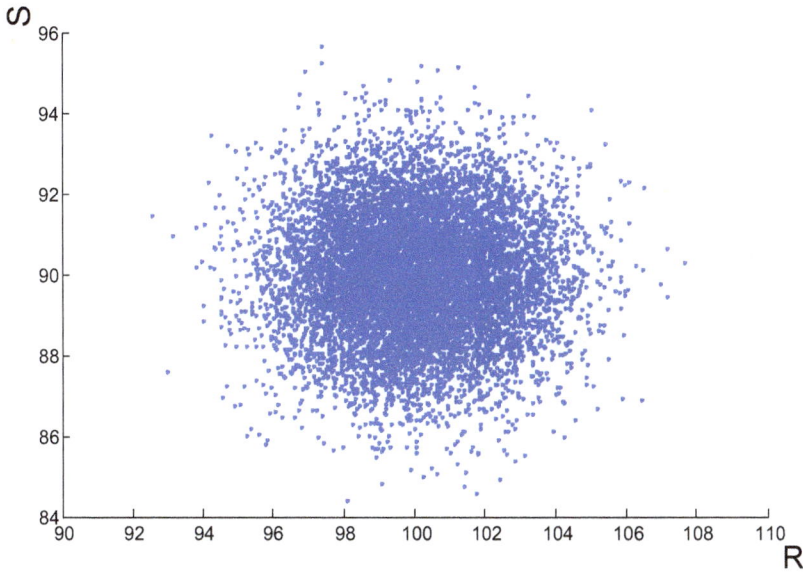

Fig. 11.19: Realization of 10,000 random samples with given parameter $\rho = 0.5$ obtained by Monte Carlo simulation.

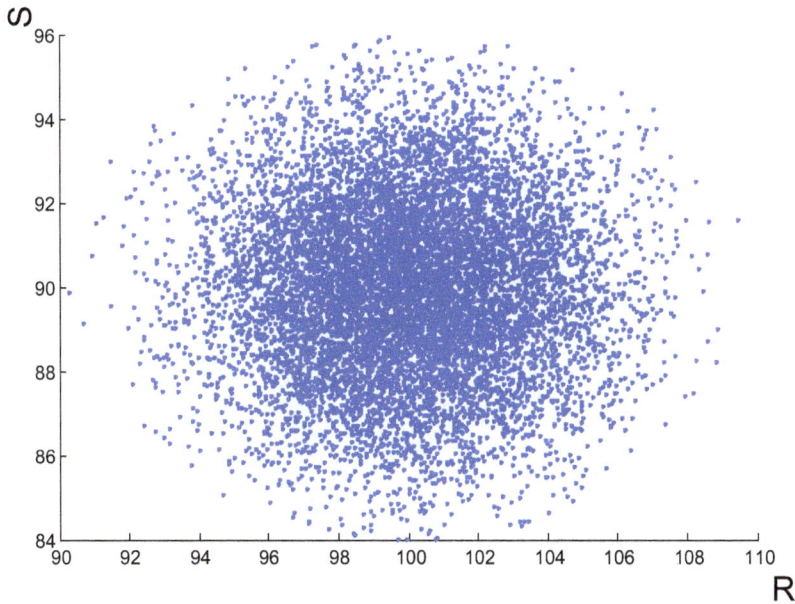

Fig. 11.20: Realization of 10,000 random samples with given parameter $\rho = 1.0$ obtained by Monte Carlo simulation.

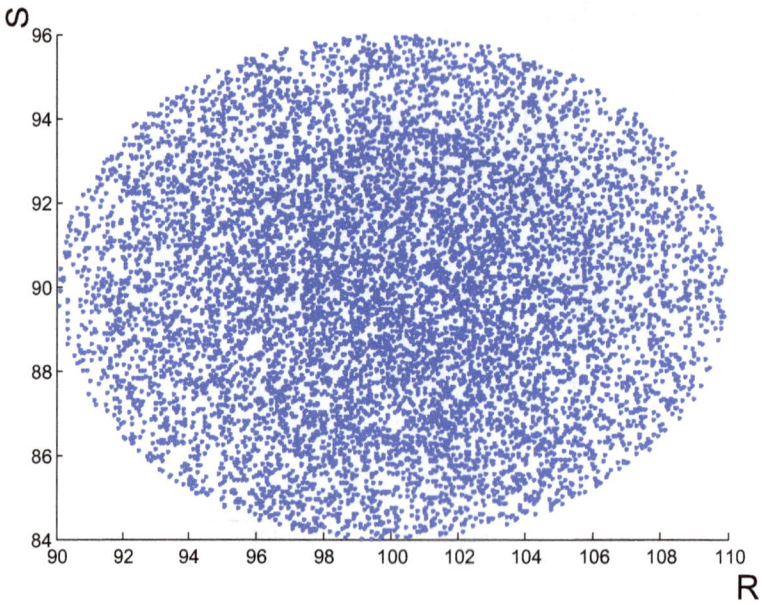

Fig. 11.21: Realization of 10,000 random samples with given parameter $p = 5.0$ obtained by Monte Carlo simulation.

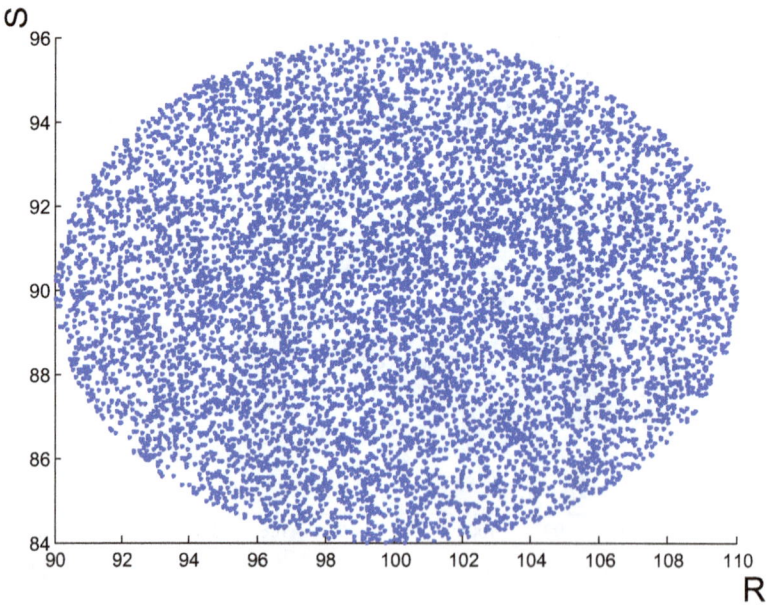

Fig. 11.22: Realization of 10,000 random samples with given parameter $p = 20.0$ obtained by Monte Carlo simulation.

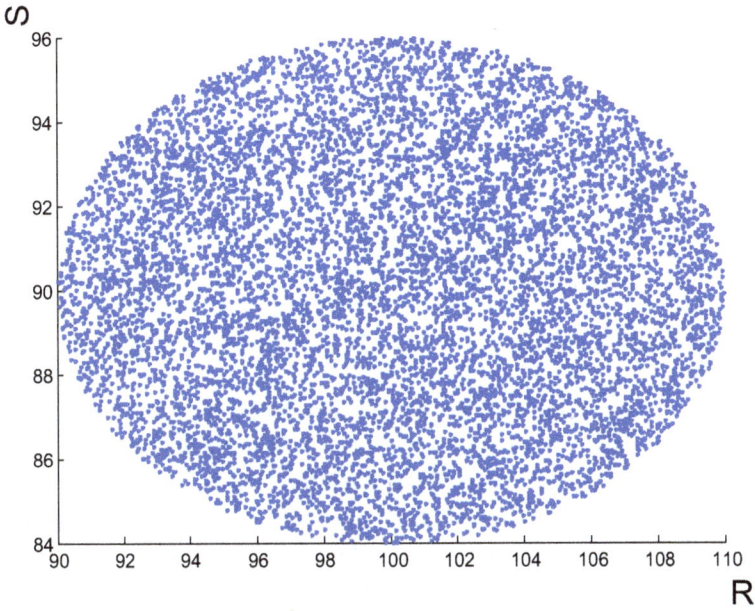

Fig. 11.23: Realization of 10,000 random samples with given parameter $p = 100.0$ obtained by Monte Carlo simulation.

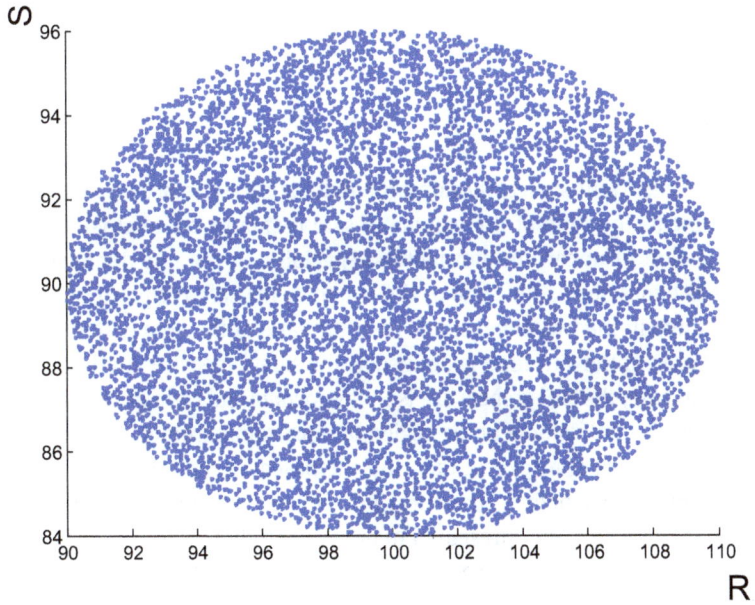

Fig. 11.24: Realization of 10,000 random samples with given parameter $p = 200.0$ obtained by Monte Carlo simulation.

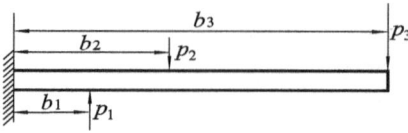

Fig. 11.25: Cantilever beam.

$$V_{failure} = \int\limits_{0.3148}^{1} \int\limits_{3-3u/\sqrt{4.69}}^{\sqrt{1-u^2}} \pi\left[(1-u^2) - 9(1-u)^2/4.69\right] du = 0.0709 \qquad (11.69)$$

is the volume of the failure region, $V_{safe} = V_{total} - V_{failure} = 1.4999$ is the volume of the safe region.

From the first example, it can be seen that the interval modeling is more appropriate than the convex modeling since the measure of non-probabilistic reliability based on interval analysis is more conservative than the measure based on convex model. The second example yields a similar conclusion. Still, by utilizing the proposed way of calculating $R_{ellipsoidal}$ is easier in implementation.

We conclude that if probabilistic information is available, it should be fully exploited for the derivation of the probabilistic reliability; however, if the information on densities is missing, one is forced to use the non-probabilistic safety measure that treats the unknown density as if it were an uniform one. The latter the way of computing $R_{ellipsoidal}$, as the ratio of the appropriate volumes, appears to be an efficient technique.

11.11 Conclusion

In this chapter, a measure of structural non-probabilistic safety is proposed within the convex modeling as the ratio of the volume of safe region to the total volume of the region of variation of uncertain parameters. Specially, in the one-dimensional case the measure is expressed as the ratio of the length of the safe region to the total length of the variation. In the two-dimensional case, the ratio of the area with the safe region to the total area of the variation is defined as the safety measure. In the three- or multidimensional case, the safety measure is in the form of the ratio of the volume of the safe region to the total volume of the region of uncertain parameters. This new measure appears to be more transparent than the one based on the "allowable maximum amount of uncertainty before structural failure" or "the minimum distance from the origin to the failure plane" presented in previous studies.

Furthermore, the interrelation between probability and convex models as well as the transition from probabilistic to convex modeling are discussed. Ignorance of the probability density is investigated. Ignorance on the probability density is modeled by a control parameter p. With the increase of p, the probabilistic reliability decreases. If

ρ is large, the random variables are nearly uniformly distributed. This result is compatible with the maximum entropy principle (Ziegler, 2010). The results of structural reliability were demonstrated to be in good agreement with the ones obtained by convex modeling.

It must be stressed that the purpose of the chapter is not to suggest replacement of the probabilistic approach by the non-probabilistic convex modeling. The latter is a viable avenue of uncertainty analysis when scarce data is available. Note that the material in this chapter was mostly drawn from the paper by Wang X-J., Wang L., Elishakoff I. and Qiu Z-P., Probability and Convexity are Not Antagonistic, *Acta Mechanica*, Vol. 219, 45–64, 2011, with permission.

Chapter 12
Conclusion: each uncertainty analysis is right

Uncertainty is the only certainty there is, and knowing how to live with Insecurity is the only security. John Allen Paulos

If you don't know where you are going, any road will get you there. Lewis Carroll

12.1 Preliminary comments

As you might know, there are those who find themselves skipping to the end of a book to know what happens. Why do they do this? Why can't they just enjoy the journey? The answer is two-fold. People who do research seem to be consumed in their own research, which appears to them most important. So, why should they read others? Nobel laureate Lev Landau (1908–1968) once quipped (or is this an apocryphal story?) that when someone tells him about his scientific problems, he simultaneously responds by the enumeration of his own problems he is devoted to. Otherwise, Landau would think about someone else's problems, and he can't afford it. Be it as it may, there are those (not yet Nobel laureates) who read the conclusion of the book rather than the entire book. To those "readers" we will elaborate the book in a nutshell.

Currently there are numerous papers on uncertainty quantification in its myriad of manifestations. The interested readers can consult definitive works – listed chronologically– by Roache (1997), Karniadakis (2002), Bae *et al.* (2003, 2004a, 2004b), Matthies (2004), Pettit (2004), Lew and Horta (2007), De Rocquigny *et al.* (2008), Adhikari *et al.* (2009), Ghanem (2009), Najm (2009), Galbally et al. (2010), Sepahvand *et al.* (2010), Biegler *et al.* (2011), Eldred *et al.* (2011), Li Y. *et al.* (2011), Liang B. and Mahadevan (2011), Adams *et al.* (2012), Ellingwood and Kinali (2009), Grigoriu (2012), Kamath (2012), Kidane et al. (2012), Lin *et al.* (2012), Scarrott and MacDonald (2012), Oladyshkin and Nowak (2012), Scarrott and MacDonald (2012), Batou and Soize (2013), Chernatynskiy *et al.* (2013), Clément *et al.* (2013), Crespo et al. (2013), Kamga *et al.* (2014), Mahadevan (2013), Nannapaneni *et al.* (2016), Papadrakakis and Stefanou (2014), Roy and Oberkampf (2011), Scarth et al. (2014), Owhadi et al. (2013), Cunha *et al.* (2014), Georgiou et al (2014), Janon *et al.* (2014), Marelli and Sudret (2014), Smith (2014), Stavroulakis *et al.* (2014), Atamturktur *et al.* (2015), Feinberg and Langtangen (2015), Han *et al.* (2015), Patelli *et al.* (2015), Sciacchitano *et al.* (2015), Sullivan (2015), Tang and Zhou (2015), Ghanem *et al.* (2017), Dey et al (2018), Zhu Y. et al. (2019), Faes and Moens (2020), Garno *et al.* (2020), Imholz *et al.* (2020), Stenger (2020), Wang and McDowell (2020), Son and Du (2021), Volodina and Challenor (2021), Zhao M.Y. *et al.* (2021), Dalbey *et al.* (2022), Farid (2022), Fröhlich *et al.* (2022), Hanea *et al.* (2022), Gray *et al.* (2022), Greś *et al.* (2022), Jain and Ramu (2022), Li *et al.* (2022), *Re-*

https://doi.org/10.1515/9783111354231-012

zaei et al. (2022), Righi *et al.* (2022), Römer *et al.* (2022), Strand *et al.* (2022), Squarcio and da Silva (2022), Wang C. *et al.* (2022), Yang *et al.* (2022), Fu *et al.* (2023), Katsidoniotaki *et al.* (2022), and Katsidoniotaki (2023).

Two quotes immediately appear to be called for in a way of describing what this chapter is about; the first one is by Alfred Freudenthal (1968) in his classic paper (by now, you would remember it by heart!) "Ignorance of the cause of variation does not make such variation random" (see also p. 24). Frendenthal did not know alternatives to randomness. This book discussion fuzziness, and boundedness (interval, ellipsoidal, super-ellipsoidal, polyhedral analyses) as alternative to randomness.

In his equally classic paper, Giuseppe Grandori (1991) resonates, as it were, with the above quote, stressing: "The probabilistic approach to structural safety is today a well-established paradigm. As to the current state of this paradigm, one can notice a dissymmetry similar to that observed by Freudenthal in the traditional approach. An overwhelming part of the research effort, in fact, has been and still is devoted to estimating failure probabilities. By contrast, only sporadic research deals with the problem of choosing an acceptable risk of failure. It is true that the adoration of the probabilistic approach is in any case a progress, even in the case when acceptable risk levels are conveniently defined, because it allows us to treat different structures with homogeneous criteria. However, the concept of structural safety will not leave the "realm of metaphysics" unless we devise a method for justifying the choice of risk acceptability levels."

As we see the first quote criticizes the deterministic approach in engineering; the other one questions the validity of a probabilistic approach. Here we are concerned with the "mother of all problems," namely, how to deal with uncertainty, which methodology to choose and why and how.

Hereinafter, author's musings on uncertainty quantification are presented, via some parables. In 1990, the present writer lucked out to introduce the notion of an uncertainty triangle. It maintains that currently there are three competing methodologies on dealing with problems involving uncertainty quantification. Those are (a) probabilistic (or stochastic) treatment of uncertain variables and functions, (b) fuzzy sets-based analysis, and (c) anti-optimization technique used either separately or in combination with optimization. Used separately it coincides then with convex modeling of uncertainty, or worst-case design. Utilized in conjunction with optimization it leads to minimizing the least favorable structural response or maximizing the minimum buckling load of the structure due to uncertainties. Its spiritual twin is the maximin model by Abraham Wald (1902–1950); for the intriguing paper on Wald, the interested reader can consult with a paper by Bill Casselman (2016) (see also the papers by Moshe Sniedovich (2007, 2016a, 2016b) on maximin principle, and by Houman Owhadi and Clint Scovel (2017) on uncertainty quantification.

This book discusses three facets of uncertainty, as manifested by stochasticity, fuzziness, and combined optimization and anti-optimization the latter also known as the min-max approach. Later, we try to answer the question about which one is preferable.

As Danish polymath Piet Hein (1906–1996) (1968) argued:

> Truth shall emerge from the interplay
>   of attitudes freely debated.
> Don't be misled by the fanatics who say
>   That only one truth should be stated:
> Truth is constructed in such a way
>   that it can't be exaggerated.

As a way of answering, we cite three parables that suggest that utility of each approach lies in the applicability to a problem at hand.

12.2 Stochasticity

Stochastic modeling, though starting from times immemorial (attributed to the ancient Egyptians), has emerged scientifically by correspondence between two French giants, Blaise Pascal (1623–1662) and Pierre Fermat (1601–1665) (see Devlin, 2008, 2010). They laid the foundations of probability that was axiomatized by one of the greatest Russian mathematicians of the Twentieth century, Kolmogorov (1903–1987) in 1933. In engineering its application has started apparently in civil engineering, specifically in the Ph. D. dissertation by Max Mayer (1926). Later it was pursued most vigorously by Alfred Freudenthal (1906–1977), first in Israel (Freudenthal, 1938) and then in the United States (Freudenthal, 1947), and Alexei Rufovich Rzhanitsyn (1949) in Russia (interested readers can also consult the French version, Rjanistsyn (1959). Their spectacular success consisted in connecting empirically known safety factors with the reliability of the structure, namely, with the probability that the structure performs its mission satisfactorily. The mission may consist of the maximum actual stress being less than the yield stress, or the buckling load to exceeding any preselected value, or that natural frequency being below or above any given value.

In the beginning, the probability theory and stochastic processes did not attract many investigators. In words of Smith (1986),

> the mathematics involved in the application of full probability theory are esoteric and, not unnaturally, most civil engineers did not welcome any move which might lead to the adaptation of such methods in their work.

Allin C. Cornell (1938–2007) describes the current consensus (Cornell, 1969):

> Probability theory provides a more accurate engineering representation of reality. Many leading civil engineers in many countries have written of the statistical nature of loads and of material properties.

According to Lovelace (1972),

> The times of straightforward structural design, when the structural engineer could afford to be fully ignorant of probabilistic approaches to analysis, are definitely over.

Currently, however, a stochastic approach is an accepted paradigm in modern engineering literature, with numerous articles, specialized journals and books devoted to it. Its popularity, however, is not an unqualified one. Freudenthal (1972) remarked:

> It is not implied that this use [of the theory of probability] is in itself sufficient to make a design more reliable or more economical, any more than that the avoidance of the probabilistic approach makes it safer.

Masanobu Shinozuka (1987) states:

> It is recognized that it is rather difficult to estimate experimentally the autocorrelation function, or in the case of weak homogeneity, the spectral density function of the stochastic viriation of material properties. In view of this, the upper bound results are particularly important, since the bounds derived . . . do not require knowledge of the autocorrelation function.

Rudolf Kalman (1930–2016), the famous discoverer of the "Kalman filter" (over 12 million hits once you type this concept on Google!) speaks of probabilistic "quasi religion" (Kalman, 1994). But let us give the podium to Kalman himself: "I see an enormous activity, seemingly aimless, I see fanatical devotion to ideas and principles, which have grown into a quasi-religion . . . Probability is an intellectual construct. It does not exist in the real world."

David Mumford (1999), the past president of the International Mathematical Union, in his article "The dawning of the age of stochasticity," claims: "My overall conclusion is that I believe stochastic methods will transform pure and applied mathematics in the beginning of the third millennium. Probability and statistics will come to be viewed as the natural tools to use in mathematical and scientific modeling."

12.3 Fuzziness

Fuzzy sets were introduced by an Australian philosopher Max Black (1909–1988) in 1937 (Black, 1937), and apparently independently by American computer scientist Lotfi Zadeh (1921–2017) who published his celebrated paper (Zadeh, 1965); according to Google Scholar, this paper received more than 100,000 citations as of October 2019. He is the one who coined the term fuzzy, with the view that with fuzzy sets a mathematical model can be made of properties that are not precisely defined, for example, human statements. His principle of incompatibility states:

> As the complexity of a system increases, our ability to make precise yet significant descriptions about its behavior diminishes until a threshold is reached beyond which precision and significance (or relevance) become almost mutually exclusive characteristics.

Ove Ditlevsen (1980) criticized fuzzy sets-based approaches: ". . . the sources of fuzzy information are non-objectivistic and non-reproductive in their very nature like the process of perception in the human brain." Likewise, William Kahan claimed that "Fuzzy theory is wrong, wrong and pernicious" (Zadeh, 2008b, p. 2753). Rudolf Kalman maintained:

> "Fuzzification" is a kind of scientific permissiveness. It tends to result in socially appealing slogans unaccompanied by the discipline of hard scientific work and patient observation.

Here is a quote on Kalman (1994) from Zadeh's (2015) paper:

> "Kalman's comments were made at the Man and Computer Conference in Bordeaux, France, 1972, at which I presented a paper describing the concept of a linguistic variable. An excerpt from his comments is reproduced below.

> "Kalman: I would like to comment briefly on Professor Zadeh's presentation. His proposals could be severely, ferociously, even brutally criticized from a technical point of view. This would be out of place here. But a blunt question remains: Is Professor Zadeh presenting important ideas or is he indulging in wishful thinking? The most serious objection to fuzzification of system analysis is that lack of methods of system analysis is not the principal scientific problem in the systems field. That problem is one of developing basic concepts and deep insight into the nature of systems, perhaps trying to find something akin to the laws of Newton. In my opinion, Professor Zadeh's suggestions have no chance to contribute to the solution of this basic problem. To take a concrete example, modern experimental research has shown that the brain, far from fuzzy, has in many areas a highly specific structure. Progress in brain research is now most rapid in anatomy where the electron microscope is the new tool clarifying regularities of structure which previously were seen only in a fuzzy way. No doubt Professor Zadeh's enthusiasm for fuzziness has been reinforced by the prevailing political climate in the U.S. – one of unprecedented permissiveness. Fuzzification is a kind of scientific permissiveness . . . I must confess that I cannot conceive of fuzzification as a viable alternative for the scientific method; I even believe that it is healthier to adhere to Hilbert's naïve optimism, *Wir wollen wissen: wir werden wissen* [We want to know, we will know].It is very unfair for Professor Zadeh to present trivial examples where fuzziness is tolerable or even comfortable and in any case irrelevant, and then imply, though not formally claim, that his vaguely outlined methodology can have an impact on deep scientific problems. In any case, if the fuzzification approach is going to solve any difficult problems, this is yet to be seen."

Zadeh responded in the following way: ". . . the skeptics will find it hard to understand why they failed to realize that fuzzy logic is a phase in a natural evolution of science – an evolution brought by the need to find an accommodation with the pervasive impression of the real world."

Bart Kosko (1990) claimed that ". . . in a hundred years from now, a thousand years from now . . . no one . . . will believe that there was a time when a concept as simple, as expressive as a "fuzzy set" met with such impersonal denial." Moreover, according to Claude Rosenthal (2004), Kosko stated: "The boat of uncertainty reasoning is being rebuilt at sea. Plank by plank fuzzy theory is beginning to gradually shape its design. Today only a few fuzzy planks have been laid. But a hundred years from now, a thousand years from now, the boat of uncertainty reasoning may little resemble the boat of today . . . Amassed fuzzy applications, hardware, and products will have broadened its sails. And no one on the boat will believe that there was a time when a concept as simple, as intuitive, as expressive as a fuzzy set met with such impassioned denial."

The interested reader can also consult with the paper by Tamir *et al.* (2015) about the progress made in the recent half century of the fuzzy sets-based analyses. The definitive books by Kosko (1994, 1997, 1999) are a must-reads.

12.4 Boundedness (Wald's min-max; guaranteed approach; convex modeling; optimization and anti-optimization under uncertainty; info-gap)

Ben-Haim (1985) introduced the convex modeling of uncertainty in the context of nuclear engineering. Ben-Haim and Elishakoff (1990) employed convex modeling in applied mechanics. This methodology is somewhat opposite to both probabilistic and fuzzy-sets-based uncertainty analyses for it does not introduce a measure, namely the probability density in a probabilistic setting or membership function in fuzzy-sets-based methodology. Rather, it deals with bounding techniques under uncertainty. This method has been known from times immemorial. For example, when scientists did not know the mathematical quantity, they would provide the lower and upper bounds for it. For example, Archimedes (circa 285–212 BCE) derived the following bounds for $\sqrt{3}$ and π, respectively,

$$265/153 < \sqrt{3} < 1351/780, \ 213/71 < \pi < 21/7$$

Naturally, the numbers $\sqrt{3}$ and π are irrational numbers, neither random variables nor fuzzy ones. The bounds, however, can be useful for various purposes. In the same manner, the convex modeling of uncertainty provides with least favorable or most favorable responses. For prudent design one then utilizes the least favorable design following the folk wisdom "better safe than sorry." Such approaches were initiated by Bulgakov (1940) in Russia and Wald in United States in 1940s, Boley (1966) in the United States in 1950s, and later were independently rediscovered by Schweppe (1968, 1973) in 1960s and Drenick in 1970s, both in the USA. For the bibliography, the interested reader can consult with the monograph by Elishakoff and Ohsaki (2010) as well as the paper by Caselman (2016) tellingly titled as *The Legend of Abraham Wald* is of interest (see also spirited discussions by Sniedovich, 2007, 2008, 2010, 2012a, 2012b,

2016a, 2016b) These methodologies have deep roots from ancient times. Roman poet Ovid (43 BCE–18 CE), for example states, in his book *Metamorphosis7:* "I see, and I desire the better: I follow the *worse.*" Likewise, many centuries later, English playwright, poet and actor William Shakespeare (1564–1616) in *Julius Caesar* (Act 5, Scene 1), instructs us:

> But, since the affairs of men rests still uncertain,
> Let's reason with the worst that may befall.

Indeed, the book by Hlavacek, Chleboun and Babuška (2004) and a paper by Yoshikawa, Elishakoff and Nakagiri (1998), contain the word *worst* in their titles. One has to stress that the present write was informed by Professor Drenick that his worst-case type methodologies were adopted neither by earthquake engineers nor by nuclear scientists because of the conservative nature of this worst-case scenario-based approache. It occurred to me that one has to combine this anti-optimization approach with optimization so as to either reduce as much as possible the maximum, least favorable, responses or increase to the maximum extent the least buckling load. This methodology that greatly enhances both convex modeling and worst-case type research was adopted in the book by Elishakoff and Ohsaki (2010).

Those who may be skeptical of the worst-case scenario approach and prefer the probabilistic one, may want to read the comment made by indefatigable pursuer of better designs and methodologies, Henri Petroski (2002, p. 16) writing about the fall of the World Trade Center towers:

> Since two hijacked airplanes loaded with jet fuel were crashed within about 15 min of each other into the two most prominent and symbolic structures of lower Manhattan, the once reassuringly low numbers generated by probabilistic risk assessment seem irrelevant. What happened in New York ceased being a hypothetical, incredible or ignorable scenario.

Is there a possibility of comparing various approaches? The answer to this nagging question is a resounding "yes." In the papers by Elishakoff, Cai and Starnes (1994) and Wang, Elishakoff, Qiu and Ma (2009) comparison is being performed between probabilistic methodology and convex modeling. Maglaras (1995), Maglaras *et al.* (1997), Chen *et al.* (1999), and others conducted a comparison between probabilistic and convex modeling.

Later, Elishakoff (1995) showed that convex modeling (Ben-Haim 1985; Ben-Haim and Elishakoff, 1990) was nothing else but a generalization of the interval algebra (McWilliam, 2001; Penmetsa and Grandhi, 2002), known from the onset of the twentieth century. A comparison of interval and stochastic analyses was conducted by Köylüoglu and Elishakoff (1998).

Here it appears instructive to discuss the history of interval mathematics. In his insightful article, Hayes (2003) writes: "Give a digital computer a problem in arithmetic, and it will grind away methodically, tirelessly, at gigahertz speed, until ultimately it produces the wrong answer. The cause of this sorry situation is not that software is full of bugs – although that is very likely true as well – nor is it that hardware is unreliable. The problem is simply that computers are discrete and finite machines, and they cannot cope with some of the continuous and infinite aspects of mathematics. Even an innocent-looking number like Ko can cause no end of trouble: In most cases, the computer cannot even read it in or print it out exactly, much less perform exact calculations with it. Errors caused by these limitations of digital machines are usually small and inconsequential, but sometimes every bit counts. On February 25, 1991, a Patriot missile battery assigned to protect a military installation at Dahrahn, Saudi Arabia, failed to intercept a Scud missile, and the malfunction was blamed on an error in computer arithmetic. The Patriot's control system kept track of time by counting tenths of a second; to convert the count into full seconds, the computer multiplied by 0.1. Mathematically, the procedure is unassailable, but computationally it was disastrous. Because the decimal fraction Ko has no exact finite representation in binary notation, the computer had to approximate. Apparently, the conversion constant stored in the program was the 24-bit binary fraction 0.00011001100110011001100, which is too small by a factor of about one ten-millionth. The discrepancy sounds tiny, but over four days it built up to about a third of a second. In combination with other peculiarities of the control software, the inaccuracy caused a miscalculation of almost 700 meters in the predicted position of the incoming missile. Twenty-eight soldiers died. Of course, it is not to be taken for granted that better arithmetic would have saved those 28 lives."

Interval mathematics was developed to answer these needs. Hayes (2003) writes about its history too: "Interval arithmetic is not a new idea. Invented and reinvented several times, it has never quite made it into the mainstream of numerical computing, and yet it has never been abandoned or forgotten either. In 1931 Rosalind Cicely Young, a recent Cam bridge Ph.D., published an "algebra of many valued quantities" that gives rules for calculating with intervals and other sets of real numbers. Of course, Young and others writing in that era did not see intervals as an aid to improving the reliability of machine computation. By 1951, however, in a textbook on linear algebra, Paul S. Dwyer of the University of Michigan was describing arithmetic with intervals (he called them "range numbers") in a way that is clearly directed to the needs of computation with digital devices. A few years later, the essential ideas of interval arithmetic were set forth independently and almost simultaneously by three mathematicians. Mieczyslaw Warmus in Poland, Teruo Sunaga in Japan, and Ramon E. Moore in the United States. Moore's version has been the most influential, in part because he emphasized solutions to problems of machine computation but also because he has continued for more than four decades to publish on interval methods and to promote their use."

Sometimes some of the parameters may be given in such a manner that stochastic analysis can be warranted, but some other parameters could better be described via alternative analyses. The combined methodology can be called for. Elishakoff and Colombi (1993a, 1993c), Elishakoff, Lin and Zhu (1994), and Meng *et al.* (2022) offered a combined stochastic-convex analysis for the scarce data applied to the Space Shuttle weather protection systems. Structural reliability and interval analysis were combined by Qiu *et al.* (2008a, 2008b). Fang, Smith and Elishakoff (1998) combined convex analysis with the fuzzy sets-based approach. Interesting applications are given in books by Chernousko (1994), Svetlitskii (2003), Ben-Haim (2006), Takewaki (2013), and Takewaki *et al.* (2013); interested readers can consult with an article by Owhadi and Scovel (2017).

Now, the natural question may arise in the mind of the keen engineer: Which of these methodologies ought to be followed? It appears that the answer be better delivered via some parables.

12.5 Mediation of the dispute between litigants: First parable

The first idea on this "fight" between the musketeers – stochasticity, fuzziness, and hybrid optimization and anti-optimization appear to be the possibility that each of these modeling techniques may be right, although applicable to different circumstances.

This reminds us of a parable of a sage who was called upon to judge two men who had a dispute (Fig. 10.1, p. 205). He decided to listen to each litigant attentively. After carefully listening to the first man telling the story from his perspective, the sage responded: "You're right." Then the sage gave an attentive ear to the second disputant, also separately. Finally, the sage ruled: "You're right." Since in ancient times, there were no court houses, the hearing took place in the judge's home. The wife of the judge was perplexed by her husband's decision-making process. She inquired: "You told each of them that he was right, though they claimed things that were contradicting to each other. How is this possible?" The judge thought for a moment and exclaimed: "You are right too!"

Can we say that each of the uncertainty modeling "is right"? Indeed, at the international conferences these three methodologies are exposed in different sessions. The sessions on stochasticity are attended overwhelmingly by the stochasticians, the fuzzy sets-based analyses are discussed in sessions on fuzziness and attended by the researchers in that field, whereas anti-optimization sessions are attended by those who embrace it or its twin brothers: guaranteed approach, or ellipsoidal modeling, or set-theoretical analysis, or worst-case scenario, or info-gap approach. Thus, everyone feels that he/she is right. Interested readers can consult papers by Hot *et al.* (2017) and Ioakimidis (2021) about the info-gap approach.

Thus, the current situation is reminding us of the Biblical story of the Babel tower (Fig. 12.1), where the builders spoke different languages, were unable to communicate with each other and thus, failing to coordinate their actions, leading to the collapse of the tower.

Specifically, we also undertook (Elishakoff, Cai and Starnes, 1994; Wang, Elishakoff, Qiu and Ma (2009)) a direct comparison between stochastic and anti-optimization procedures.

Fig. 12.1: Miscommunication between the proponents of stochasticity, fuzziness, and convexity analysts reminds us of the story of Babel Tower (courtesy of Wikimedia Commons).

When we presented results at some conference (Ben-Haim and Natke, 1997), one of its organizers exclaimed: "How is it possible to contrast these two totally different methodologies?!" Our response was as follows: "Assume that there is a firm with considerable means. It orders analysis of some uncertain phenomenon that it is concerned with, to two different organizations with different schools of thought, providing them with the same input information. The two organizations, which are unaware of the other's involvement, deliver the results to the company. It is natural then for the company's associates to compare the results delivered by the two above organizations. Therefore, even if the organizations themselves are not engaged in comparing their results with other approaches, such comparison may take place by a third party."

12.6 Which analysis is right: Second parable

This parable represents a modification of one due to Gotthold Ephraim Lessing (1729–1781) written in 1779: "In the Orient in ancient times there lived a man who possessed a ring of inestimable worth. Its stone was an opal that emitted a hundred colors, but its real value lay in its ability to make its wearer beloved of God and man. The ring passed from father to most favored son for many generations, until finally, its owner was a father with three sons, all equally deserving. Unable to decide which of the three sons most worthy was, the father commissioned a master artisan to make exact copies of the ring, then gave each son a ring, and each son believed that he alone had inherited the original true ring (Fig. 12.2)."

Fig. 12.2: Parable of three rings by Ephraim Lessing (courtesy of Wikimedia Commons).

But instead of harmony, the father's plan brought only discord to his heirs. Shortly after the father died, each son claimed to be the sole ruler of the father's house, each basing his claim to authority on the ring given to him by the father. The discord grew even stronger and more hurtful when a close examination of the rings failed to disclose any differences.

The dispute among the brothers grew until their case was finally brought before a judge. After hearing the history of the original ring and its meticulous powers, the judge pronounced his conclusion: "the authentic ring," he said, "had the power to make its owner beloved of God and man, but each of your rings has brought only hatred and strife. None of you is loved by others; each love only himself. Therefore, I must conclude that none of you has the original ring. Your father must have lost it, then attempted to hide his loss by having three counterfeits made, and these are the rings that cause you so much grief." The judge continued: "Or it may be that your father, weary of the tyranny of a single ring, made duplicates, which he gave to you. Let each of you demonstrate his belief in the power of this ring by conducting his life in such manner that he fully merits – as anciently promised – the love of God and man." It should be noted that the story is older and earlier variants are known from Giovanni Boccaccio, in his classic book The Decameron (written between 1350 and 1335), day 1, tale 3 (borrowing the ideas from Jewish / Islamic authors) (Ashliman, 2004, p.53).

12.7 Third parable: Blind men and an elephant

Wikipedia tells us: "The parable of the blind men and an elephant (Fig. 12.3) originated in the ancient Indian subcontinent, from where it has been widely diffused. It is a story of a group of blind men who have never come across an elephant before and who learn and conceptualize what the elephant is like by touching it. Each blind man feels a different part of the elephant's body, but only one part, such as the side or the tusk. They then describe the elephant based on their limited experience and their descriptions of the elephant are different from each other. In some versions, they come to suspect that the other person is dishonest, and they come to blows. The moral of the parable is that humans tend to claim absolute truth based on their limited, subjective experience as they ignore other people's limited, subjective experiences which may be equally true." One of the most famous versions of the nineteenth century was the poem "The Blind Men and the Elephant" by John Godfrey Saxe (1816–1887) (see Saxe, 2016).

> And so these men of Indostan
> Disputed loud and long,
> Each in his own opinion
> Exceeding stiff and strong,
> Though each was partly in the right
> And all were in the wrong!

The poem starts with the description of the problem at hand:

> It was six men of Indostan
> To learning much inclined,
> Who went to see the Elephant
> (Though all of them were blind),
> That each by observation
> Might satisfy his mind

Fig. 12.3: Blind men and elephant (freely available on Wikimedia Commons).

Each in his own opinion maintains that the elephant is like a wall, snake, spear, tree, fan or rope, depending upon where they had touched. Their polarized debate leads to a conclusion:

> The disputants, I ween,
> Rail on in utter ignorance
> Of what each other mean,
> And prate about an Elephant
> Not one of them has seen!

Hopefully, someone will explain to our disputants that the elephant is like a wall, snake, spear, tree, fan, or rope, that the elephant exhibits all the above properties. Likewise, uncertainty is like randomness or like fuzziness or like boundedness or a combination thereof. Our approach, hopefully, allows one to "see the elephant," to borrow from the title of Paranjape's (2022) paper albeit on another topic. Note that this parable was resorted to in another book (Raizer and Elishakoff, 2022).

12.8 Picking and choosing an uncertainty quantification methodology

Naturally, having these three possible modeling opportunities provides one with a choice depending on the amount of information provided and the degree of precision demanded. If only scarce information is provided one cannot come up with either probability assessment, not with determination of the fuzzy sets based and levels of confidence. If scarce information is given, one can only try to get scarce information on the output: normally it is maximum and/or minimum. In other words, in such a case one can furnish only the least and most favorable responses.

If only qualitative information is provided, possibly then the only feasible approach is a fuzzy sets-based analysis.

However, if one has a massive event in mind, with events occurring under macroscopically identical conditions, one can apply the stochastic approach.

Not everyone may be satisfied with the comment that several possible models of uncertainty is a good thing. In this insightful look, tellingly titled *The Paradox of Choice: Why More is Less*, Barry Schwartz (2004) notes:

> As choices proliferate, people have a harder and harder time making decisions. And they end up less satisfied with the decisions they make. They are filled with regret over those that turned out well but might have been better.

Still, it appears preferable, in the humble opinion of this writer to try to construct bridges between different approaches rather than to ignite a fight between them. In this spirit, the paper by Wang, Wang, Elishakoff and Qiu (2011b) is titled *"Probability and convexity are not antagonistic* as an answer, as it were to the questions posed in this writer's (Elishakoff, 1999a) paper titled "Are the Probabilistic and Anti optimization Methods Interrelated?" After all, according to the wise paper by Elizabeth Paté-Cornell (1996), there are six levels of uncertainty description (it appears to be not random that the paper had not less than 298 citations at the time of re-reading it.) Instead of having the Babel Tower artificially erected between these treatments it appears advisable to look for the grain of truth in each of them, depending on circumstances and available data.

12.9 Where are we going: The Rashomon effect?

One morning, as the famous scholar approached the town square, the policeman walked up to him and asked, "Sir, may I know where you are going?"

The scholar replied, "I don't know."

The policeman seized on this and said, "Old man, you are lying to me. I know you are going to that pub over there. I see you every day. I'm going to arrest you for lying to a member of the police force."

The policeman took the scholar to the nearest police station and put him in one of the cells. As he was locking the door, the policeman proudly remarked, "Now you foolish man you will realize never again to lie to me."

The scholar replied, "My son, I have no idea why you claim I lied to you. I told you I didn't know where I was going. Indeed, I did not – I thought I was going to synagogue but, as you can see, it turned out I was going to jail."

It is more than a story; it is a parable of our lives. It is a parable on research about uncertainty. In the next piece I will write where we should be going to.

Very often the so-called *Rashomon effect* takes place. It stems from the classic 1950 Japanese movie *Rashomon* (Fig. 12.4) by director Akira Kurosawa. It vividly shows that the same event can be described in several almost contradictory ways by different people. In the beginning of the movie, the woodcutter and the Shinto priest find themselves thrown together one dark day, huddling by the gates of a city, trying to remain dry during the heavy afternoon rainstorm. They are soon joined by a common wayfarer, to whom each man describes a lurid event they both saw earlier.

The woodcutter maintains he located the body of a murdered samurai while he was out in a forest cutting wood. The priest claims he remembers seeing the samurai journeying with his wife through the forest on the day he was murdered. These two men are summoned by authorities to testify in court as witnesses. There they are joined by the third man who was a bandit. The latter was captured and accused of the samurai's murder. The viewer hears several testimonies of the murder recounted in court. The stories of the woodcutter, the bandit, the samurai's widow, and even – hold the breath – of the murdered samurai himself, are chillingly told via the medium. The movie seemingly propagates the idea of the nonexistence of objective reality, and that there are often several alternative, subjective, sometimes self-serving, and often totally contradictory ways of seeing the same situation or event. This movie introduced the so-called "Rashomon effect." According to Anderson (2016), The Rashomon effect is defined in the modern academic context as "The naming of an epistemological framework – or ways of thinking, knowing, and remembering – required for understanding complex and ambiguous situations."

It appears apropos to reproduce the quote of the author unknown to the present writer: "When Columbus started his journey in 1492, he didn't know where he was going. When he got there, he did not know where he was. When he returned to Spain, he did not know where in the world he has been." This naturally brings us to the quip by the former Defense Minister Donald Rumsfeld: ". . . as we know, there are known knowns; there are things we know we know. We also know there are known unknowns; that is to say, we know there are some things we do not know. But there are also unknown unknowns – the ones we don't know we don't know." (see also p. VII) Likewise, we embrace the statements of Robert Greene (2013) "The need for certainty is the greatest disease the mind faces," and of Ursula K. Le Guin (2016) "The only thing that makes life possible is permanent, intolerable uncertainty: not knowing what comes next."

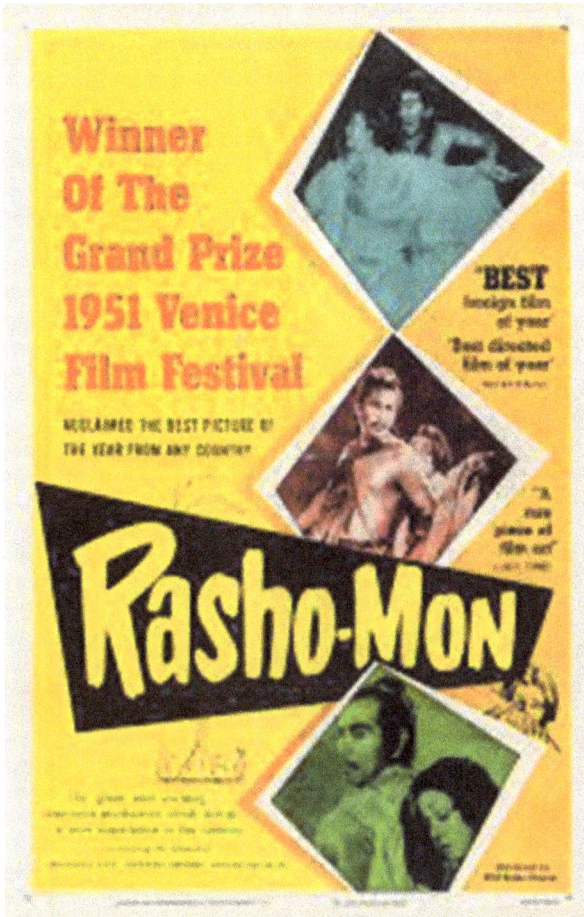

Fig. 12.4: "Rashomon" effect is often discussed in the literature as the ways of thinking, knowing, and remembering for understanding complex and ambiguous situations (courtesy of Wikimedia Commons).

12.10 Conclusion

At this juncture it appears apropos to paraphrase from the Frank Sinatra's song "My Way":

> "And now, the end [of this book- I.E.] is near
> And so I face the final curtain
> My friend, I'll say it clear
> I'll state my case, of which I'm certain,"

specifically, the concept of randomness and attendant probabilistic analysis is not the only game in town of uncertainty, despite its spectacular success. Likewise, fuzziness does not uniformly reign, and concept of boundedness, via interval, ellipsoidal, super-ellipsoidal or polyhedral analyses, though simple and attractive, does not eliminate the other two possible approaches. This book presents three alternative tools as parts of "uncertainty's multi-tooled Swiss army knife," as it were. Each must be used in proper circumstances, depending on availability and character of the experimental data, and the preference of the engineer.

And to those who embrace, in words of philosopher John Dewey, "quest for certainty" and have a "comforting feeling of certainty", in terminology of Maggie Jackson (2023), we humbly recommend to follow the sage advice of the latter author of the book *Uncertain: The Wisdom and Wonder of Being Unsure*: "We need not fear the indefinite. For that is where we find the better solution and path of hope. This is uncertainty's edge."

Bibliography

Abaoui-Ismaili M. and Bernard P., Asymptotic Analysis and Linearization of the Randomly Perturbed Two-Wells Duffing Oscillator, *Probabilistic Engineering Mechanics*, Vol. 12, 171–178, 1997.

Abdel-Rahman, A. Y. A., Matrix analysis of wave propagation in periodic structures, University of Southampton, 1979.

Abramowitz M. and Stegun I. (eds.), *Handbook of Mathematical Functions*, New York: Dover, p. 949, 1965.

Acar E., Bayrak G., Jung Y., Lee I., Ramu P. and Ravichandran S.S., Modeling, Analysis, and Optimization under Uncertainties: A Review, *Structural and Multidisciplinary Optimization*, Vol. 64(5), 2909–2945, 2021.

Adams B.M., Bohnhoff W.J., Dalbey K.R., Eddy J.P., Eldred M.S., Gay D.M., Haskell K., Hough P.D. and Swiler L.P., DAKOTA, A Multilevel Parallel Object-Oriented Framework for Design Optimization, Parameter Estimation, Uncertainty Quantification, and Sensitivity Analysis: Version 5.0 User's Manual, Sandia National Laboratories, Tech. Rep. SAND2010-2183, 2009.

Adams M., Lashgari A., Li B., McKerns M., Mihaly J., Ortiz M., Owhadi H., Rosakis A.J., Stalzer M. and Sullivan T.J., Rigorous Model-Based Uncertainty Quantification with Application to Terminal Ballistics, Part II: Systems with Uncontollable Inputs and Large Scatter, *Journal of the Mechanics and Physics of Solids*, Vol. 60(5), 1002–1010, 2012.

Adelman H.M. and Haftka R.T., Sensitivity Analysis for Discrete Structural Systems – A Survey, NASA Technical Memorandum 86333, NASA Langley Research Center, 1984.

Adhikari S., Friswell M.I., Lonkar K. and Sarkar A., Experimental Case Studies for Uncertainty Quantification in Structural Dynamics, *Probabilistic Engineering Mechanics*, Vol. 24(4), 473–492, 2009.

Afshari S.S., Enayatollahi F., Xu X. and Liang X., Machine Learning-Based Methods in Structural Reliability Analysis: A Review, *Reliability Engineering and System Safety*, Vol. 219, article 108223, 2022.

Agarwal H., Renaud J.E., Preston E.L. and Padmanabhan D., Uncertainty Quantification Using Evidence Theory in Multidisciplinary Design Optimization, *Reliability Engineering & System Safety*, Vol. 85(1–3), 281–294, 2004.

Aldosary M., Wang J. and Li C., Structural Reliability and Stochastic Finite Element Methods: State-of-the-Art Review and Evidence-Based Comparison, *Engineering Computations*, Vol. 35, 2165–2214, 2018.

Alefled G. and Claudio D., The Basic Properties of Interval Arithmetic, Its Software Realizations and Some Applications, *Computers and Structures*, Vol. 67, 3–8, 1998.

Alefeld G. and Herzberger J., *Introduction to Interval Computations*, New York: Academic Press, 1983.

Ali T., Boruah H. and Dutta P., Modeling Uncertainty in Risk Assessment Using Double Monte Carlo Method, *International Journal of Engineering Innovation and Technology*, Vol. 1(4), 114–118, 2012.

Alvin K.F., Oberkampf W.L., Diegert K.V. and Rutherford B.M., Uncertainty Quantification in Computational Structural Dynamics: A New Paradigm for Model Validation, *Society for Experimental Mechanics, Inc, 16th International Modal Analysis Conference*, Vol. 2, 1191–1198, 1998.

Amazigo J.C., Buckling of Stochastically Imperfect Structures, in *Buckling of Structures* (B. Budiansky, ed.), Berlin: Springer, pp. 172–182, 1976.

Amer K.B., A New Philosophy of Structural Reliability, Fail Safe Vs Safe Life – The 1988 Alexander A. Nikolsky Lecture, *Journal of American Society*, Vol. 34(1), 1989.

Amyotte P., Margeson A., Chiasson A. and Khan F.I., There Is No Such Thing as a Black Swan Process Incident, *Hazards 24*, Vol. 159, 12–21, 2014.

Anderson R., The Rashomon Effect and Communication, *Canadian Journal of Communication*, Vol. 41(2), 250–265, 2016.

Andrianov I.B. and Manevich L.I., *Asymptotology: Ideas, Methods, Results*, Moscow: Aslan Publishers, 1994 (in Russian).

Ang A.H.S., An Intuitive Basis of the Probability Density Evolution Method (PDEM) for Stochastic Dynamics, in *Risk and Reliability Analysis: Theory and Applications*, Cham: Springer, pp. 99–108, 2017.

https://doi.org/10.1515/9783111354231-013

Ang A.H.-S. and Tang W.H., *Probability Concepts in Engineering, Planning and Design*, New York: Wiley, 1984.

Ang A.H.-S. and Tang W.H., *Probability Concepts in Engineering: Emphasis on Applications in Civil and Environmental Engineering*, New York: Wiley, 2007.

Anh N.D., Private Communication to I.E., May 25, 2006a.

Anh N.D., Private Communication to I.E., May 24, 2006b.

Anh N.D. and Di Paola M., Some Extensions of Gaussian Equivalent Linearization, *International Confernce on Nonlinear Stochastic Dynamics*, Hanoi, Vietnam, pp. 5–16, Dec. 7–10, 1995.

Anh N.D. and Schiehlen W.A., Technique for Obtaining Approximate Solutions in Gaussian Equivalent Linearization, *Computer Methods in Applied Mechanics and Engineering*, Vol. 168, 113–119, 1998.

Anh N.D., Krause R. and Schiehlen W., Statistical Linearization and Large Excitation of Nonlinear Stochastic Mechanical Systems, in *Nonlinear Stochastic Mechanics* (N. Bellomo and F. Casciati, eds.), Berlin: Springer, pp. 1–12, 1992.

Arbocz J., Present and Future of Shell Stability Analysis, *Zeitschrift fuer Flugwissenschaften und Weltraumforschung*, Vol. 5, 335–348, 1991.

Arbocz J., Future Directions and Challenges in Shell Stability, *The 38th AIAA/ASME/ASCE/AHS/ASC Structures, Structural Dynamics, and Materials Conference and Exhibit*, AIAA Paper 97-1077, pp. 1949–1962, 1997.

Arbocz J. and Hol J.M.A.M., Collapse of Axially Compressed Cylindrical Shells with Random Imperfections, *AIAA Journal*, Vol. 29, 2247–2256, 1991.

Argyris J., Papadrakakis M. and Stefanou G., Stochastic Finite Element Analysis of Shells, *Computer Methods in Applied Mechanics and Engineering*, Vol. 191(4142), 4781–4804, 2002.

Aronov J. and Papic L., *Reliability and Safety Management of Engineering Systems through the Prism of Black Swan Theory, System Reliability Management, Solutions and Technologies*, Boca Raton: CRC Press, pp. 103–112, 2018.

Arora J.S., *Introduction to Optimum Design*, New York: McGraw Hill, 1989.

Arregui-Mena J.D., Margetts L. and Mummery P.M., Practical Application of the Stochastic Finite Element Method, *Archives of Computational Methods in Engineering*, Vol. 23(1), 171–190, 2016.

Ashliman D.L., *Folk and Fairy Tales: A Handbook*, Westport, CT: Greenwood Press, 2004.

Atalik T.S. and Utku S., Stochastic Linearization of Multi-Degree-of-Freedom Non-Linear Systems, *Earthquake Engineering and Structural Dynamics*, Vol. 4, 411–420, 1976.

Atamturktur H.S., Moaveni B., Papadimitriou C. and Schoenherr T., *Model Validation and Uncertainty Quantification*, Vol. 3, Berlin: Springer, 2015.

Atanassov E. and Dimov I.T., What Monte Carlo Models Can Do and Cannot Do Efficiently?, *Applied Mathematical Modelling*, Vol. 32(8), 1477–1500, 2008.

Au F.T., Cheng Y.S., Tham L.G. and Zeng G.W., Robust Design of Structures Using Convex Models, *Computers and Structures*, Vol. 81(28–29), 2611–2619, 2003.

Au S.-K., Connecting Bayesian and Frequentist Quantification of Parameter Uncertainty in System Identification, *Mechanical Systems and Signal Processing*, Vol. 29, 328–342, 2012.

Au S.-K., *Operational Modal Analysis: Modeling, Bayesian Inference, Uncertainty, Laws*, Singapore: Springer, 2017.

Au S.-K. and Beck J.L., Estimation of Small Failure Probabilities in High Dimensions by Subset Simulation, *Probabilistic Engineering Mechanics*, Vol. 16(4), 263–277, 2001.

Au S.-K. and Wang Y., *Engineering Risk Assessment with Subset Simulation*, New York: Wiley, 2014.

Auciello N.M., Transverse Vibrations of a Linearly Tapered Cantilever Beam with Tip Mass of Rotary Inertia and Eccentricity, *Journal of Sound and Vibration*, Vol. 194(1), 25–34, 1996.

Augusti G., Baratta A. and Casciati F., *Probabilistic Methods in Structural Engineering*, London: Chapman and Hall, 1984.

Augustin T., Optimal Decisions under Complex Uncertainty – Basic Notions and a General Algorithm for Data-Based Decision Making with Partial Prior Knowledge Described by Interval Probability, *ZAMM – Zeitschrift für Angewandte Mathematik und Mechanik*, Vol. 84(10–11), 1–10, 2004.

Augustin T., Coolen F.P., De Cooman G. and Troffaes M.C., *Introduction to Imprecise Probabilities*, New York: John Wiley, 2014.

Aven T., Implications of Black Swans to the Foundations and Practice of Risk Assessment and Management, *Reliability Engineering & System Safety*, Vol. 134, 83–91, 2015.

Ayyub B.M., *Uncertainty Modeling and Analysis in Civil Engineering*, Boca Raton: CRC Press, 1997.

Ayyub B.M., *Elicitation of Expert Opinions for Uncertainty and Risks*, Boca Raton: CRC Press, 2001.

Ayyub B.M., *Risk Analysis in Engineering and Economics*, Boca Raton: Chapman and Hall/CRC, 2003.

Ayyub B.M. and Klir G.J., *Uncertainty Modeling and Analysis in Engineering and the Sciences*, Boca Raton: Chapman and Hall, 2006.

Ayyub B.M. and Lai K.-L., Structural Reliability Assessment with Ambiguity and Vagueness in Failure, *Naval Engineering Journal*, Vol. 104(3), 21–35, 1992.

Ayyub B.M. and McCuen R.H., *Probability, Statistics, and Reliability for Engineers and Scientists*, Boca Raton: CRC Press, 2011.

Ayyub B.M. and McCuen R.H., *Probability, Statistics, and Reliability for Engineers and Scientists*, Second Edition, Boca Raton: CRC Press, 2016.

Babuška I., Tempone R. and Nobile F., Worst-Case Scenario Analysis for Elliptic PDE's with Uncertainty, in *Structural Dynamics EURODYN 2005* (C. Soize and G.I. Schuëller, eds.), Rotterdam: Millpress, pp. 889–894, 2005.

Bae H.R., Grandhi R.V. and Canfield R.A., Uncertainty Quantification Using Evidence Theory with a Cost-Effective Algorithm, in *Computational Fluid and Solid Mechanics 2003*, Elsevier Science Ltd, pp. 2197–2200, 2003.

Bae H.R., Grandhi R.V. and Canfield R.A., An Approximation Approach for Uncertainty Quantification Using Evidence Theory, *Reliability Engineering & System Safety*, Vol. 86(3), 215–225, 2004a.

Bae H.R., Grandhi R.V. and Canfield R.A., Epistemic Uncertainty Quantification Techniques Including Evidence Theory for Large-Scale Structures, *Computers & Structures*, Vol. 82(13–14), 1101–1112, 2004b.

Bakhshian B.Z., Nazirov R.R. and Eliasberg P.E., *Determination and Correlation of Motion: Guaranteed Approach*, Moscow: "Nauka" Publishing House, 1980 (in Russian).

Ballent W., Corotis R.B. and Torres-Machi C., Representing Uncertainty in Natural Hazard Risk Assessment with Dempster Shafer (Evidence) Theory, *Sustainable and Resilient Infrastructure*, Vol. 4(4), 137–151, 2019.

Banichuk N.V. and Neittaanmäki P., *Structural Optimization with Uncertainties*, Berlin: Springer, 2010.

Barantsev R.G., Asymptotology, *Procedure of Leningrad University*, Vol. 1, 69–76, 1976 (in Russian).

Barantsev R.G., Asymptotic versus Classical Mathematics, *Topics in Mathematical Analysis*, Vol. 11, 49–64, 1989.

Barmish B.R. and Lagoa C.M., The Uniform Distribution: A Rigorous Justification for Its Use in Robustness Analysis, *Mathematics of Control, Signals and Systems*, Vol. 10(3), 203–222, 1997.

Barmish B.R., Corless M. and Leitmann G., A New Class of Stabilizing Controllers for Uncertain Dynamical Systems, *SIAM Journal of Continuous Optimization*, Vol. 21, 246–255, 1983.

Baroth J., Breysse D. and Schoefs F. (eds.), *Construction Reliabilit: Safety, Variability and Sustainability*, London: ISTE–Wiley, 2013.

Barthelemy J.F. and Haftka R.T., Approximation Concepts for Optimum Structural Design – A Review, *Structural Optimization*, Vol. 5(3), 129–144, 1993.

Basseville M., Benvenisle A. and Wilsky A.S., Multiscale Autoregressive Processes – Part I: Schur-Lcvinson Parametrizations, *IEEE Transactions on Signal Processing*, Vol. 40(8), 1915–1934, 1992a.

Basseville M., Benvenisle A. and Wilsky A.S., Multiscale Autoregressive Proceses – Part II: Lattice Structure for Whitening and Modeling, *IEEE Transactions on Signal Processing*, Vol. 40(8), 1935–1954, 1992.

Batou A. and Soize C., Uncertainty Quantification in Low-Frequency Dynamics of Complex Beam-Like Structures Having a High-Modal Density, *International Journal for Uncertainty Quantification*, Vol. 3(6), 475–485, 2013.

Bažant Z.P., Speech upon Conferral of the Alfred Freudenthal Medal at EMI Conference at MIT, Cambridge, May 31, 2018.

Bažant Z.P. and Le J.-L., *Probabilistic Mechanics of Quasibrittle Structures: Strength, Lifetime, and Size Effect*, Cambridge, UK: Cambridge University Press, 2017.

Beaman J.J., Accuracy of Statistical Linearization, in *New Approaches to Nonlinear Problems in Dynamics* (P.J. Holmes, ed.), Philadelphia: SIAM, pp. 195–207, 1980.

Beaman J.J. and Hedrick J.K., Improved Statistical Linearization for Analysis and Control of Nonlinear Stochastic Systems, *Journal of Dynamical Systems, Measurement, and Control*, Vol. 103, 22–27, 1981.

Bedrosian B., Barbela M., Drenick R.F. and Tsirk A., Critical Excitation for Calculating Earthquake Effects on Nuclear Plant Structures: An Assessment Study, *NUREG/CR-1673 RD*, Oradell, NJ: Burns and Roe, 1980.

Beer M., Engineering Quantification of Inconsistent Information, *International, Journal of Reliability and Safety*, Vol. 3(1–3), 174–200, 2009.

Beer M., Ferson S. and Kreinovich V., Imprecise Probabilities in Engineering Analyses, *Mechanical Systems and Signal Processing*, Vol. 37(1–2), 4–29, 2013.

Beer M., Kougioumtzoglou I.A. and Patelli E., Emerging Concepts and Approaches for Efficient and Realistic Uncertainty Quantification, in *Maintenance and Safety of Aging Infrastructure: Structures and Infrastructures*, pp. 121–162, 2014.

Behera D. and Chakraverty S., Solving the Nondeterministic Static Governing Equations of Structures Subjected to Various Forces under Fuzzy and Interval Uncertainty, *International Journal of Approximate Reasoning*, Vol. 116, 43–61, 2020.

Bell E.T., *Mathematics: Queen and Servant of Science*, New York: McGraw-Hill, p. 306, 1951.

Bell K.J., Some Observations in the Teaching and Practices of Pessimization, *Heat Transfer Engineering*, Vol. 13(1), 5–6, 1992.

Bellizi S. and Bouc R., Analysis of Multi-Degree of Freedom Non-Linear Mechanical Systems with Random Input. Part II: Equivalent Linear Systems with Random Matrices and Power Spectral Density Matrix, *Probabilistic Engineering Mechanics*, Vol. 14, 245–256, 1999.

Bellman R., The Roles of the Mathematician in Applied Mechanics, in *U.S. National Congress in Applied Mechanics*, New York: ASME Press, pp. 195–204, 1962.

Benaroya H., Han S.M. and Nagurka M., *Probability Models in Engineering and Science*, Boca Raton: CRC Press, 2005.

Benaroya H. and Rehak M., Finite Element Methods in Probabilistic Structural Analysis: A Selective Review, *Applied Mechanics Reviews*, Vol. 41(5), 201–213, 1988.

Ben-Haim Y., *The Essay of Spatially Random Material*, Dordrecht: Kluwer Academic Publishers, 1985.

Ben-Haim Y., Optimizing Multihypothesis Diagnosis of Control-Actuator Failures in Linear Systems, *Journal of Guidance, Control, and Dynamics*, Vol. 13(4), 744–750, 1990a.

Ben-Haim Y., Detecting Unknown Lateral Forces on a Bar by Vibration Measurement, *Journal of Sound Vibration*, Vol. 140, 13–29, 1990b.

Ben-Haim Y., Failure of an Axially Compressed Beam with Uncertain Initial Deflection of Bounded Strain Energy, *International Journal of Engineering Science*, Vol. 31, 989–1001, 1993.

Ben-Haim Y., Convex Models for Uncertainty in Radial Pulse Buckling of Shells, *Journal of Applied Mechanics*, Vol. 15, 1–30, 1993.

Ben-Haim Y., Convex Models of Uncertainty: Applications and Implications, *Erkenntnis: An International Journal of Analytic Philosophy*, Vol. 41, 139–156, 1994.

Ben-Haim Y., A Non-Probabilistic Concept of Uncertainty, *Structural Safety*, Vol. 14, 227–245, 1994.

Ben-Haim Y., Must Reliability be Probabilistic?, *Journal of Statistics and Mathematical Simulation*, Vol. 55(3), 263–265, 1996a.

Ben-Haim Y., *Robust Reliability in the Mechanical Sciences*, Berlin: Springer Verlag, 1996b.

Ben-Haim Y., Robust Reliability of Structures, *Advances in Applied Mechanics*, Vol. 33, 1–41, 1997.

Ben-Haim Y., *Information Gap Decision Theory*, London: Academic Press, 2001.

Ben-Haim Y., *Info-Gap Decision Theory: Decisions Under Severe Uncertainty*, Second Edition, Amsterdam: Academic Press, 2006.

Ben-Haim Y. and Elishakoff I., Dynamics and Failure of a Thin Bar with Unknown but Bounded Imperfections, in *Recent Advances on Impact Dynamics of Engineering Structures* (D. Hui and N. Jones, eds.), New York: ASME Press, pp. 89–96, and-Vol. 105, AD-Vol. 17, 1989.

Ben-Haim Y. and Elishakoff I., Non-Probabilistic Models of Uncertainty in the Non-Linear Buckling of Shells with General Imperfections: Theoretical Estimates of the Knockdown Factor, *Journal of Applied Mechanics*, Vol. 111, 403–410, 1989.

Ben-Haim Y. and Elishakoff I., *Convex Models of Uncertainty in Applied Mechanics*, Amsterdam: Elsevier Science Publishers, 1990.

Ben-Haim Y. and Elishakoff I., Convex Models of Vehicle Response to Uncertain but Bounded Terrain, *Journal of Applied Mechanics*, Vol. 16, 90–99, 1990.

Ben-Haim Y. and Natke G. (eds.), *Uncertainty: Models and Measures*, Berlin: Akademie Verlag, 1997.

Benjamin J.R. and Cornell C.A., *Probability, Statistics and Decision for Civil Engineers*, New York: McGraw Hill, 1970.

Benov D.M., The Manhattan Project, the First Electronic Computer and the Monte Carlo Method, *Monte Carlo Methods and Applications*, Vol. 22(1), 73–79, 2016.

Ben-Tal A. and Nemirovski A., Robust Optimization–Methodology and Applications, *Mathematical Programming*, Vol. 92(3), 453–480, 2002.

Ben-Tal A., El Ghaoui L. and Nemirovski A., *Robust Optimization*, Princeton, NJ: Princeton University Press, 2009.

Beran M.Y., Mason T.A., Adams B.L. and Olson T., Bounding Elastic Constants of an Orthotropic Polycrystal Using Measurements of the Microstructure, *Journal of Mechanics of Physics and Solids*, Vol. 44, 1543–1563, 1996.

Bergman L.A., Shinozuka M., Bucher C.G., Sobczyk K., Dasgupta G., Spanos P.D., Deodatis G., Spencer B.F., Ghanem R.G., Sutoh A. and Grigoriu M., A State-of-the-Art Report on Computational Stochastic Mechanics, *Probabilistic Engineering Mechanics*, Vol. 12(4), 197–321, 1997.

Bernard P., About Stochastic Linearization, in *Nonlinear Stochastic Mechanics* (N. Bellomo and F. Casciati, eds.), Berlin: Springer Verlag, pp. 61–70, 1992.

Bernard P., Stochastic Linearization: What is Available and What is Not, *Computers and Structures*, Vol. 67, 9–18, 1998.

Bernard P. and Wu L., Stochastic Linearization: The Theory, *Journal of Applied Probability*, Vol. 35, 718–730, 1998.

Bernardini A., Fuzzy Measures in the Knowledge-Based Diagnosis of Seismic Vulnerability of Masonry Buildings, in *Probabilistic Mechanics and Structural and Geotechnical Reliability* (Y.K. Lin, ed.), New York: ASCE Press, pp. 25–28, 1992.

Bernardini A., A Fuzzy Set Approach to the Response Evaluation of Uncertain Mechanical Systems, in *Proceedings of the Third International Conference on Stochastic Structural Dynamics* (D.H. Soffara, ed.), San Juan, Porto Rico, pp. 30–39, 1997.

Bernardini A., What are the Random and Fuzzy Sets and How to Use Them for Uncertainty Modelling in Engineering Systems?, in *Whys and Hows in Uncertainty Modelling* (I. Elishakoff, ed.), Vienna: Springer, pp. 63–125, 1999.

Bernardini A., Upper and Lower Probabilities of Events from Random Sets and Fuzzy Sets, in *Seguridad en Ingenieria* (A.J. Bignoli, ed.), Buenos Aires: Academia Nacional de Ingenieria, 2000.

Bernardini A. and Tonon F., A Combined Fuzzy and Random-Set Approach to the Multiobjective Optimization of Uncertain Systems, in *Proceedings Seventh Specialty ASCE Conference on Advances in Probabilistic Mechanics and Structural Reliability*, 7–9 August 1996, Worcester, MA, New York: ASCE Press, pp. 314–317, 1996.

Bernardini A. and Modena C., Applications of the Fuzzy Sets Theory to the Reliability Evaluations of Structural Systems, in *Proceedings of Fuzzy Systems and Knowledge Engineering*, Guangdong Higher Education Publ House, pp. 541–548, 1987.

Bernardini A. and Tonon F., *Bounding Uncertainty in Civil Engineering: Theoretical Background*, Berlin: Springer, 2010.

Bertsimas D., Brown D.B. and Caramanis C., Theory and Applications of Robust Optimization, *SIAM Review*, Vol. 53(3), 464–501, 2011.

Berveiller M., Sudret B. and Lemaire M., Stochastic Finite Element: A Non-Intrusive Approach by Regression, *European Journal of Computational Mechanics*, Vol. 15, 81–92, 2006.

Besseling J.F., Laws of Physics and Variational Principles, in *Variational Methods in Engineering* (C.A. Brebbia, ed.), Berlin: Springer, pp. 63–72, 1985.

Bhat R.B., Plate Deflections using Orthogonal Polynomials, *Journal of Engineering Mechanics*, Vol. 111(11), 1301–1309, 1985.

Bickford W.B., A Consistent Higher Order Beam Theory, *Developments in Theoretical and Applied Mechanics*, Vol. 11, 137–151, 1982.

Biegler L., Biros G., Ghattas O., Heinkenschloss M., Keyes D., Mallick B., Tenorio L., Van Bloemen Waanders B., Willcox K. and Marzouk Y. (eds.), *Large-Scale Inverse Problems and Quantification of Uncertainty*, Vol. 712, New York: Wiley, 2011.

Bielajew A.F., History of Monte Carlo, in *Monte Carlo Techniques in Radiation Therapy*, Boca Raton: CRC Press, pp. 3–15, 2021.

Bieneck M.P., Fan E.C. and Lackman L.M., Dynamic Stability of Cylindrical Shells, *AIAA Journal*, Vol. 4(3), 495–500, 1966.

Bignoli A.J., Assessment of Proneness to Failure and Structural Risk with Fuzzy Sets, in *Structural Failure, Product Liability and Technological Inference* (H.P. Rossmanith, ed.), Amsterdam: Elsevier Science, pp. 505–512, 1993.

Black M., Vagueness: Exercise in Logical Analysis, *Philosophy of Science*, Vol. 4, 427–455, 1937.

Bland D.R., *The Theory of Viscoelasticity*, New York: Academic Press, 1975.

Blekhman I.I., Myshkis A.D. and Panovko Y.G., *Mechanics and Applied Mathematics: Logics and Specifics of Applications of Mathematics*, Moscow: "Nauka" Publishers, pp. 205–209, 1983 (in Russian).

Blekhman I.I., Myshkis A.D. and Panovko Ya. G., *Mechanics and Applied Mathematics: Logic and Features of the Applications of Mathematics*, Moscow: "Nauka" Publishing House, 1990 (in Russian).

Blockley D.I., *The Nature of Structural Design and Safety*, Chichester, U.K: Ellis Horwood, 1980.

Blockley D.I., *The Nature of Structural Design*, Chichester: Ellis Horwood, 1980.

Blockley D.I., *Bridges: The Science and Art of the World's Most Inspiring Structures*, Oxford University Press, 2010.

Blyth M., Coping with the Black Swan: The Unsettling World of Nassim Taleb, *Critical Review*, Vol. 21(4), 447–465, 2009.

Boley B.A., Bounds on the Maximum Thermoelastic Stress and Deflection in a Beam and Plate, *Journal of Applied Mechanics*, Vol. 33(4), 881–887, 1966.

Bolotin V.V., *Statistical Methods in Structural Mechanics*, Moscow: State Publishing House in Civil Engineering Architecture and Building Materials, pp. 9–10, 1961 (in Russian).

Bolotin V.V., *Application of the Methods of the Theory of Probability and the Theory of Reliability to Analysis of Structures*, Moscow: State Publishing House for Buildings, 1971 (in Russian). English translation: FTD-MT-24-771-73, Foreign Technology Div., Wright Patterson AFB, Ohio, 1974.

Bolotin V.V., *Non-Conservative Problems of the Theory of Elastic Stability*, Oxford: Pergamon, 1963.

Bolotin V.V., *Statistical Methods in Structural Mechanics*, San Francisco: Holden-Day, 1969.

Bolotin V.V., *Wahrscheinlichkeitsmethoden zur Berechnung von Konstruktionen*, Berlin: VEB Verlag für Bauwesen, 1981 (in German).

Bolotin V.V., *Random Vibration of Elastic Bodies*, Dordrecht: Kluwer, 1984.

Bolotin V.V., Statistical Methods in the Nonlinear Theory of Elastic Shells, *Izestiya Akademii Nauk SSSR, Otdelenie Tekhnicheskikh Nauk*, Vol. 3, 1958 (in Rusian). English Translation, NASA TTF-85, 1962.

Bolotin V.V., *Prognosis of the Lifetime of Machines and Structures*, Moscow: "Mashinostroenie" Publishers, p. 84, 1984 (in Russian).

Bolotin V.V., *Random Vibrations of Elastic Systems*, The Hague: Martinus Nijhoff Publishers, pp. 240–292, 1984.

Bolotin V.V. and Elishakoff I., Random Vibrations of Elastic Shells Containing an Acoustic Medium, *Mechanics of Solids*, Vol. 6, 99–107, 1971.

Bonstrom H., Corotis R. and Porter K., The Role of Uncertainty in the Political Process for Risk Investment Decisions, *Applications of Statistics and Probability in Civil Engineering*, pp. 2753–2760, 2011.

Booton R.C., The Analysis of Nonlinear Control Systems with Random Inputs, *Proceedings Symposium on Nonlinear Circuit Analysis*, Vol. 2, 341–344, 1953.

Booton R.C., Nonlinear Control Systems with Random Inputs, *IRE Transactions Circuit Theory*, Vol. 1, 32–34, 1954.

Borgonovo E. and Plischke E., Sensitivity Analysis: A Review of Recent Advances, *European Journal of Operational Research*, Vol. 248(3), 869–887, 2016.

Borkar V.S., The Birth of MCMC, *Resonance*, Vol. 27(7), 1105–1106, 2022a.

Borkar V.S., Markov Chain Monte Carlo, *Resonance*, Vol. 27(7), 1107–1115, 2022b.

Bouc R. and Defelippi M., Multi-modal Nonlinear Spectral Response of a Beam with Impact under Random Load, *Probabilistic Engineering Analysis*, Vol. 12, 163–170, 1997.

Bourinet J.M., Mattrand C. and Dubourg V., A Review of Recent Features and Improvements Added to FERUM Software, in *Proceedings of the 10th International Conference on Structural Safety and Reliability* (ICOSSAR'09), 2009.

Brandimarte P., *Handbook in Monte Carlo Simulation: Applications in Financial Engineering, Risk Management, and Economics*, Hoboken, NJ: Wiley, 2014.

Branover H., Reconciling Conflicts between Science and the Torah, *Chabad Magazine*, 10–11, 1996.

Bras R.L. and Rodriguez-Iturbe I., *Random Functions and Hydrology*, New York: Dover, 1985.

Breitung K., A Criticism of Statistical Methods in Probabilistic Models in Structural Reliability, in *Probabilistic Mechanics and Structural and Geotechnical Reliability* (Y.K. Lin, ed.), New York: ASCE Press, pp. 236–239, 1992.

Breitung K., The Lindley Paradox, Information and Generalized Functions, in *Proceedings of 3rd International Symposium on Uncertainty Modeling and Analysis and Annual Conference of the North American Fuzzy Information Processing Society*, IEEE Press, pp. 720–723, 1995.

Breitung K.W., *Asymptotic Approximations for Probability Integrals*, Berlin: Springer, 2006.

Breitung K., 40 Years FORM: Some New Aspects?, *Probabilistic Engineering Mechanics*, Vol. 42, 71–77, 2015.

Breitung K., FORM/SORM, SS and MCMC: A Mathematical Analysis of Methods for Calculating Failure Probabilities, in *International Probabilistic Workshop 2021*, Cham: Springer, pp. 353–367, 2021.

Brooks S., Gelman A., Jones G. and Meng X.L. (eds.), *Handbook of Markov Chain Monte Carlo*, Boca Raton: CRC Press, 2011.

Brown C.B., Fuzzy Safety Measure, *Journal of Engineering Mechanics*, Vol. 105, 855–872, 1979.

Brown F.B., Recent Advances and Future Prospects for Monte Carlo, *Progress in Nuclear Science and Technology*, Vol. 2, 1–4, 2011.

Brown R. and Chua L., Clarifying Chaos: Examples and Counterexamples, *International Journal of Bifurcation and Chaos*, Vol. 6(2), 219–249, 1996.

Bucher C., *Computational Analysis of Randomness in Structural Mechanics*, London: CRC Press, 2009.

Budiansky B. and Hutchinson J.W., Dynamic Buckling of Imperfection-Sensitive Structures, in *Proceedings of 11th International Congress of Applied Mechanics* (H. Görtler, ed.), Berlin: Springer, pp. 636–651, 1964.

Budiansky B. and Hutchinson J.W., Buckling: Progress and Challenge, in *Trends in Solid Mechanics* (J.F. Besseling and A.M.A. van der Heijden, eds.), Alphen aan den Rijn: Sijthoff and Noorhoff, pp. 93–116, 1979.

Bulgakov B.V., Fehleranhäufung bei Kreiselapparaten, *Ingenieur-Archiv*, Vol. 11(6), 461–469, 1940 (in German).

Burezyriski T. and Skrzypczyk J., *Fuzzy Boundary Element Method: A New Methodology*, Gliwice: Silesian Technical University, 1996.

Burezyriski T. and Skrzypczyk J., Fuzzy Aspects of the BEM, *Engineering Analysis with Boundary Elements*, Vol. 19, 209–216, 1997.

Busby H.R., Jr. and Weingarten V.I., Response of Nonlinear Beam to Random Excitation, *Journal of Engineering Mechanics Division*, Vol. 99, 55–68, 1973.

Butler R., Dodwell T.J., Haftka R.T., Kim N.H., Kim T., Kynaston S. and Scheichl R., Uncertainty Quantification of Composite Structures with Defects Using Multilevel Monte Carlo Simulations, in *17th AIAA Non-Deterministic Approaches Conference*, pp. 1598–1612, 2015.

Cacuci D.G. and Ionescu-Bujor M., A Comparative Review of Sensitivity and Uncertainty Analysis of Large-Scale Systems – II: Statistical Methods, *Nuclear Science and Engineering*, Vol. 147(3), 204–217, 2004.

Cadzow J.A., High Performance Spectral Estimation – A New ARMA Method, *IEEE Transactions on Acoustics, Speech and Signal Processing*, Vol. 28(5).

Cadzow J.A., Spectral Estimation: An Overdetermined Rational Model Equation Approach, *Proceedings of IEEE*, Vol. 70(2), 907–939, 1982.

Cafeo J.A. and Thacker B.H., Concepts and Terminology of Validation for Computational Solid Mechanics Models, *SAE Transactions*, Vol. 113, 155–161, 2004.

Cai G-Q. and Zhu W.-Q., *Elements of Stochastic Dynamics*, Singapore: World Scientific, 2016.

Cai K.-Y., *Introduction to Fuzzy Reliability*, Boston: Kluwer Academic, 1996.

Cai K.Y., Wen C. and Zhang M., Fuzzy Variables as a Basis for the Theory of Fuzzy Reliability in the Possibility Context, *Fuzzy Sets and Systems*, Vol. 42, 145–172, 1991.

Callens R.R., Faess M.G. and Moens D., Multilevel Quasi-Monte Carlo for Interval Analysis, *International Journal for Uncertainty Quantification*, Vol. 12(4), 2022.

Cai K.-Y., Wen C.-Y. and Zhang M.-L., Reliability Behavior of Typical Systems with Two Types of Failure, *Fuzzy Sets and Systems*, Vol. 43, 17–32, 1991.

Cambell C.W., Monte Carlo Turbulence Simulation Using Rational Approximations to von Karman Spectra, *AIAA Journal*, Vol. 24(1), 62–66, 1986.

Cambou B., Application of First-Order Uncertainty Analysis in the Finite Element Method in Linear Elasticity, *Proceedings of 2nd International Conference on Applied Statistics and Structural Engineering*, pp. 67–87, 1975.

Cao H.J. and Duan B.Y., An Approach on the Non-Probabilistic Reliability of Structures Based on Uncertainty Convex Models, *Chinese Journal of Computational Mechanics*, Vol. 22(5), 546–549, 2005.

Cao L., Liu J., Xie L., Jiang C. and Bi R., Non-Probabilistic Polygonal Convex Set Model for Structural Uncertainty Quantification, *Applied Mathematical Modelling*, Vol. 89, 504–518, 2021.

Cappelle B. and Kerre E.E., Issues in Possibilistic Reliability Theory, in *Reliability and Safety Analyses Under Fuzziness* (T. Onisawa and J. Kacprzyk, eds.), Heidelberg: Springer, pp. 61–80, 1995.

Casciati F., Equivalent Linearization Technique in the Analysis of Seismic Excited Structures, *Proceedings, Euro-China Joint Seminar of Earthquake Engineering*, Beijing, pp. 398–410, 1986.

Casciati F., Faravelli L. and Hasofer A.M., New Philosophy for Stochastic Equivalent Linearization, *Probabilistic Engineering Mechanics*, Vol. 8, 179–185, 1993.

Casselman B., The Legend of Abraham Wald, *American Mathematical Society*, 2016, available at http://www.ams.org/publicoutreach/feature-column/fc-2016-06 (accessed on 31 December 2020).

Caughey T.K., Response of Nonlinear Systems to Random Excitation, Lecture Note, California Institute of Technology, 1953 (unpublished, quoted in Caughey T.K., 1963).

Caughey T.K., Equivalent Linearization Techniques, *Journal of Acoustical Society of America*, Vol. 35(11), 1706–1711, 1963.

Caughey T.K., July 27, and an undated one. Personal Communications to I.E., 1998.

Caughey T.K. and Dienes J.K., Analysis Nonlinear First-Order System with a White Noise Input, *Journal of Applied Physics*, Vol. 23, 2476–2479, 1961.

Ceberio M., Kreinovich V., Pownuk A. and Bede B., From Interval Computations to Constraint-Related Set Computations: Towards Faster Estimation of Statistics and ODEs under Interval, p-Box, and Fuzzy Uncertainty, in *Novel Developments in Granular Computing: Applications for Advanced Human Reasoning and Soft Computation* (J.T. Yao, ed.), IGI Global Publisher, pp. 131–147, 2010.

Ceberio M. and Kreinovich V. (eds.), *How Uncertainty-Related Ideas Can Provide Theoretical Explanation for Empirical Dependencies*, Berlin: Springer, 2021.

Cederbaum G., Elishakoff I., Aboudi J. and Librescu L., *Random Vibration and Reliability of Composite Structures*, Lancaster, Philadelphia: Technomic, 1992.

Chabridon V., Balesdent M., Bourinet J.-M., Morio J. and Gayton N., Evaluation of Failure Probability Under Parameter Epistemic Uncertainty: Application to Aerospace System Reliability Assessment, *Aerospace Science and Technology*, Vol. 69, 526–537, 2017.

Chase L., *Uncertainty Quantification: Advances in Research and Applications*, Nova Science Publishers Inc., 2019.

Chaing W.L., Dong W.M. and Wong F.-S., Dynamic Response of Structures with Uncertain Parameters: A Comparative Study of Probabilistic and Fuzzy Set Models, *Probabilistic Engineering Mechanics*, Vol. 2, 82–91, 1987.

Chakraverty S. (ed.), *Mathematics of Uncertainty Modeling in the Analysis of Engineering and Science Problems*, IGI Global, 2014.

Chakraverty S., *New Paradigm in Computational Modeling Ad Its Applications*, London: Academic Press, 2021.

Chakraverty S. and Rout S., *Affine Arithmetic Based Solution of Uncertain Static and Dynamic Problems*, Morgan and Claypool Publishers, 2020.

Chakraverty S., Tapaswini S. and Behera D., *Fuzzy Differential Equations and Applications for Engineers and Scientists*, Boca Raton: CRC Press, 2016.

Chamis C.C., Probabilistic Structural Analysis Methods for Space Propulsion System Components, *NASA TM-88861*, 1986.

Champneys A.R., Dodwell T.J., Groh R.M., Hunt G.W., Neville R.M., Pirrera A., Sakhaei A.H., Schenk M. and Wadee M.A., Happy Catastrophe: Recent Progress in Analysis and Exploitation of Elastic Instability, *Frontiers in Applied Mathematics and Statistics*, Vol. 5, article 34, 2019.

Chamis C.C., Probabilistic Structural Analysis Methods for Space Propulsion System Components, *Probabilistic Engineering Mechanics*, Vol. 2, 100–110, 1987.

Chang R.J., Non-Gaussian Linearization Method for Stochastic Parametrically and Externally Excited Nonlinear Systems, *Journal of Dynamic Systems, Measurement, and Control*, Vol. 114, 20–26, 1992.

Chang R.J. and Young G.E., Methods and Gaussian Criterion for Statistical Linearization of Stochastic Parametrically and Externally Excited Nonlinear Systems, *Journal of Applied Mechanics*, Vol. 56, 179–185, 1989.

Chao R.Y. and Ayyub B.M., Finite Element Analysis with Fuzzy Variables, in *ASCE Engineering Congress, Proceedings*, Vol. 1, New York:: ASCE Press, pp. 643–650, 1996.

Chaudhuri A., Haftka R.T., Ifju P., Chang K., Tyler C. and Schmitz T., Experimental Flapping Wing Optimization and Uncertainty Quantification Using Limited Samples, *Structural and Multidisciplinary Optimization*, Vol. 51(4), 957–970, 2015.

Chaudhuri A., Waycaster G., Price N., Matsumura T. and Haftka R.T., NASA Uncertainty Quantification Challenge: An Optimization-Based Methodology and Validation, *Journal of Aerospace Information Systems*, Vol. 12(1), 10–34, 2015.

Chen G., Cascade Linearization of Nonlinear Systems Subjected to Gaussian Excitations, in *Stochastic Structural Dynamics* (B.F. Spencer, Jr. and E. Johnson, eds.), Rotterdam: Balkema, pp. 69–76, 1999.

Chen S., Nikolaidis E., Cudney H., Rosca R. and Haftka R., Comparison of Probabilistic and Fuzzy Set Methods for Designing under Uncertainty, in *40th Structures, Structural Dynamics, and Materials Conference and Exhibit*, pp. 1579–1789, 1999.

Chernatynskiy A., Phillpot S.R. and LeSar R., Uncertainty Quantification in Multiscale Simulation of Materials: A Prospective, *Annual Review of Materials Research*, Vol. 43, 157–182, 2013.

Chernousko F.L., Optimal Guaranteed Estimates of Indeterminacies with the Aid of Ellipsoids, *Technical Cybernetics, Issue*, Vol. 18(3), 1–9, 1981 (in Russian).

Chernousko F.L., Ellipsoidal Bounds for Sets of Attainability and Uncertainty in Control Problems, *Optimization Control Applications and Methods*, Vol. 3(2), 187–202, 1982.

Chernousko F.L., On Equations of Ellipsoids Approximating Reachable Sets, *Probabilistc Control and Information Theory*, Vol. 12(2), 97–110, 1983.

Chernousko F.L., *State Estimation for Dynamic Systems*, Boca Raton, FL: CRC Press, 1994.

Chernousko F.L., What is Ellipsoidal Modelling and How to Use It for Control and State Estimation?, in *Whys and Hows in Uncertainty Modelling* (I. Elishakoff, ed.), Vienna: Springer, pp. 127–188, 1999.

Chernousko F.L., Properties of Optimal Ellipsoids Approximating Reachable Sets of Uncertain Systems, *Mathematical and Computer Modelling of Dynamical Systems*, Vol. 11(2), 135–147, 2005.

Chibaro A., Lawrence L.S., Corso J.M., Julian O.G. and Gierasch A.R., Discussion of "Safety and the Probability of Structural Failure", *Transactions of the American Society of Civil Engineers*, Vol. 121(1), 1376–1393, 1956.

Choi S.-K., Grandhi R. and Canfield R.A., *Reliability-Based Structural Design*, London: Springer, 2007.

Chojaczyk A., Teixeira A., Neves L., Cardoso J. and Soares C.G., Review and Application Artificial Neural Networks Models in Reliability Analysis of Steel Structures, *Structural Safety*, Vol. 52, 78–89, 2015.

Chou K.C., McIntosh C. and Corotis R.B., Observations on Structural System Reliability and the Role of Modal Correlations, *Structural Safety*, Vol. 1(3), 189–198, 1982.

Chua L.O., Yao Y. and Yang Q., Generating Randomness from Chaos with Desired Randomness, *International Journal of Circuit Theory and Applications*, Vol. 18(3), 215–240, 1990.

Cicirello A. and Giunta F., Machine Learning Based Optimization for Interval Uncertainty Propagation, *Mechanical Systems and Signal Processing*, Vol. 170, article 108619, 2022.

Cicirello A. and Langley R.S., Probabilistic Assessment of Performance under Uncertain Information Using a Generalized Maximum Entropy Principle, *Probabilistic Engineering Mechanics*, Vol. 53, 143–153, 2018.

Civanlar M.R. and Trussell H.J., Constructing Membership Functions Using Statistical Data, *Fuzzy Sets and Systems*, Vol. 18(1), 1–13, 1986.

Clarkson B.L., Pope R.J. and Ranky M.F., Experimental Work to Evaluate Parameters Required in the Statistical Energy Analysis Prediction Method, Final Report, ESA Contract No. 4100/79/NL/PP, Rider 1, 1981.

Clarkson B.L., Can the Statistical Energy Analysis (SEA) Method be Used to Estimate the Response of Structures to Random Forces, Seminar, Florida Atlantic University, October 20, 1994.

Clément A., Soize C. and Yvonnet J., Uncertainty Quantification in Computational Stochastic Multiscale Analysis of Nonlinear Elastic Materials, *Computer Methods in Applied Mechanics and Engineering*, Vol. 254, 61–82, 2013.

Clough R.W. and Penzien J., *Dynamics of Structures*, Auckland: McGraw-Hill, p. 547, 1975.

Colajanni I. and Elishakoff I., A Subtle Error in the Stochastic Linearization Technique, *Chaos, Solitons and Fractals*, Vol. 9, 479–491, 1998a.

Colajanni P. and Elishakoff I., A New Look at the Stochastic Linearization Technique for Hyperbolic Tangent Oscillator, *Chaos, Solitons and Fractals*, Vol. 9, 1611–1623, 1998b.

Comba J.L.D. and Stolfi J., Affine Arithmetic and Its Applications to Computer Graphics, *Anais do VI Simposio Brasileiro de Computaao Grafica e Processamento de Imagens* (SIBGRAPI'93), Recife (Brazil), 1993.

Constantinou M.C., Vibration Statistics of the Duffing Oscillator, *Soil Dynamics and Earthquake Engineering*, Vol. 4, 221–223, 1985.

Constantinou M.C. and Tadjbakhsh I.G., Response of a Sliding Structure to Filtered Random Excitation, *Journal of Structural Mechanics*, Vol. 12, 401–418, 1984.

Conte J.P., Pister K.S. and Mahin S.A., Nonstationary ARMA Modeling of Seismic Motions, *Soil Dynamics and Earthquake Engineering*, Vol. 11, 411–426, 1992.

Cooper N.G., Eckhardt R. and Shera N., *From Cardinals to Chaos: Reflections on the Life and Legacy of Stanislaw Ulam*, New York: Cmabridge University Press, 1989.

Coppa A.P. and Nash W.A., Dynamic Buckling of Shell Structures Subject to Longitudinal Impact, *FDL-TDR -64-65*, Philadelphia, PA: General Electric Corporation, 1964.

Coppe A., Haftka R.T., Kim N.H. and Yuan F.G., Uncertainty Reduction of Damage Growth Properties using Structural Health Monitoring, *Journal of Aircraft*, Vol. 47(6), 2030–2038, 2010.

Cornell C.A., Probability-Based Structural Code, *ACI Journal*, Vol. 66, 974–985, 1969.

Cornell C.A., Structural Safety: Some Historical Evidence that It is a Healthy Adolescent, in *Proceedings of ICOSSAR' 81*, Trondheim, Norway, pp. 19–31, 1981.

Corless M.Y. and Leitmann G., Continuous State Feedback Guaranteeing Uniform Ultimate Boundedness for Uncertain Dynamic Systems, *IEEE Transactions on Automatic Control*, Vol. 26(5), 1139–1144, 1981.

Corless M., Control of Uncertain Nonlinear Systems, *Journal of Dynamic Systematic Measurement Continuity*, Vol. 115, 362–373, 1993.

Corless M., Goodall P.D., Leitmann G. and Ryan E.P., Model-Following Controls for a Class of Uncertain Dynamical Systems, *Procedures IFAC Identification and Systems Parameter Estimation Conference*, pp. 1895–1899, New York, 1985.

Corless M., Letimann G. and Ryan E.P., Tracking in the Presence of Bounded Uncertainties, in *Proceedings, Fourth International Conference Control Theory*, New York: Academic Press, 1985.

Cornell C.A., Probability-Based Structural Code, *ACI Journal*, Vol. 66, 974–985, 1969.

Cornell C.A., Structural Safety: Some Historical Evidence that It Is a Healthy Adolescent, in *Structural Safety and Reliability* (T. Moan and M. Shinozuka, eds.), Amsterdam: Elsevier Scientific Publishing Company, pp. 19–29, 1981.

Cornell C.A., Structural Safety: Some Historical Evidence that It Is Healthy Adolescent, in *Structural Safety and Reliability* (T. Moan and M. Shinozuka, eds.), Amsterdam: Elsevier, pp. 19–29, 1981.

Corotis R.B., Socially Relevant Structural Safety, in *Proceeding of ICASP 9, Applications of Statistics and Probability in Civil Engineering*, San Francisco, CA, pp. 15–24, 2003a.

Corotis R.B., Risk and Uncertainty, in *Proceeding of ICASP 9, Applications of Statistics and Probability in Civil Engineering*, San Francisco, CA, 2003b.

Corotis R.B., Risk-Setting Policy Strategies for Hazards, in *Life-Cycle Performance of Deteriorating Structures: Assessment, Design and Management*, pp. 1–8, 2004.

Corotis R.B., Reliability, Risk or Reality, in *Advances in Reliability and Optimization of Structural Systems: Proceedings 12th IFIP Working Conference on Reliability and Optimization of Structural Systems*, Aalborg, Denmark, 22–25 May 2005.

Corotis R.B., Risk and Risk Perception for Low Probability, High Consequence Events in the Built Environment, in *Recent Developments In Reliability-Based Civil Engineering*, (A. Haldar, ed.) pp. 1–20. 2006.

Corotis R.B., Risk Communication with Generalized Uncertainty and Linguistics, *Structural Safety*, Vol. 31(2), 113–117, 2009.

Corotis R.B., An Overview of Uncertainty Concepts Related to Mechanical and Civil Engineering, *ASCE-ASME Journal of Risk and Uncertainty in Engineering Systems Part B: Mechanical Engineering*, Vol. 1(4), article 040801, 2015.

Corotis R.B., Societal Issues in Adopting Life-Cycle Concepts within the Political System, *Structures & Infrastructure Engineering*, Vol. 5(1), 59–65, 2019.

Corotis R.B. and Nafday A.M., Application of Mathematical Programming to System Reliability, *Structural Safety*, Vol. 7(2–4), 149–154, 1990.

Coveyou R.R., Random Number Generation is too Important to be Left to Chance, in *A Collection of Papers Presented by Invitation at the Symposia on Applied Probability and Monte Carlo Methods and Modern Aspects of Dynamics*, sponsored by the Air Force Office of Scientific Research at the 1967 National Meeting of SIAM in Washington, D.C. (B.R. Agins and M.H. Kalos, eds.), *Studies in Applied Mathematics*, Vol. 3, pp. 70–11, 1969.

Cozarelli F.A. and Huang W.N., Effect of Random Material Parameters on Nonlinear Steady Creep Solution, *International Journal of Solids and Structures*, Vol. 2, 1477–1454, 1971.

Crandall S.H., Random Vibration of One- and Two-Dimensional Structures, in *Developments in Statistics* (R.P. Krishnaiah, ed.), Vol. 2, Academic Press, pp. 1–82, 1979.

Crandall S.H., Wide-Band Random Vibration of Structures, in *Proc of 7th US National Congress of Applied Mechanics*, New York: ASME Press, pp. 131–138, 1974.

Crandall S.H., Structured Response Patterns Due to Wide-Band Random Excitation, in *Stochastic Problems in Dynamics* (B.L. Clarkson, ed.), London: Pitman Press, pp. 366–389, 1977.

Crandall S.H, Random Vibration of One- and Two-Dimensional Structures, in *Developments in Statistics* (RP Krishnaiah, ed.), vol 2, New York: Academic, pp 1-82, 1979.

Crandall S.H., Localized Response Reductions in Wide-Band Random Vibration of Uniform Vibration of Uniform Structures, *Ingenieur-Archiv*, Vol. 49, 347–359, 1980.

Crandall S.H., Spatial Correlation in Structural Response to Wideband Excitation, in *Stochastic Structural Dynamics: Progress in Theory and Applications* (S.T. Ariaratnam, G.I. Schuëller and I. Elishakoff, eds.), London: Elsevier Applied Science, pp. 21–36, 1988.

Crandall S.H., Is Stochastic Equivalent Linearization a Subtly Flawed Procedure?, *Probabilistic Engineering Mechanics*, Vol. 16, 169–176, 2001.

Crandall S.H., On Using Non-Gaussian Distributions to Perform Statistical Linearization, in *Advances in Stochastic Structural Dynamics* (W.Q. Zhu, G.Q. Cai and R.C. Zhang, eds.), Boca Raton: CRC Press, pp. 49–62, 2003.

Crandall S.H., On the Statistical Linearization Methods of I. E. Kazakov, *Problems of Mechanical Engineering and Automation*, (Moscow), (4), 41–74, 2004a.

Crandall S.H., On Using Non-Gaussian Distributions to Perform Statistical Linearization, *International Journal of Non-Linear Mechanics*, Vol. 39, 1395–1406, 2004b.

Crandall S.H., Fifty Years of Stochastic Equivalent Linearization, *Seminar*, Center for Applied Stochastics Research, College of Engineering, Florida Atlantic University, 12 February 2004.

Crandall S.H., A Half-Century of Stochastic Equivalent Linearization, *Structural Control and Health Monitoring*, Vol. 13, 27–40, 2006.

Crandall S.H. and Kulvets A.P., Source Correlation Effects on Structural Response, in *Application in Statistics* (P.R. Krishnaiah, ed.), North-Holland, pp. 163–182, 1977.

Crandall S.H. and Kulvets A.P., Random Vibration of Mechanisms on Plates, in *Proc of 5th World Congress on Theory of Machines and Mech*, New York: ASME Press, pp. 1568–1571, 1979.

Crandall S.H. and Wittig L., Chladni's Patterns for Random Vibration of a Plate, in *Dynamic Response of Structures* (G. Herrmann and N. Perrone, eds.), New York: Pergamon, pp. 55–71, 1972.

Crandall S.H. and Yildiz A., Random Vibrations of Beams, *Journal of Applied Mechanics*, Vol. 31, 267–275, 1962.

Crandall S.H. and Zhu W.Q., Wide-Band Random Excitation of Square Plates, in *Random Vibrations and Reliability* (K. Henning, ed.), Berlin: Akademie – Verlag, pp. 231–244, 1983a.

Crandall S.H. and Zhu W.Q., Random Vibration: A Survey of Recent Developments, *Journal of Applied Mechanics*, Vol. 50, 953–962, 1986.

Crandall S.H. and Zhu W.Q. Wide-Band Random Vibration of Equilateral Triangular Plate, in *Random Vibrations* AMD – vol 65 (T.C. Huang and P.D. Spanos, eds.), New York: ASME Press, pp. 35–50, 1984.

Crandall S.H., Nonlinearities in Structural Dynamics, *The Shock and Vibration Digest*, Vol. 6(8), 1–13, 1974.

Crandall S.H., On Statistical Linearization for Nonlinear Oscillators, in *Problems of the Asymptotic Theory of Nonlinear Oscillators*, Academy of Sciences of the Ukrainian SSR, Naukova Dumka, pp. 115–122, Reprinted, 1980, *Nonlinear System Analysis and Synthesis*, Vol. 2, *Techniques and Applications*, (R.V. Ramnath, J.K. Hedrick and H.M. Paynter, eds.) ASME, New York, pp. 199–209, 1977.

Crespo L.G., Kenny S.P. and Giesy D.P., The NASA Langley Multidisciplinary Uncertainty Quantification Challenge, http://uqtools.larc.nasa.gov/nda-uq-challenge-problem-2014/, 2013.

Crocker M.J. (ed.), *Encyclopedia of Acoustics*, Vol. 4, New York: Wiley, 1997.

Crocker M.J. (ed.), *Handbook of Acoustics*, New York: Wiley, 1998.

Cui T., Marzouk Y.M. and Willcox K.E., Data-Driven Model Reduction for the Bayesian Solution of Inverse Problems, *International, Journal for Numerical Methods in Engineering*, Vol. 102(5), 966–990, 2015.

Cunha A.A.M.F., The Role of the Stochastic Equivalent Linearization Method in the Analysis of the Nonlinear Seismic Response of Building Structures, *Earthquake Engineering and Structural Dynamics*, Vol. 23, 837–857, 1994.

Cunha A., Jr., Nasser R., Sampaio R., Lopes H. and Breitman K., Uncertainty Quantification through Monte Carlo Method in a Cloud Computing Setting, *Computer Physics Communications*, Vol. 185, 1355–1363, 2014.

Cunningham P.R. and White R.G., A Review of Analytical Methods for Aircraft Structures Subjected to High-Intensity Random Acoustic Loads, *Proceedings of the Institution of Mechanical Engineers, Part G: Journal of Aerospace Engineering*, Vol. 218, 231–242, 2004.

da Silva Á.C.R., Jr., Squarcio R.M.F., Cavichiol J.L. and da Neto J.M., Application of the Neumann-Monte Carlo Methodology for Quantification of the Uncertainty of the Problem of Stochastic Bending of Kirchhoff Plates, *International Journal for Uncertainty Quantification*, Vol. 11(3), 85–97, 2021.

Dahlberg T., Modal Cross-Spectral Terms May Be Important and an Alternative Method of Analysis be Preferable, *Journal of Sound and Vibration*, Vol. 84, 503–508, 1982.

Dalbey K., Eldred M., Geraci G., Jakeman J., Maupin K., Monschke J.A., Seidl D., Tran A., Menhorn F., and Zeng X., Dakota, A Multilevel Parallel Object-Oriented Framework for Design Optimization, Parameter Estimation, Uncertainty Quantification, and Sensitivity Analysis, Version 6.16 Theory Manual (No. SAND2022-6172). Sandia National Lab.(SNL-NM), Albuquerque, NM, 2022.

Das S. and Ghanem R., Hybrid Representations for Complex Dynamical Stochastic Systems: Coupled Non-Parametric and Parametric Models, in *Structural Dynamics EURODYN 2005* (C. Soize and G.I. Schuëller, eds.), Rotterdam: Millpress, pp. 53–60, 2005.

Davies P., The Analysis of Vibration and Acoustic Data Using Time Domain Methods, *Ph.D. Thesis*, Unversity of Southhampton, 1985.

Davies P. and Hammond J.K., The Use of Signal Envelopes to Describe the Bandlimited Response of a Class of Systems and to Define and Alternative Shock Spectrum, *Journal of Sound Vibration*, Vol. 111, 93–114, 1986.

Davies P. and Hammond J.K., The Envelope Response of a Nonlinear System and It Relationship to Upper Bound of the Response of the Equivalent Linear System, in *Proceedings, 6th International Modal Analysis Conference*, pp. 1466–1470, Kissimmee, FL, 1988.

Davies R.E., Random Pressure Excitation of Shells and Statistical Dependence Effect of Normal Mode Response, *NASA CR-311*, 1965.

Davis J.P. and Hall J.W., A Software-Supported Process for Assembling Evidence and Handling Uncertainty in Decision-Making, *Decision Support Systems*, Vol. 35(3), 415–433, 2003.

Davood H., Interval Mathematics as a Potential Weapon against Uncertainty, in *Mathematics of Uncertainty Modeling in the Analysis of Engineering and Science Problems* (S. Chakraverty, ed.), Hershey, Pennsylvania: IGI Global, 2014.

De Buffon G.C., *Essai d'arithmétique morale*, Suppl. L'Histoire Naturalle, 1777 (in French).

De Cooman G., Modeling Probabilistic Uncertainty in Two-state Reliability Theory, *Fuzzy Sets Syst*, Vol. 83, 215–238, 1996.

De Cursi E.S., *Uncertainty Quantification Using R*, Berlin: Springer, 2023.

De Cursi E.S. and Sampao R., *Uncertainty Quantification and Stochastic Mechanics with MATLAB*, London: ISTE Press, 2015.

De Finetti B., *Theory of Probability*, Vol. 1, New York: John Wiley and Sons, 1974.

De Rocquigny E., Devictor N. and Tarantola S. (eds.), *Uncertainty in Industrial Practice: A Guide to Quantitative Uncertainty Management*, New York: Wiley, 2008.

Degrauwe D., Lombaert G. and De Roeck G., Improving Interval Analysis in Finite Element Calculations by Means of Affine Arithmetic, *Computers & Structures*, Vol. 88(3), 247–254, 2010.

Der Kiureghian A., *Structural and System Reliability*, Cambridge, UK: Cambridge University Press, 2022.

Dempster A.P., Upper and Lower Probability Induced Multivalued Mapping, *The Annals of Mathematical Statistics*, Vol. 38(2), 325–339, 1967.

Dempster A.P., A Generalization of Bayesian Inference, *Journal of the Royal Statistical Society: Series B (Methodological)*, Vol. 30(2), 205–232, 1968.

Dempster A.P., Upper and Lower Probabilities Induced by a Multivalued Mapping, in *Classic Works of the Dempster-Shafer Theory of Belief Functions*, Berlin: Springer, pp. 57–72, 2008.

Deodatis G., Simulation of Non-Stationary Stochastic Vector Process, in *Proceedings of I0th ASCE Engineering Mechanics Conference*, Boulder, CO, pp. 806–809, 1995.

Deodatis, G., Non-Stationary Stochastic Vector Processes: Seismic Ground Motion Applications, *Probabilistic Engineering Mechanics*, Vol. 11, 149–168, 1996.

Deodatis, G., Simulation of Ergodic Multi-Variate Stochastic Processes, *Engineering Mechanics*, Vol. 122(8), 778–787, 1996.

Deodatis G. and Graham L., Weighted Integral Method and the Variability Response Function as Part of an SFEM Formulation, in *Uncertainty Modeling in Finite Element, Fatigue and Stability of Systems* (A. Haldar, A. Guran and B.M. Ayyub, eds.), Singapore: World Scientific, pp. 71–116, 1997.

Deodatis G. and Shinozuka M., Bounds on Response Variability of Stochastic Systems, *Journal of Engineering Mechanics*, Vol. 115, 2543–2563, 1989.

Deodatis G. and Shinozuka M., Simulation of Seismic Ground Motion Using Stochastic Waves, *Engineering Mechanics*, Vol. 115(2), 2723–2737, 1989.

Depina I., Le T., Fenton G. and Eiksund G., Reliability Analysis with Meta Model Line Sampling, *Structural Safety*, Vol. 60, 1–15, 2016.

Dessombz O., Thouverez F., Laîné J.P. and Jézéquel L., Analysis of Mechanical Systems Using Interval Computations Applied to Finite Element Methods, *Journal of Sound and Vibration*, Vol. 239(5), 949–968, 2001.

Devlin R., *The Unfinished Game: Pascal, Fermat, and the Seventeenth-Century Letter that Made the World Modern*, New York: Basic Books, 2008.

Devlin K., The Pascal-Fermat Correspondence: How Mathematics is Really Done, *The Mathematics Teacher*, Vol. 103(8), 579–582, 2010.

Devroye L., Nonuniform Random Variate Generation, *Handbooks in Operations Research and Management Science*, Vol. 13, 83–121, 2006.

Dey S., Mukhopadhyay T. and Adhikari S., *Uncertainty Quantification in Laminated Composites: A Meta-Model-Based Approach*, Boca Raton: CRC Press, 2018.

Dimarogonas A.D., Interval Rotor Dynamics, in *Procedure of International Conference of Vibration Engineering* (Z.C. Zheng, ed.), Beijing: International Academic Publishing, pp. 745–758, 1994.

Dimentberg M., *Statistical Dynamics of Nonlinear and Time-Varying Systems*, New York: Wiley, 1988.

Daubechies I., *Ten Lectures on Wavelets*, Philadelphia: SIAM, 1992.

Dijkerman R.W. and Mazumdar R.R., Wavelet Representation of Stochastic Processes and Multiresolution Stochastic Models, *IEEE Transactions on Signal Processing*, Vol. 42(7), 1640–1652, 1994.

Di Paola M. and Elishakoff I., Non-Stationary Response of Linear Systems under Stochastic Gaussian and Non-Gaussian Excitation: A Brief Overview of Recent Results, *Chaos, Solitons & Fractals*, Vol. 7(7), 961–971, 1996.

Di Paola M., Failla G., Pirrotta A., Sofi A. and Zingales M., The Mechanically Based Non-Local Elasticity: An Overview of Main Results and Future Challenges, *Philosophical Transactions of the Royal Society A: Mathematical, Physical and Engineering Sciences*, Vol. 371, 1993, article 20120433, 2013.

Di Paola M., Failla G. and Sumelka W., New Prospects in Non-Conventional Modelling of Solids and Structures, *Meccanica*, Vol. Vol. 57, 751–755, 2022.

Di Paola M. and Pisano A.A., Multivariate Stochastic Wave Generation, *Applied Ocean Research*, Vol. 18, 361–365, 1997.

Ditlevsen O., Formal and Real Structural Safety: Influence of Gross Errors, in *Lectures on Structural Reliability (Institute of Building Technology and Structural Engineering)* (P. Thoft-Christensen, ed.), Lyngby, Denmark, pp. 121–147, 1980.

Ditlevsen O., *Uncertainty Modelling with Applications to Multidimensional Civil Engineering Systems*, New York: McGraw-Hill, 1981.

Ditlevsen O., The Fake of Reliability Measures as Absolutes, *Nuclear Engineering and Design*, Vol. 71, 439–440, 1982.

Ditlevsen O., *Structural Reliability Methods*, SBI, 1990 (in Danish).

Dixit U.S. and Dixit P.M., A Finite Element Analysis of Flat Rolling and Application of Fuzzy Set Theory, *International Journal of Machine Tools and Manufacture*, Vol. 36(8), 947–969, 1996.

Do B. and Ohsaki M., A Random Search for Discrete Robust Design Optimization of Linear-Elastic Steel Frames under Interval Parametric Uncertainty, *Computers & Structures*, Vol. 249, article 106506, 2021.

Dobson S., Noori M., Hou Z. and Dimentberg M., Direct Implementation of Stochastic Linearization for SDOF Systems with General Hysteresis, *Structural Engineering and Mechanics*, Vol. 6, 473–484, 1998.

Dodwell T.J., Kynaston S., Butler R., Haftka R.T., Kim N.H. and Scheichl R., Multilevel Monte Carlo Simulations of Composite Structures with Uncertain Manufacturing Defects, *Probabilistic Engineering Mechanics*, Vol. 63, article 103116, 2021.

Dogan V. and Vaicaitis R., Nonlinear Response of Cylindrical Shells to Random Excitation, *Nonlinear Dynamics*, Vol. 20, 33–53, 1999.

Donley M.G. and Spanos P.D., *Dynamic Analysis of Non-Linear Structures by the Method of Statistical Quadratization*, New York: Springer, 1990.

Donley M.G. and Spanos P.D., Equivalent Statistical Quadratization for Multi-Degree-of- Freedom Nonlinear Systems, in *Nonlinear Stochastic Mechanics* (N. Bellomo and F. Casciati, eds.), Berlin: Springer, pp. 185–200, 1992.

Doucet A., Freitas N. and Gordon N. (eds.), *Sequential Monte Carlo Methods in Practice*, Springer Science + Business Media, 2001.

Dowell E.H., Comments on "Turbulent-Induced Plate Vibrations: An Evaluation of Finite – and Infinite-Plate Models", *Journal of Acoustical Society of America*, Vol. 47, 376, 1971.

Drenick R.F., Functional Analysis of Effects of Earthquakes, in *2nd Joint United States-Japan Seminar on Applied Stochastics*, Washington, DC, 1968.

Drenick R.F., Model-Free Design of Aseismic Structures, *Journal of Engineering Mechanics Division*, Vol. 96, 483–493, 1970.

Drenick R.F., On a Class of Non-Robust Problems in Stochastic Dynamics, in *Stochastic Problems in Dynamics* (B.L. Clarkson, ed.), London: Pitman, pp. 237–255, 1977.

Drenick R.F., Novemestky F. and Bagchi G., Critical Excitation of Structures, in *Wind and Seismic Effects, Procedure 12th Joint UJNR Penel Conference*, U.S. National Bureau of Standards Special Publication, pp. 133–142, 1984.

Drenick R.F. and Yun C.B., Reliability of Seismic Resistance Prediction, *Journal of Structural Division*, Vol. 105, 1879–1891, 1979.

Du L., Choi K.K., Youn B.D. and Gorsich D., Possibility-Based Design Optimization Method Design Problems with Both Statistical and Fuzzy Input Data, *Journal of Mechanical Design*, Vol. 128(4), 928–935, 2006.

Du X.P., Interval Reliability Analysis, in *International Design Engineering Technical Conferences and Computers and Information in Engineering Conference*, Vol. 48078, pp. 1103–1109, 2007.

Du X.P., Venigella P.K. and Liu D.S., Robust Mechanism Synthesis with Random and Interval Variables, *Mechanism and Machine Theory*, Vol. 44, 1321–1337, 2009.

Dubois D. and Prade H., *Fuzzy Sets and Systems: Theory and Applications*, New York: Academic Press, 1980.

Dubois D. and Prade H., On Several Representations of an Uncertain Body of Evidence, *Fuzzy Information and Decision Processes* (M.M. Gupta and E. Sanchez, eds.), Amsterdam: North-Holland, pp. 167–181, 1982.

Dubois D. and Prade H., Evidence Measures Based on Fuzzy Information, *Automatica*, Vol. 21, 547–562, 1985.

Dubois D. and Prade H., A Set-Theoretic View of Belief Functions. Logical Operations and Approximations by Fuzzy Sets, *International Journal. of General Systems*, Vol. 12, 193–226, 1986.

Dubois D. and Prade H., *Possibility Theory – An Approach to Computerized Processing of Uncertainty*, New York: Plenum Press, 1988.

Dubois D. and Prade H., Representation and Combination of Uncertainty with Belief Functions and Possibility Measures, *Computational Intelligence*, Vol. 4, 244–264, 1988.

Dubois D. and Prade H., Fuzzy Sets, Probability and Measurement, *European Journal of Operational Research*, Vol. 40, 135–154, 1989.

Dubois D. and Prade H., Consonant Approximations of Belief Functions, *International Journal of Approximate Reasoning*, Vol. 4, 491–449, 1990.

Dubois D. and Prade H., Random Sets and Fuzzy Interval Analysis, *Fuzzy Sets and Systems*, Vol. 42, 87–101, 1991.

Dubourg V., Sudret B. and Deheeger F., Metamodel-Based Importance Sampling for Structural Reliability Analysis, *Probabilistic Engineering Mechanics*, Vol. 33, 47–57, 2013.

Dworetzky A., Stochastic Approximation, *Proceedings of 3rd Berkeley Symposium on Mathematical Statistics and Probability*, Vol. 1, 1969.

Dyne S.J.C., Hammond J.K. and Davies P., A Method of Finding an Upper Bound for the Response of Structures to Blast Waves, in *Recent Advances in Structural Dynamics* (M. Petyt, H.F. Wolfe and C. Mei, eds.), Southampton, UK: Institute of Sound and Vibration Research, pp. 551–559, 1988.

Echard B., Gayton N. and Lemaire M., AK-MCS: An Active Learning Reliability Method Combining Kriging and Monte Carlo Simulation, *Structural Safety*, Vol. 33(2), 145–154, 2011.

Echard B., Gayton N., Lemaire M. and Relun N., A Combined Importance Sampling and Kriging Reliability Method for Small Failure Probabilities with Time-Demanding Numerical Models, *Reliability Engineering & System Safety*, Vol. 111, 232–240, 2013.

Eckhardt R., Stan Ulam, John von Neumann, and the Monte Carlo Method, *Los Alamos Science*, Vol. 15, 131–137, 1987.

Ekeland I., *The Broken Dice and Other Mathematical Tales of Chance*, The University of Chicago Press, pp. 142, 1993.

Ekimov V.V., *Probabilistic Methods in the Structural Mechanics of Ships*, Leningrad: "Sudostroenie" Publishing House, 1966 (in Russian).

Eldred M.S. and Elman H.C., Design under Uncertainty Employing Stochastic Expansion Methods, *International Journal for Uncertainty Quantification*, Vol. 1(2), 119–146, 2011.

Eldred M.S., Swiler L.P. and Tang G., Mixed Aleatory-Epistemic Uncertainty Quantification with Stochastic Expansions and Optimization-Based Interval Estimation, *Reliability Engineering & System Safety*, Vol. 96(9), 1092–1113, 2011.

Eliasberg P.E., *Determination of Motion through Results of Measurement*, Moscow: Nauka Publishing House, 1976.

Elishakoff I., Vibration Fields in Circular Cylindrical Shells, Subjected to Random Loads, *PhD Dissertation, Department of Dynamics and Strength of Machines, Moscow Power Engineering Insitutte and State University*, 1971 (in Russian).

Elishakoff I., On the Role of Cross-Correlations in the Random Vibrations of Shells, *Journal of Sound Vibration*, Vol. 50, 239–252, 1977.

Elishakoff I., Axial Impact Buckling of a Column with Random Initial Imperfections, *Journal of Applied Mechanics*, Vol. 45, 361–365, 1978.

Elishakoff I., Impact Buckling of Thin Bar via Monte Carlo Method, *Journal of Applied Mechanics*, Vol. 45, 586–590, 1978.

Elishakoff I., Buckling of a Stochastically Imperfect Finite Column on a Nonlinear Elastic Foundation: A Reliability Study, *Journal of Applied Mechanics*, Vol. 46, 411–416, 1979a.

Elishakoff I., Simulation of Space-Random Fields for Solution of Stochastic Boundary-Value Problems, *Journal of the Acoustical Society of America*, Vol. 61, 399–403, 1979b.

Elishakoff I., Hoff's Problem in a Probabilistic Setting, *Journal of Applied Mechanics*, Vol. 47, 403–408, 1980.

Elishakoff I., Stochastic Simulation of an Initial Imperfection Data Bank for Isotropic Shells with General Imperfections, in *Buckling of Structures – Theory and Experiment* (I. Elishakoff, J. Arbocz, A. Libai and J. Babcock, eds.), Amsterdam: Elsevier Science, pp. 195–209, 1982.

Elishakoff I., How to Introduce the Imperfection Sensitivity Concept into Design, in *Collapse: Buckling of Structures in Theory and Practice* (J.M.T. Thompson and G.W. Hunt, eds.), Cambridge, UK: Cambridge University Press, pp. 345–357, 1983a.

Elishakoff I., A Simple Model Explaining Some Recent Random Vibration Results, in *Procedure 4th International Conference Applications of Statistics and Probability in Soil and Structural Engineering*, Bologna: Pitagora Press, pp. 493–507, 1983b.

Elishakoff I., *Probabilistic Methods in the Theory of Structures*, New York: Wiley, 1983.

Elishakoff I., *Probabilistic Theory of Structures*, second edition, New York: Dover, 1999a.

Elishakoff I., *Probabilistic Methods in the Theory of Structures: Strength of Materials, Random Vibrations, and Random Buckling*, Singapore: World Scientific, 2017.

Elishakoff I., A Model Elucidating Significance of Cross-Correlation in Random Vibration Analysis, in *Random Vibration-Status and Recent Developments* (I. Elishakoff and R.H. Lyon, eds.), Amsterdam: Elsevier Science Publishers, pp. 101–112, 1986.

Elishakoff I., Reliability Approach to the Random Imperfection Sensitivity of Columns, *Acta Mechanica*, Vol. 85, 151–170, 1986.

Elishakoff I., Random Vibration of Multi-Degree-of-Freedom Systems with Associated Effect of Cross-Correlations, in *Analysis and Estimation of Stochastic Mechanical Systems* (W. Schiehlen and W. Wedig, eds.), Vienna: Springer, pp. 22–31, 1988a.

Elishakoff I., Wide-Band Random Vibration of Continuous Structures with Associated Effect of Cross-Correlations, in *Analysis and Estimation of Stochastic Mechanical Systems* (W. Schiehlen and W. Wedig, eds.), Vienna: Springer, pp. 32–42, 1988b.

Elishakoff I., An Idea of the Uncertainty Triangle, *Shock Vibration Digest*, Vol. 22(10), 1, 1990.

Elishakoff I., Convex Versus Probabilistic Models of Uncertainty in Structural Dynamics, Opening Keynote Lecture, in *Structural Dynamics: Recent Advances* (M. Petyt, H.F. Wolfe and C. Mei, eds.), London: Elsevier Applied Science, pp. 3–21, 1991a.

Elishakoff I., Essay on Probability, Convexity and Earthquake Engineering (Keynote Lecture), in *Procedure 5th Italian National Conference Earthquake Engineering*. L'ingegneria Sismica in Italia 2, Vol. 2, pp. 907–935, 1991b.

Elishakoff I., Method of Stochastic Linearization Revisited and Improved, in *Computational Stochastic Mechanics* (P.D. Spanos and C.A. Brebbia, eds.), London: Elsevier Applied Science, pp. 101–111, 1991c.

Elishakoff I., Method of Stochastic Linearization Revisited and Improved, in *Computational Stochastic Mechanics* (P.D. Spanos and C.A. Brebbia, eds.), Southampton: Computational Mechanics Publications, pp. 101–111, 1991d.

Elishakoff I., Essay on Probability, Convexity and Earthquake Engineering, *Proceedings of the 5th Italian National Conference on Earthquake Engineering, "L'Ingegneria Sismica in Italia,"* Vol. 2, pp. 907–935, 1991e.

Elishakoff I., Essay on Reliability Index, Probabilistic Interpretation of Safety Factor, and Convex Models of Uncertainty, in *Reliability Problems: General Principles and Applications in Mechanics of Solids & Structures* (F. Casciati and J.B. Roberts, eds.), Vienna: Springer, pp. 237–271, 1992.

Elishakoff I., *Essay on Uncertainty in Structures, International Forum for Safety Engineering and Science,* Tokyo, Japan, pp. 38–65, 1994.

Elishakoff I., Essay on Uncertainties in Elastic and Viscoelastic Structures: From A.M. Freudenthal's Criticisms to Modern Convex Modeling, *Computers and Structures,* Vol. 56(6), 871–895, 1995a.

Elishakoff I., Convex Modeling – A Generalization of Interval Analysis for Non-Probabilistic Treatment of Uncertainty, (proc. of the Int workshop on Application of Interval Competition (APIC '95)), Supplement, 76–79, 1995b.

Elishakoff I., Some Results in Stochastic Linearization of Nonlinear Systems, in *Nonlinear Dynamics and Stochastic Mechanics* (W. Kliemann and N. Sri Namachchivaya, eds.), Boca Raton: CRC Press, pp. 259–281, 1995.

Elishakoff I., Discussion on the Paper: "A Non-Probabilistic Concept of Reliability, *Structural Safety,* Vol. 17(3), 195–199, 1995.

Elishakoff I., Random Vibration of Structures: A Personal Perspective, *Applied Mechanics Reviews,* Vol. 48(12), 809–825, 1995.

Elishakoff I., Fuzzy Finite Element Approach for Analysis of Imprecisely Defined Systems: Comment on Paper by S.S. Rao and J.P. Sawyer, *AIAA Journal,* Vol. 35, 403, 1997.

Elishakoff I., How to Introduce the Imperfection-Sensitivity Concept in Design – 2, in *Stability Analysis of Plates and Shells: A Collection of Papers in Honor of Dr Manuel Stein* (N.F. Knight Jr. and M.P. Nemeth, eds.), NASA SP-1998-206280, pp. 273–267, 1998.

Elishakoff I., Three Versions of the Finite Element Method Based on Concepts of either Stochasticity, Fuzziness, or Anti-Optimization, *Applied Mechanics Reviews,* Vol. 51(3), 209–218, 1998.

Elishakoff I., Are the Probabilistic and Antioptimization Methods Interrelated ?, in *Whys and Hows in Uncertainty Modeling,* Vienna: Springer Verlag, pp. 285–318, 1999a.

Elishakoff I., Are Probabilistic and Anti-Optimization Approaches Compatible?, in *Whys and Hows in Uncertainty Modelling* (I. Elishakoff, ed.), Vienna: Springer, pp. 263–355, 1999b.

Elishakoff I., What May Go Wrong with Probabilistic Methods, in *Whys and Hows in Uncertainty Modeling* (I. Elishakoff, ed.), Vienna : Springer, pp. 263–284, 1999c.

Elishakoff I., Are the Probabilistic and Antioptimization Methods Interrelated, in *Whys and Hows in Uncertainty Modeling* (I. Elishakoff, ed.), Vienna : Springer, pp. 285–318, 1999d.

Elishakoff I., *Probabilistic Theory of Structures,* Mineola: Dover Publications, 1999e.

Elishakoff I., Possible Limitations of Probabilistic Methods in Engineering, *Applied Mechanics Reviews,* Vol. 53(2), 19–25, 2000a.

Elishakoff I., Multiple Combinations of the Stochastic Linearization Criteria by the Moment Method, *Journal of Sound and Vibration,* Vol. 237, 550–559, 2000b.

Elishakoff I., Stochastic Linearization Technique: A New Interpretation and A Selective Review, *The Shock and Vibration Digest,* Vol. 32(3), 179–188, 2000c.

Elishakoff I., Review Of Marek P., Brozzetti J. and Gustar M., Probabilistic Assessment of Structures Using Monte Carlo Simulations, *Applied Mechanics Reviews,* Vol. 55(2), B31–B32, 2002.

Elishakoff I., Notes on Philosophy of the Monte Carlo Method, *International Applied Mechanics,* Vol. 39(7), 753–762, 2003.

Elishakoff I., *Safety Factors and Reliability: Friends or Foes?*, Dordrecht, Netherlands: Kluwer Academic Publishers, 2004.

Elishakoff I., Probabilistic Resolution of the Twentieth Century Conundrum in Elastic Stability, *Thin-Walled Structures*, Vol. 59, 35–57, 2012.

Elishakoff I., *Resolution of Twentieth Century Conundrum in Elastic Stability*, Singapore: World Scientific/Imperial College Press, 2014.

Elishakoff I. and Ohsaki M., *Optimization and Anti-Optimization of Structures under Uncertainty*, London: Imperial College Press, 2010.

Elishakoff I., Whys and Hows of the Parameterized Interval Analyses : A Guide for the Perplexed, *International Journal for Computational Methods in Engineering Science and Mechanics*, Vol. 14, 495–504, 2013.

Elishakoff I., *Dramatic Effect of Cross-Correlations in Random Vibrations of Discrete Systems, Beams, Plates, and Shells*, Berlin: Springer, 2020a.

Elishakoff I., *Handbook on Timoshenko-Ehrenfest Beam and Uflyand-Mindlin Plate Theories*, Singapore: World Scientific, 2020b.

Elishakoff I. and Andriamasy L., Nonclassical Linearization Criteria in Nonlinear Stochastic Dynamics, *Journal of Applied Mechanics*, Vol. 77(4), paper 04501, 2010.

Elishakoff I. and Andriamasy L., The Tale of Stochastic Linearization Technique: Over Half A Century of Progress, in *Nondeterministic Mechanics* (I. Elishakoff and C. Soize, eds.), pp. 115–189, 2012.

Elishakoff I., Andriamasy L. and Dolley M., Application and Extension of the Stochastic Linearization by Anh and Di Paola, *Acta Mechanica*, Vol. 204, 89–98, 2009.

Elishakoff I. and Arbocz J., Reliability of Axially Compressed Cylindrical Shells with Random Axisymmetric Imperfections, *International Journal of Solids and Structures*, Vol. 18, 563–583, 1982.

Elishakoff I. and Arbocz J., Reliability of Axially Compressed Cylindrical Shells with General Nonsymmetric Imperfections, *Journal of Applied Mechanics*, Vol. 52, 122–128, 1985.

Elishakoff I. and Archaud E., Modified Monte Carlo Method for Buckling Analysis of Nonlinear Imperfect Structure, *Archive of Applied Mechanics*, Vol. 83, 1327–1339, 2013.

Elishakoff I., Baruch M., Zhu L.P. and Caimi R., Random Vibration of Space Shuttle Weather Protection Systems, *Journal of Shock and Vibration*, Vol. 2, 111–118, 1995.

Elishakoff I. and Bekel Y., Application of Lamé's Super Ellipsoids to Model Initial Imperfections, *Journal of Applied Mechanics*, Vol. 80, article 061006, 2013.

Elishakoff I. and Ben-Haim Y., Dynamics of a Thin Cylindrical Shell under Impact with Limited Deterministic Information on Its Initial Imperfections, *Journal of Structural Safety*, Vol. 8(1–4), 103–112, 1990.

Elishakoff I. and Bert C.W., Complementary Energy Criterion in Nonlinear Stochastic Dynamics, in *Applications of Statistics and Probability*, Rotterdam: Balkema Publishers, pp. 821–825, 2000.

Elishakoff I., Bi R.-G., Jiang C., Han X. and Long X.Y., Structural Reliability in View of Principle of Indifference, in *Advances in Computational Engineering Science* (M. Liu, X. Han, Y.-T. Gu and Z. Li, eds.), Mason, OH: ScienTech Publishers, pp. 83–99, 2017.

Elishakoff I., Cai G.Q. and Starnes J.H. Jr., Probabilistic and Convex Models of Uncertainty in Buckling of Structures, in *Structural Safety and Reliability* (G.I. Schuëller, M. Shinozuka and J.T.P. Yao, eds.), Vol. 2, Rotterdam: Balkema, pp. 761–766, 1993.

Elishakoff I., Cai G.Q. and Starnes J.H. Jr, Nonlinear Buckling of a Column with Initial Imperfection via Stochastic and Non-Stochastic, Convex Models, *International Journal of Non-Linear Mechanics*, Vol. 29, 65–70, 1994.

Elishakoff I., Cederbaum G. and Librescu L., Response of Moderately Thick Laminated Cross-Ply Composite Shells Subjected to Random Excitation, *AIAA Journal*, Vol. 27, 975–981, 1989.

Elishakoff I. and Colajanni P., Stochastic Linearization Revisited, *Chaos, Solitons & Fractals*, Vol. 8, 1957–1972, 1997.

Elishakoff I. and Colajanni P., Booton's Problem Critically Re-Examined, *Journal of Sound and Vibration*, Vol. 210, 683–691, 1998.

Elishakoff I. and Colombi P., Combination of Probabilistic and Convex Models of Uncertainty when Scarce Knowledge is Present on Acoustic Excitation Parameters, *Computer Methods in Applied Mechanics and Engineering*, Vol. 104(2), 187–209, 1993.

Elishakoff I. and Colombi P., Successful Combination of the Stochastic Linearization and Monte-Carlo Methods, *Journal of Sound and Vibration*, Vol. 160(3), 554–558, 1993b.

Elishakoff I. and Colombi P., Ideas of Probability and Convexity Combined for Analyzing Parameter Uncertainty, in *Structural Safety and Reliability* (G.I. Schüeller and M. Shinozuka, eds.), *ICOSSAR'93*, pp. 109–113, 1993c.

Elishakoff I. and Crandall S.H., Sixty Years of Stochastic Linearization Technique, *Meccanica*, Vol. 52, 299–305, 2017.

Elishakoff I. and Ducreux B., Modified Interval Analysis for Structures with Uncertain Boundary Conditions, in *Structural Mechanics and Building Constructions : Collection of Papers Dedicated to A.V. Perelmuter's 80th Birth Anniversary*, Moscow: "SKAD SOFT" Publishers, pp. 154–172, 2013.

Elishakoff I. and Ducreux B., Dramatic Effect of Cross Correlations in Random Vibration of Point-Driven Spherically Curved Panel, *Archive of Applied Mechanics*, Vol. 84, 473–490, 2014.

Elishakoff I., Elisseeff P. and Glegg S.A.L., Non-Probabilistic, Convex-Theoretic Modeling of Scatter in Material Properties, *AIAA Journal*, Vol. 32, 843–849, 1994.

Elishakoff I. and Falsone G., Some Recent Developments in Stochastic Linearization Technique, in *Computational Stochastic Mechanics* (A.H.-D. Cheng and C.Y. Yang, eds.), Southampton: Computation Mechanics Publications, pp. 175–194, 1993.

Elishakoff I. and Ferracuti B., Fuzzy Sets Based Interpretation of the Safety Factor, *Fuzzy Sets and Systems*, Vol. 157(18), 2495–2512, 2006a.

Elishakoff I. and Ferracuti B., Four Alternative Definitions of the Fuzzy Safety Factor, *Journal of Aerospace Engineering*, Vol. 19(4), 281–287, 2006b.

Elishakoff I., Gabriele S. and Wang Y., Generalized Galileo Galilei Problem in Interval Setting for Functionally Related Loads, *Archive of Applied Mechanics*, Vol. 86(7), 1203–1217, 2016.

Elishakoff I., Gana-Shvili Y. and Givoli D., Treatment of Uncertain Imperfections as a Convex Optimization Problem, in *Proceedings of the Sixth International Conference on Applications of Statistics and Probability in Civil Engineering* (L. Esteva and S.E. Ruiz, eds.), Vol. 1, Mexico City, pp. 150–157, 1991.

Elishakoff I. and Hasofer A.M., On the Accuracy of Hasofer-Lind Reliability Index, *Proceedings of ICOSSAR-85, International Conference on Structural Safety and Reliability*, Kobe, Japan, Vol. 1, pp. 229–239, 1985.

Elishakoff I. and Hasofer A.M., Exact versus Approximate Analyses in Structural Reliability, *Applied Scientific Research, International Journal of Thermal, Mechanical and Electromagnetic Phenomena in Continua*, 44, 303–312, 1987.

Elishakoff I. and Hasofer A.M., Effect of Human Error on Reliability of Structures, *Proceedings of the 33rd AAIA/ASME/ASCE/AMS/ASC Structures Structural Dynamics and Material Conference*, pp. 3222–3237, Dallas, TX, 1992.

Elishakoff I. and Hasofer A.M., Detrimental and Serendipitous Effect of Human Error on Reliability of Structures, *Computational Methods of Applied Mechanical Engineering*, Vol. 129, 1–7, 1996.

Elishakoff I. and Hollkamp J., Computerized Symbolic Solution for a Non-Conservative System in Which Instability Occurs by Flutter in One Range of the Parameter and by Divergence in the Other, *Computer Methods in Applied Mechanics and Engineering*, Vol. 62, 27–46, 1987.

Elishakoff I., Haftka R.T. and Fang J.J., Structural Design under Bounded Uncertainty: Optimization with Anti-Optimization, *Computers and Structures*, Vol. 53, 1401–1405, 1994.

Elishakoff I. and Li Q., How to Combine Probabilistic and Antioptimization Methods, in *Whys and Hows in Uncertainty Modeling* (I. Elishakoff, ed.), Vienna: Springer, pp. 319–340, 1999.

Elishakoff I., Lin Y.-K. and Zhu L.P., *Probabilistic and Convex Modeling of Acoustically Excited Structures*, Amsterdam: Elsevier Science Publishers, 1994.

Elishakoff I., Li Y.W. and Starnes J.H. Jr., *Non-Classical Problems in the Theory of Elastic Stability*, Cambridge University Press, 2001.

Elishakoff I., Li Y.W. and Starnes J.H. Jr, A Deterministic Method to Predict the Buckling Load Variability Due to Uncertain Moduli, *Computer Methods of Applied Mechanics and Engineering*, Vol. 111, 155–167, 1994.

Elishakoff I., Lin Y.K. and Zhu L.P., *Probabilistic and Convex Modeling of Acoustically Excited Structures*, Amsterdam: Elsevier, 1994.

Elishakoff I. and Livshits D., Some Closed-Form Solutions in Random Vibration of Timoshenko Beams, *Recent Advances in Structural Dynamics* (M. Petyt, ed.), Southampton University Press, pp. 639–648, 1984.

Elishakoff I. and Lottati I., Divergence and Flutter of Non-Conservative Systems with Intermediate Support, *Computer Methods in Applied Mechanics and Engineering*, Vol. 66, 241–250, 1988.

Elishakoff I. and Lubliner E., Random Vibrations of a Structure via Classical and Nonclassical Theories, in *Probabilistic Methods in Mechanics, and Structures* (S. Eggwertz, ed.), Berlin: Springer, pp. 455–468, 1985.

Elishakoff I. and Miglis Y., Novel Parameterized Intervals May Lead to Sharp Bounds, *Mechanics Research Communications*, Vol. 44, 1–8, 2012a.

Elishakoff I. and Miglis Y., Overestimation-Free Computational Version of Interval Analysis, *International Journal of Computational Methods in Engineering Science and Mechanics*, Vol. 13, 319–328, 2012b.

Elishakoff I. and Nordstrand T., Probabilistic Analysis of Uncertain Eccentricities in a Model Structure, in *Proceedings of the Sixth International Conference Applications of Statistics and Probability in Civil Engineering* (L. Esteva and S.E. Ruiz, eds.), Vol. 1, Mexico City, pp. 250–256, 1991.

Elishakoff I. and Ohsaki M., *Optimization and Anti-Optimization of Structures under Uncertainty*, London: Imperial College Press, 2010.

Elishakoff I. and Pletner B., Analysis of Base Excitation on an Uncertain Function with Specified Bounds on It and Its Derivatives, in *Structural Vibration and Acoustics* (T.C. Huang, *et al.*, ed.), New York: ASME Press, pp. 177–184, 1991.

Elishakoff I. and Ren Y., Finite Element Method for Stochastic Structures Based on Inverse of Stiffness Matrix, in *Uncertainty Modeling in Finite Element, Fatigue and Stability of Systems* (A. Haldar, A. Guran and B.M. Ayyub, eds.), Singapore: World Scientific, pp. 51–70, 1997.

Elishakoff I. and Ren Y-J., The Bird's Eye View on Finite Element Method for Stochastic Structures, *Computer Methods in Applied Mechanics and Engineering*, Vol. 168, 51–61, 1999.

Elishakoff I. and Ren Y.J., *Finite Element Methods for Structures with Large Stochastic Variations*, Oxford, UK: Oxford University Press, 2003.

Elishakoff I., Ren Y.J. and Shinozuka M., Some Exact Solutions for Bending Beams with Spatially Stochastic Stiffness, *International Journal of Solids Structures*, Vol. 32, 2315–2327, 1995.

Elishakoff I., Ren Y.J. and Shinozuka M., Some Thoughts and Attendant New Results in Finite Element Method for Stochastic Structures, *Chaos, Solitons and Fractals*, Vol. 7, 597–609, 1996.

Elishakoff I., Ren Y.J. and Shinozuka M., Critical Observations and Attendant New Results in the FEM for Stochastic Problems, *Chaos, Solitons, Fractals*, Vol. 7, 597–609, 1996.

Elishakoff I. and Santoro R., Error in the Finite Difference Based Probabilistic Dynamic Analysis: Analytical Evaluation, *Journal of Sound and Vibration*, Vol. 281(3–5), 1195–1206, 2005.

Elishakoff I. and Santoro R., Error in Finite Difference Based Probabilistic Dynamic Analysis: Analytical Evaluation, *Journal of Sound and Vibration*, Vol. 281(3–5), 1195–1206, 2005.

Elishakoff I. and Starnes J.H. Jr., Safety Factor and the Non-Deterministic Approaches, AIAA Paper 99-1614, *Proceedings, AIAA/ASME/ASCE/AHS/ASC Structures, Structural Dynamics and Materials Conference*, St. Louis, pp. 3084–3099, 1999.

Elishakoff I., van Manen S., Vermeulen P.G. and Arbocz J., First Order Second Moment Analysis of the Buckling of Shells with Random Imperfections, *American Institute of Aeronautics and Astronautics (AIAA) Journal*, Vol. 25, 1113–1117, 1987.

Elishakoff I., van Zanten A.T. and Crandall S.H., Wide-Band Random Axisymmetric Vibration of Cylindrical Shells, *Journal of Applied Mechanics*, Vol. 46, 417–422, 1979.

Elishakoff I., Verhaeghe W. and Moens D., Probabilistic and Interval Analyses Contrasted in Impact Buckling of a Clamped Column, *Journal of Applied Mechanics*, Vol. 80, paper 011022, 2013.

Elishakoff I., Wang X.-J., Hu J. and Qiu Z.-P., Minimization of the Least Favorable Static Response of a Two-Span Beam Subjected to Uncertain Loading, *Thin-Walled Structures*, Vol. 70, 49–56, 2013.

Elishakoff I., Wang X.-J., Li Y.-L., Hu J. and Qiu Z.-P., Regulating the Dynamic Behavior of a Column with Uncertain Initial Imperfections by Support Placing, *International Journal of Solids and Structures*, Vol. 50(2), 396–402, 2012.

Elishakoff I. and Zhang R., Comparison of the New Energy-Based Versions of the Stochastic Linearization Technique, in *Nonlinear Stochastic Mechanics* (N. Bellomo and F. Casciati, eds.), Berlin: Springer Verlag, pp. 201–212, 1992b.

Elishakoff I. and Zhang R., Comparison of the New Energy-Based Versions of the Stochastic Linearization Technique, in *Nonlinear Stochastic Mechanics* (N. Bellomo and F. Casciati, eds.), Berlin: Springer, pp. 201–212, 1992.

Elishakoff I. and Zhang X.T., An Appraisal of Different Stochastic Linearization Techniques, *Journal of Sound and Vibration*, Vol. 153, 370–375, 1992.

Ellingwood B.R., Acceptable Risk Bases for Design of Structures, *Progress in Structural Engineering and Materials*, Vol. 3(2), 170–179, 2001.

Ellingwood B.R. and Kinali K., Quantifying and Communicating Uncertainty in Seismic Risk Assessment, *Structural Safety*, Vol. 31, 179–187, 2009.

Elseifi M.A., Gurdal Z. and Nikolaidis E., Convex/Probabilistic Models of Uncertainties in Geometric Imperfections of Stiffened Composite Panels, *AIAA Journal*, Vol. 37(4), 468–474, 1999.

Emam H.H., Pradlwarter H.J. and Schüeller G.I., A Computational Procedure for the Implementation of Equivalent Linearization on Finite Element Analysis, *Earthquake Engineering and Structural Dynamics*, Vol. 29, 1–17, 2000.

Ermakov S.M., *Monte Carlo Methods and Mixed Problems*, Moscow: "Nauka" Publishing House, 1985 (in Russian).

Faber M.H., On the Treatment of Uncertainties and Probabilities in Engineering Decision Analysis, *Journal of Offshore Mechanics and Arctic Engineering*, Vol. 127, 243–248, 2005.

Faber M.H., Vrouwenvelder T. and Zilch K. (eds.), *Aspects of Structural Reliability: In Honor of R. Rackwitz*, Munich: Herbert Utz Publishing, 2007.

Faes M., Daub M., Marelli S., Patelli E. and Beer M., Engineering Analysis with Probability Boxes: A Review of Computational Methods, *Structural Safety*, Vol. 93, article 102092, 2021.

Faes M. and Moens D., Recent Trends in the Modeling and Quantification of Non-Probabilistic Uncertainty, *Archives of Computational Methods in Engineering*, Vol. 27(3), 633–671, 2020.

Fahy F.J., Statistical Energy Analysis – A Critical Review, *Shock Vibration Digest*, Vol. 6, 14–33, 1974.

Fahy F.J., Statistical Energy Analysis: A Wolf in Sheep's Clothing, *Proceedings of Inter-Noise 93*, pp. 13–26, Leuven, Belgium, 1993.

Failla G., Spanos P.D. and Di Paola M., Response Power Spectrum of Multi-Degree-of-Freedom Nonlinear Systems by a Galerkin Technique, *Journal of Applied Mechanics*, Vol. 70, 708–723, 2003.

Falsone G., Stochastic Linearization of MDoF Systems under Parametric Excitation, *International Journal Non-Linear Mechanics*, Vol. 27, 1025–1037, 1992.

Falsone G., Stochastic Linearization for the Response of MDOF Systems Subjected to External and Parametric Gaussian Excitations, in *Computational Stochastic Mechanics* (P.D. Spanos and C.A. Brebbia, eds.), Southampton: Computational Mechanics Publications, 1992a.

Falsone G., Stochastic Linearization of MDOF Systems under Parametric Excitations, *International Journal of Non-Linear Mechanics*, Vol. 27, 1025–1037, 1992b.

Falsone G. and Elishakoff I., Modified Stochastic Linearization Technique for Colored Noise Excitation of Duffing Oscillator, *International Journal of Non-Linear Mechanics*, Vol. 29, 65–69, 1994.

Falsone G. and Ricciardi G., Stochastic Linearization: Classical Approach and New Developments, in *Recent Research Developments in Structural Dynamics* (A. Luongo, ed.), pp. 81–106, 2003.

Fan F.-G. and Ahmadi G., On Loss of Accuracy and Non-Uniqueness of Solutions Generated by Equivalent Linearization and Cumulant-Neglect Methods, *Journal of Sound and Vibration*, Vol. 137, 385–401, 1990.

Fang J.J., Smith S.M. and Elishakoff I., Hybrid Anti-Optimization and Fuzzy-Set-Based Analyses for Structural Optimization under Uncertainty, *Mathematical Problems in Engineering*, Vol. 4, 187–200, 1998.

Fang J. and Fang T.S., A Weighted Equivalent Linearization Method in Random Vibration, *Chinese Journal of Applied Mechanics*, Vol. 8(3), pp. 114–120, 1991 (in Chinese).

Fang J., Elishakoff I. and Caimi R., Nonlinear Response of a Beam under Stationary Random Excitation by Improved Stochastic Linearization Method, *Applied Mathematical Modeling*, Vol. 19, 106–111, 1995.

Fang J.J. and Elishakoff I., Nonlinear Response of a Beam under Stationary Random Excitation by Improved Stochastic Linearization Method, *Applied Mathematical Modelling*, Vol. 19, 106–111, 1995.

Fang T., Zhang T.S. and Fang J., Nonstationary Response of Nonlinear Systems under Random Excitation, *Proceedings, Third National Conference on Random Vibrations*, Taiwan, pp. 147–155, 1991 (in Chinese).

Farid M., Data-Driven Method for Real-Time Prediction and Uncertainty Quantification of Fatigue Failure under Stochastic Loading Using Artificial Neural Networks and Gaussian Process Regression, *International Journal of Fatigue*, Vol. 155, article 106415, 2022.

Farsangi E.N., Noori M., Gardoni P., Takewaki I., Varum H. and Bogdanovic A. (eds.), *Reliability-Based Analysis and Design of Structures and Infrastructure*, Boca Raton: CRC Press, 2021.

Feinberg J. and Langtangen H.P., Chaospy: An Open-Source Tool for Designing Methods of Uncertainty Quantification, *Journal of Computational Science*, Vol. 11, 46–57, 2015.

Fellin W., Lessmann H., Oberguggenberger M. and Vieider R. (eds.), *Analyzing Uncertainty in Civil Engineering*, Berlin: Springer, 2005.

Fenton G.A., Error Evaluation of Three Random-Field Generators, *Journal of Engineering Mechanics*, Vol. 120(12), 2478–2497, 1994.

Fenton G.A. and Vanmarcke E.H., Simulation of Random Fields via Local Average Subdivision, *Engineering Mechanics*, Vol. 116(8), 1733–1749, 1990.

Fernández-Godino M.G., Panda N., O'Malley D., Larkin K., Hunter A., Haftka R.T. and Srinivasan G., Accelerating High-Strain Continuum-Scale Brittle Fracture Simulations with Machine Learning, *Computational Materials Science, Vol*, Vol. 186, article 109959, 2021.

Fernández-Godino M.G., Park C., Kim N.H. and Haftka R., Review of Multi-Fidelity Models, arXiv preprint arXiv:1609.07196, 2016 Sep 23.

Ferry Borges J. and Castanheta M., *Structural Safety*, Second Edition, Lisbon: National Civil Engineering Laboratory, 1971.

Ferson S., What Monte Carlo Methods Cannot Do, *Human and Ecological Risk Assessment: An International Journal*, Vol. 2(4), 990–1007, 1996.

Ferson S., *RAMAS Risk Calc 4.0 Software: Risk Assessment with Uncertain Numbers*, Boca Raton: Lewis Publishers, 2002.

Ferson S., Estimating Rare-Event Probabilities without Data, in *Proceedings of the Twelfth International Conference on Structural Safety & Reliability (ICOSSAR)*, Vienna, Austria, p. 1, 2017.

Ferson S. and Ginzburg L.R., Different Methods are Needed to Propagate Ignorance and Variability, *Reliability Engineering & System Safety*, Vol. 54(2–3), 133–144, 1996.

Ferson S. and Hajagos J.G., Arithmetic with Uncertain Numbers: Rigorous and (Often) Best Possible Answers, *Reliability Engineering & System Safety*, Vol. 85(1–3), 135–152, 2004.

Ferson S., Joslyn C.A., Helton J.C., Oberkampf W.L. and Sentz K., Summary from the Epistemic Uncertainty Workshop: Consensus Amid Diversity, *Reliability Engineering & System Safety*, Vol. 85(1–3), 355–369, 2004.

Ferson S., Kreinovich V., Hajagos J., Oberkampf W. and Ginzburg L., Experimental Uncertainty Estimation and Statistics for Data Having Interval Uncertainty, Sandia National Laboratories, Report SAND2007-0939, 2007.

Ferson S., Nelsen R.B., Hajagos J., Berleant D.J., Zhang J., Tucker W.T., Ginzburg L.V. and Oberkampf W.L., *Dependence in Probabilistic Modeling, Dempster-Shafer Theory, and Probability Bounds Analysis*, SAND REPORT 2004-3072, Albuquerque and Livermore, Sandia National Laboratories, 2004.

Ferson S., O'Rawe J. and Balch M., Computing with Confidence: Imprecise Posteriors and Predictive Distributions, *Vulnerability, Uncertainty, and Risks: Quantification, Mitigation, and Management*, pp. 895–904, 2014.

Fetz T., Efficient Computation of Upper Probabilities of Failure, in *Safety, Reliability, Risk, Resilience and Sustainability of Structures and Infrastructure* (C. Bucher and B.R. Ellingwood, eds.), Vienna: TU-Verlag, pp. 493–502, 2017.

Fetz T. and Oberguggenberger M., Imprecise Random Variables, Random Sets, and Monte Carlo Simulation, *International Journal of Approximate Reasoning*, Vol. 78, 252–264, 2016.

Feynman R.P., *What Do You Care What Other People Think?*, New York: Bantam Books, 1988.

Fishman G.S., *Monte Carlo Concepts: Algorithms and Applications*, Berlin: Springer, 1995.

Flügge W., *Theory of Viscoelasticity*, New York: Academic Press, 1975.

Fortier M., Rebba R., Koch P., Karl A., Broggi M. and Wright L., Challenge in Uncertainty Quantification, *Benchmark, The International Magazine for Engineering Designers & Analysis from NAFEMS, Medical Modelling*, 40–43, 2014.

Fournier A., Fussel D. and Carpenter L., Computer Rendering of Stochastic Models, *Communications ACM*, Vol. 25(6), 371–384, 1982.

Frangopol D.M., Structural Optimization Using Reliability Concepts, *Journal of Structural Engineering*, Vol. 111(11), 2288–2301, 1985.

Frangopol D.M., Probabilistic Structural Optimization, *Progress in Structural Engineering and Materials*, Vol. 1(2), 223–230, 1998.

Frangopol D.M., Life-Cycle Performance, Management, and Optimisation of Structural Systems under Uncertainty: Accomplishments and Challenges, *Structure and Infrastructure Engineering*, Vol. 7(6), 389–413, 2011.

Frangopol D.M. and Corotis R.B., Reliability-Based Structural System Optimization: State-of-the-Art versus State-of-the-Practice, in *Analysis and Computation*, New York: ASCE Press, pp. 67–78, 1996.

Frangopol D.M. and Hong K., Probabilistic-Fuzzy Model for Seismic Hazard, in *Reliability and Safety Analyses under Fuzziness* (T. Onisawa and J. Kacprzyk, eds.), Heidelberg: Springer, pp. 302–325, 1995.

Frangopol D.M., Kallen M.J. and Noortwijk J.M., Probabilistic Models for Life-Cycle Performance of Deteriorating Structures: Review and Future Directions, *Progress in Structural Engineering and Materials*, Vol. 6(4), 197–212, 2004.

Frangopol D.M. and Kim S., *Life-Cycle of Structures under Uncertainty: Emphasis on Fatigue-Sensitive Civil and Marine Structures*, Boca Raton: CRC Press, 2019.

Fraser J.R., *Applied Linear Programming*, Englewood Cliffs, NJ: Prentice Hall, 1918.

Freidlin M.I. and Wentzel A.D., *Random Perturbations of Dynamical Systems*, New York: Springer, 1984.

Freudenthal A.M., Allowable Stresses and Safety of Structures, *Journal of Association of Engineers in Palestine/Eretz Israel*, Vol. 1, 149–153, 1938 (in Hebrew).

Freudenthal A.M., Safety of Structures, *Transnational ASCE*, Vol. 112, 125–180, 1947.

Freudenthal A.M., Safety and Probability of Structural Failure, *Transnational ASCE*, Vol. 121, 1337–1375, 1956.

Freudenthal A.M., Fatigue Sensitivity and Reliability of Mehcanical Systems, Especially Aircraft Structures, WADD Technical Report 61-53, Wright Patterson, AFB, Ohio, 1961.

Freudenthal A.M., Critical Appraisal of Safety Criteria and Their Basic Concepts, in *Preliminary Publication, 8th Congress, International Association of Bridge and Structural Engineers*, New York, pp. 13–28, 1968.

Freudenthal A.M., Introductory Remarks, in *International Conference on Structural Safety and Reliability* (A.M. Freudenthal, ed.), New York: Pergamon Press, pp. 5–6, 1972a.

Freudenthal A.M., Introductory Remarks, *Proceedings, International Conference on Structural Safety and Reliability*, (A.M. Freudenthan, ed.), Oxford: Pergamen Press, pp. 5–6, 1972b.

Frey M., Finite Point Process Model of Shot Noise, in *Compuiational Stochastic Mechanics* (P.D. Spanos, ed.), Rotterdam: Balkema, pp. 29–37, 1995.

Friedlander B. and Porat B., Modificd Yulo-Walker Method of ARMA Spectral Estimation, *IEEE Transactions on Aerospace Elecron Systems*, Vol. 20(2).

Friot E. and Bouc R., Fast Synthesis of ARMA Models for the Recursive Simulation of Scalar Random Process with a Given Target Spectum, Journal of Sound and Vibration, Vol. 170(3), 415–421, 1994.

Fröhlich B., Hose D., Dieterich O., Hanss M. and Eberhard P., Uncertainty Quantification of Large-Scale Dynamical Systems Using Parametric Model Order Reduction, *Mechanical Systems and Signal Processing*, Vol. 171, article 108855, 2022.

C. F., Sinou J.J., Zhu W., Lu K. and Yang Y., A State-of-the-Art Review on Uncertainty Analysis of Rotor Systems, *Mechanical Systems and Signal Processing*, Vol. 183, article 109619, 2023.

Fuchs M., Clouds, p-Boxes, Fuzzy Sets, and Other Uncertainty Representations in Higher Dimensions, *Acta Cybernetica*, Vol. 19(1), 61–92, 2009.

Galambos Y., Lechner J. and Simiu E. (eds.), *Extreme Value Theory and Applications*, Vol. 1, Dortrecht: Kluwer Academic Publ, Preface, IX, pp. 1–60, 1994.

Galbally D., Fidkowski K., Willcox K. and Ghattas O., Non-Linear Model Reduction for Uncertainty Quantification in Large-Scale Inverse Problems, *International Journal for Numerical Methods in Engineering*, Vol. 81(12), 1581–1608, 2010.

Gangadharan S., Nikolaidis E., Lee K. and Haftka R., The Use of Antioptimization to Compare Alternative Structural Models, Proceedings, *34th Structures, Structural Dynamics and Materials Conference*, pp. 1355–1367, 1993.

Gangadharan S.N., Nikolaidis E., Lee K., Haftka R.T. and Burdisso R., Antioptimization for Comparison of Alternative Structural Models and Damage Detection, *AIAA Journal*, Vol. 37(7), 857–864, 1999.

Ganzerli S. and Pantelides C.P., Load and Resistance Convex Models for Optimum Design, *Structural Optimization*, Vol. 17(4), 259–268, 1999.

Ganzerli S. and Pantelides C.P., Optimum Structural Design via Convex Model Superposition, *Computers and Structures*, Vol. 74, 639–647, 2000.

Gao J., Yao J. and Chen L., The Statistical Methods of Membership Function in Structural Serviceability Failure Criterion, *KSCE Journal of Civil Engineering*, Vol. 25(11), 4314–4321, 2021.

Gao S., Niemann H.J. and Ou K., Analysis of Stochastic Response of Rotational Shell Considering Its Constructive Nonlinearity, *Journal of Beijing Institute of Technology*, Vol. 6, 180–186, 1997.

Garno J., Ouellet F., Bae S., Jackson T.L., Kim N.H., Haftka R., Hughes K.T. and Balachandar S., Calibration of Reactive Burn and Jones-Wilkins-Lee Parameters for Simulations of a Detonation-Driven Flow Experiment with Uncertainty Quantification, *Physical Review Fluids*, Vol. 5(12), article 123201, 2020.

Garofalo F. and Leitmann G., Nonlinear Composite Control of Nominally Linear Singularity Perturbed Uncertain System, In *Procedure 12th IMACS World Congress*, Paris, 1988.

Garrelick J.M. and Chayes L., Comment on the Contribution of Cross-Modal Terms to the Local Response of Structures to Random Pressure Fields, *Journal of Acoustics Society of America*, Vol. 63, 1626–1628, 1978.

Gayton N., Bourinet J.M. and Lemaire M., CQ2RS: A New Statistical Approach to the Response Surface Method for Reliability Analysis, *Structural Safety, Vol*, Vol. 25(1), 99–121, 2003.

Gayton N., Mohamed A., Sorensen J.D., Pendola M. and Lemaire M., Calibration Methods for Reliability-Based Design Codes, *Structural Safety*, Vol. 26(1), 91–121, 2004.

Gell-Mann M., From Renormalizability to Calculability?, in *Shelter Island II: Proc of 1983 Shelter Island Conf on Quantum Field Theory and the Fundamental Problems of Physics* (R. Jackie, N.N. Khuri, S. Weinberg and E. Witten, eds.), Cambridge, MA: MIT Press, pp. 3–23, 1985.

Georgiou G., Khodaparast H.H. and Cooper J.E., Uncertainty Quantification of Aeroelastic Stability, in *Mathematics of Uncertainty Modeling in the Analysis of Engineering and Science Problems*, IGI Global, pp. 329–356, 2014.

Georgescu I., The Early Days of Monte Carlo Method, *Nature Review Physics*, Vol. 5, 372, 2023.

Gersch W. and Foutch D.A., Least Square Estimation of Structural System Parameters Using Covariance Function Data, *IEEE Transactions on Automatic Control*, Vol. 19(6), 898–903, 1974.

Gersch W. and Yonemoto J. Synthesis of Multivariate Random Vibration Systems: A Two-Stage Least Squares ARMA Model Approach, *Journal of Sound and Vibration*, Vol. 52(4), 553–565, 1977.

Ghanem R.G., Uncertainty Quantification in Computational and Prediction Science, *International Journal for Numerical Methods in Engineering*, Vol. 80(6-7), 671–672, 2009.

Ghanem R., Higdon D. and Owhadi H. (eds.), *Handbook of Uncertainty Quantification*, New York: Springer, 2017.

Ghanem R. and Sarkar A., Reduced Models for the Medium-Frequency Dynamics of Stochastic Systems, The, *Journal of the Acoustical Society of America*, Vol. 113(2), 834–846, 2003.

Ghanem R. and Spanos P.D., Galerkin-Based Response Surface Approach for Reliability Analysis, *Proceedings of the Fifth International Conference on Structural Safety and Reliability*, San Francisco, CA, Vol. II, pp. 1081–1088, 1989.

Ghanem R. and Spanos P.D., *Stochastic Finite Elements: A Spectral Approach*, New York: Springer, 1991 (Second edition, Mineola, N.Y.: Dover, 2003).

Ghanem R. and Spanos P.D., A Stochastic Galerkin Expansion for Nonlinear Random Vibration Analysis, *Probabilistic Engineering Mechanics*, Vol. 8, 255–264, 1993.

Ghanem R. and Spanos P.D., Spectral Techniques for Stochastic Finite Elements, *Architecture Computation of Methods in Engineering*, Vol. 4, 63–100, 1997.

Ghanem R. and Spanos P.D., *Stochastic Finite Elements: A Spectral Approach*, second edition, Courier Corporation, 2003.

Ghanem R, Higdon D and Owhadi H (eds.), *Handbook of Uncertainty Quantification*, New York: Springer, 2017.

Ghanem R., Yadegaran I., Thimmisetty C., Keshavarzzadeh V., Masri S., Red-Horse J., Moser R., Oliver T., Spanos P. and Aldraihem O.J., Probabilistic Approach to NASA Langley Research Center Multidisciplinary Uncertainty Quantification Challenge Problem, *Journal of Aerospace Information Systems*, Vol. 12(1), 170–188, 2015.

Giles M.B., Multilevel Monte Carlo Path Simulation, *Operations Research*, Vol. 56(3), 607–617, 2008.

Giles M.B., Multilevel Monte Carlo Methods, in *Monte Carlo and Quasi-Monte Carlo Methods*, Berlin: Springer, pp. 83–103, 2012.

Giles M.B., Multilevel Monte Carlo Methods, *Acta Numerica*, Vol. 24, 259–328, 2015.

Ghiocel D. and Lungu D., *Wind, Snow and Temperature Effects on Structures Based on Probability*, Turnbridge Wells, Kent, UK: Abacus Press, 1975.

Giannini R. and Pinto P.E., Stochastic Analysis Methods, *Computer Analysis and Design of Earthquake Resistant Structures*, Vol. 3, 103–152, 1997.

Givoli D. and Elishakoff I., Stress Concentration at a Nearly Circular Hole with Uncertain Irregularities, *Journal of Applied Mechanics*, Vol. 59, 5.65–5.71, 1992.

Goel R.P., Transverse Vibrations of Tapered Beams, *Journal of Sound and Vibration*, Vol. 47(1), 1–7, 1976.

Gogu C., Qiu Y., Segonds S. and Bes C., Optimization Based Algorithms for Uncertainty Propagation through Functions with Multidimensional Output within Evidence Theory, *Journal of Mechanical Design*, Vol. 134(10), article 100914, 2012.

Goldszmidt M. and Pearl J., Reasoning with Qualitative Probabilities Can be Tractable, in *Uncertainty in Artificial Intelligence* (D. Dubois and B. D'Ambrosio, eds.), Morgan Kaufmann, pp. 112–120, 1992.

Good I.J., Reliability Always Depends on Probability, *Journal of Statistical Computation and Simulation*, Vol. 52, 192–193, 1995.

Goodman I.R., Fuzzy Sets as Equivalence Classes of Random Sets, in *Fuzzy Sets and Possibility Theory* (R. Yager, ed.), pp. 327–343, 1982.

Grandori G., Paradigms and Falsification in Earthquake Engineering, *Meccanica*, Vol. 26, 17–21, 1991.

Grandori G., Guagenti E. and Tagliani A., Seismic Hazard Analysis: How to Measure Uncertainty?, *Computers & Structures*, Vol. 67, 47–51, 1998.

Graupe D., Krause D.J. and Moore J.B., Identification of Autoregressive Moving-Average Parameters of Time Series, *IEEE Transactions on Automatic Cantrol*, Vol. 20, 104–107, 1975.

Gray A., Wimbush A., De Angelis M., Hristov P.O., Calleja D., Miralles-Dolz E. and Rocchetta R., From Inference to Design: A Comprehensive Framework for Uncertainty Quantification in Engineering with Limited Information, *Mechanical Systems and Signal Processing*, Vol. 165, 108210, 2022.

Greene R., *Mastery*, New York: Penguin Books, 2013.

Grenander U. and Szego G., *Toepliz Forms and Their Applications*, Berkeley CA: University of California Press, 1958.

Greś S., Döhler M., Jacobsen N.J. and Mevel L., Uncertainty Quantification of Input Matrices and Transfer Function in Input/Output Subspace System Identification, *Mechanical Systems and Signal Processing*, Vol. 167, article 108581, 2022.

Grigoriu M., Spectral Representation Method in Simulation, *Probabilistic Engineering Mechanics*, Vol. 8(2), 75–90, 1993.

Grigoriu M., *Applied Non-Gaussian Processes*, Englewood Cliffs, N.J: Prentice Hall, 1995.

Grigoriu M., Sample Properties of Some Equivalent Non-Gaussian Processes, *13th ASCE Engineering Mechanics Conference*, The Johns Hopkins University, Baltimore, June 13–16, 1999.

Grigoriu M., *Stochastic Calculus: Applications in Science and Engineering*, Boston, MA: Birkhäuser, 2002.

Grigoriu M., *Stochastic Systems: Uncertainty Quantification and Propagation*, London: Springer, 2012.

Grigoriu M. and Ariaratnam S.T., Response of a Linear Systems to Polynomials of Gaussian Processes, *Applied Mechanics*, Vol. 55, 905–910, 1988.

Grossi R.O. and Bhat R.B., A Note on Vibrating Tapered Beams, *Journal of Sound Vibration*, Vol. 147(1), 174–178, 1991.

Grunddmann H., Hartmann Ch. and Waubke H., Structures Subjected to Stationary Stochastic Loadings. Preliminary Assessment by Statistical Linearization Combined with an Evolutionary Algorithm, *Computers and Structures*, Vol. 67, 53–64, 1998.

Gumbel E.J., Bivariate Exponential Distributions, *American Statistical Association Journal*, Vol. 55(292), 698–707, 1960.

Guo J. and Du X., Sensitivity Analysis with Mixture of Epistemic and Aleatory Uncertainties, *AIAA Journal*, Vol. 45(9), 2337–2349, 2007.

Guo J. and Du X.P., Reliability Sensitivity Analysis with Random and Interval Variables, *International Journal for Numerical Methods in Engineering*, Vol. 78, 1585–1617, 2009.

Guo S.X. and Lu Z.Z., Procedure for Analyzing the Fuzzy Reliability of Mechanical Structures When Parameters of Probabilistic Models are Fuzzy, *Journal of Mechanical Strength*, Vol. 25(5), pp. 527–529, 2003 (in Chinese).

Guo S.X., Lu Z.Z. and Feng Y.S., A Non-Probabilistic Model of Structural Reliability Based on Interval Analysis, *Chinese Journal of Computational Mechanics*, Vol. 18(1), pp. 56–60, 2001 (in Chinese).

Guo S.X. and Lu Z.Z., Interval Arithmetic and Static Interval Finite Element Method, *Applied Mathematics and Mechanics (English Edition)*, Vol. 20(12), 1390–1396, 2001.

Gürdal Z., Haftka R.T. and Hajela P., *Design and Optimization of Laminated Composite Materials*, New York: Wiley, 1999.

Gustafson J.L., *The End of Error: Unum Computing*, Boca Raton: Chapman and Hall, 2017.

Gutman S. and Leitmann G., Stabilizing Control for Linear Systems with Bounded Parameter and Input Uncertainty, *Proceedings, 7th IFIP Conference Optimization Techniques*, Berlin: Springer, 1976.

Gutman S., Uncertain Dynamical Systems – A Lyapunov Min-max Approach, *IEEE Transactions on Automatic Control*, Vol. 24, 437–443, 1979.

Haftka R., Review of Analysis and Optimization of Prismatic and Axisymmetric Shell Structures: Theory, Practice and Software, *AIAA Journal*, Vol. 42(2), 433–434, 2004.

Haftka R.T. and Adelman H.M., Recent Developments in Structural Sensitivity Analysis, *Structural Optimization*, Vol. 1(3), 137–151, 1989.

Haftka R.T. and Gürdal Z., *Elements of Structural Optimization*, Third revised and expanded edition,Berlin, springer Science+Business Media B.V, 2012.

Haftka R.T., Villanueva D. and Chaudhuri A., Parallel Surrogate-Assisted Global Optimization with Expensive Functions–A Survey, *Structural and Multidisciplinary Optimization*, Vol. 54(1), 3–13, 2016.

Hagedorn P. and Nascimiento N., Structural Vibrations Excited by Random Field Processes: Effect of the Cross-Correlation of the Exitation, in *Procedural of International Conference on Structural Safety and Reliability*, pp. 27–29, 1985.

Haldar A. and Reddy R.K., Random-Fuzzy Analysis of Existing Structures, *Fuzzy Sets Systems*, Vol. 48, 201–210, 1992.

Haldar A. and Mahadevan S., *Probability, Reliability, and Statistical Methods in Engineering*, New York: Wiley, 2000a.

Haldar A. and Mahadevan S., *Reliability Assessment Using Stochastic Finite Element Analysis*, New York: Wiley, 2000b.

Hall J.W., Blockley D.I. and Davis J.P., Uncertain Inference Using Interval Probability Theory, *International Journal of Approximate Reasoning*, Vol. 19, 247–264, 1998.

Hall J.W. and Lawry J., Generation, Combination and Extension of Random Set Approximations to Coherent Lower and Upper Probabilities, *Reliability Engineering & System Safety,Vol*, Vol. 85(1–3), 89–101, 2004.

Hamann D., Walz N.-P., Fischer A., Hanss M.F. and Eberhard P., Fuzzy Arithmetical Stability Analysis of Uncertain Machining Systems, *Mechanical Systems and Signal Processing*, Vol. 98, 534–547, 2018.

Hammersley J.M. and Handscomb D.C., *Monte Carlo Methods*, London: Mathuen, 1964.

Han S., Tao M., Topcu U., Owhadi H. and Murray R.M., Convex Optimal Uncertainty Quantification, *SIAM Journal on Optimization*, Vol. 25, 1368–1387, 2015.

Hanea A.M., Hemming V. and Nane G.F., Uncertainty Quantification with Experts: Present Status and Research Needs, *Risk Analysis*, Vol. 42(2), 254–263, 2022.

Hansen E.R., *A Generalized Interval Arithmetic, in Interval Mathematics* (K. Nickel ed.), Berlin: Springer, pp. 7–18, 1975.

Hansen E. and Walster G.W., *Global Optimization Using Interval Analysis*, New York: Marcel Dekker, 2003.

Hanss M., The Transformation Method for the Simulation and Analysis of Systems with Uncertain Parameters, *Fuzzy Sets and Systems*, Vol. 130(3), 277–289, 2002.

Hanss M., *Applied Fuzzy Arithmetic: An Introduction with Engineering Applications*, Berlin: Springer, 2005.

Hart G.C., *Uncertainty Analysis, Loads, and Safety in Structural Engineering*, Englewood Cliffs, N.J: Prentice Hall, 1982.

Haruyama S. and Barsky B.A., Using Stochastic Modeling for Texture Generation, *IEEE Transactions on Computer Graphics Applications*, Vol. 4(3), 7–19, 1984.

Hasofer A.M., Objective Probabilities for Unique Objects, in *Risk, Structural Engineering and Human Error* (M. Grigoriu, ed.), pp.1–16, 1984.

Hasofer A.M. and Lind N.C., Exact and Invariant Second-Moment Code Format, *Journal of the Engineering Mechanics Division*, Vol. 100, 111–121, 1974.

Hasuike T., Katagiri H. and Tsubaki H., An Interactive Algorithm to Construct an Appropriate Nonlinear Membership Function Using Information Theory and Statistical Method, *Procedia, Computer Science*, Vol. 61, 32–37, 2015.

Haugen E.B., *Probabilistic Mechanical Design*, New York: Wiley-Interscience, 1980.

Hauseux P., Hale J.S. and Bordas S.P., Calculating the Malliavin Derivative of Some Stochastic Mechanics Problems, *PloS One*, Vol. 12(12), article e0189994, 2017.

Hayes B., Computing Science: A Lucid Interval, *American Scientist*, Vol. 91(6), 484–488, 2003.

Heath T.L., *Works of Archimedes*, Cambridge Univ. Press, (Second Edition by Dover Publications 1953), 1897.

Hein P., The Overdoers, in *Grooks 2*, Garden City, N.Y: Doubleday & Company Inc, pp. 28, 1968.

Helton J.C. and Burmaster D.E., (eds.), Special Issue on the Treatment of Aleatory and Epistemic Uncertainty, in *Reliability Engineering and System Safety*, Vol. 54, 1996.

Helton J.C., Johnson J.D. and Oberkampf W.L., An Exploration of Alternative Approaches to the Representation of Uncertainty in Model Predictions, *Reliability Engineering and System Safety*, Vol. 85(1–3), 39–71, 2004.

Herrmann G., Stability of Equilibrium of Elastic Systems Subjected to Non-Conservative Forces, *Applied Mechanics Reviews*, Vol. 20, 103–108, 1967.

Hess P.E., Bruchman D., Assakkaf I.A. and Ayyub B.M., Uncertainties in Material and Geometric Strength and Load Variables, *Naval, Engineers Journal*, Vol. 114(2), 139–166, 2002.

Hessling J.P., (ed.), *Uncertainty Quantification and Model Calibration*, Rijeka, Croatia: InTech. 2017.

Heindl G., Kreinovich V. and Rifqi M., In Case of Interval (or More General) Uncertainty, No Algorithm Can Choose the Simplest Representative, *Reliable Computing*, Vol. 8(3), 213–227, 2002.

Hickey J. and Langley R., Alternative Metrics for Design Decisions Based on Separating Aleatory and Epistemic Probabilistic Uncertainties, *Mechanical Systems and Signal Processing*, Vol. 181, 109532, 2022.

Hilton H.H., Hsu J. and Kirby J.S., Linear Viscoelastic Analysis with Random Material Properties, in *Stochastic Structural Dynamics I – Now Theoretical Developments* (Y.K. Lin and I. Elishakoff eds.), Berlin: Springe, pp.83–110. 1991.

Hirota K., Fuzzy Concept Very Clear in Japan, *Asashi Evening News*, Aug 20, 1991.

Hlavacek I., Chleboun J. and Babuška I., *Uncertain Input Data Problems and the Worst Scenario Method*, Amsterdam: Elsevier, 2004.

Hoff N.J., Buckling and Stability (Forty-first Wilbur Wright Memorial Lecture, *Journal of Research of Aeronautical Society*, Vol. 58, 3–52, 1954.

Hoff N.J., Dynamic Stability of Structures, in *Dynamic Stability of Structures* (G. Hermann ed.), New York: Pergamon Press, pp. 7–77, 1965.

Hoff N.J., The Dynamics of the Buckling of Elastic Columns, *Journal of Applied Mechanics*, Vol. 18, 68–74, 1951.

Hoff N.J., Nardo S.V. and Erickson B., An Experimental Investigation of the Process of Buckling of Columns, *Procedure of Society Express Stress Analysis*, Vol. 13, 201–208, 1955.

Hoff N.J., Nardo S.V. and Erickson B., The Maximum Load Supported by an Elastic Column in a Rapid Compression Test, *Procedure of National Congress of Applied Mechanics*, New York: ASME Press, pp. 419–423, 1951.

Hofmann A., Hanss M. and Eberhard P. Possibilistic Investigation of Mechanical Control Systems under Uncertainty, in *IFToMM World Congress on Mechanism and Machine Science*, Springer, pp. 147–154, 2019.

Hohenbichler M., Gollwitzer S., Kruse W. and Rackwitz R., New Light on First-and Second-Order Reliability Methods, *Structural Safety*, Vol. 4(4), 267–284, 1987.

Holnicky M., *Reliability Analysis for Structural Design*, Stellenbosch: Sun Press, 2009.

Hose D. and Hanss M., Fuzzy Linear Least Squares for the Identification of Possibilistic Regression Models, *Fuzzy Sets and Systens*, Vol. 367, 82–95, 2019.

Hose D., Mäck M. and Hanss M., Robust Optimization in Possibility Theory, *Journal of Risk and Uncertainty in Engineering Systems, Part B: Mechanical Engineering*, Vol. 5(4), article 041001, 2019.

Hoshiya M., *Conditional Simulation of a Stochastic Field, Structural Safety and Reliability* (G.I. Schuëller, M. Shinozuka and J.T.P. Yao eds.), Vol. 1, Balkema: Rotterdam: pp. 349–353, 1994.

Hot A., Weisser T. and Cogan S., An Info-Gap Application to Robust Design of a Prestressed Space Structure under Epistemic Uncertainties, *Mechanical Systems and Signal Processing*, Vol. 91, 1-9, 2017.

Høyland A. and Raussand M., *System Reliability Theory*, New York: Wiley-Interscience, pp. 13, 1984.

Hu X., Chen X., Parks G.T. and Yao W., Review of Improved Monte Carlo Methods in Uncertainty-Based Design Optimization for Aerospace Vehicles, *Progress in Aerospace Sciences*, Vol. 86, 20–27, 2016.

Huang N.C., Nachbar W. and Nemat-Nasser S., On Willems' Experimental Verification of the Critical Load in Beck's Problem, *Journal of Applied Mechanics*, Vol. 34, 243–245, 1967.

Huang W.N. and Cozarelli F.A., Steady Creep on a Beam with Random Material Parameters, *Journal of Franklin Institution*, Vol. 294, 323–338, 1972.

Huang W.N. and Cozarelli F.A., Damped Lateral Vibration in an Axially Creeping Beam with Random Material Parameters, *International Journal of Solids Structure*, Vol. 7, 765–788, 1973.

Hull T. and Dobell A., Random Number Generators, *SIAM Review*, Vol. 4, 230–254, 1962.

Hunt G.W., Reflections and Symmetries in Space and Time, *IMA Journal of Applied Mathematics*, Vol. 76, 2–26, 2011.

Hunt M., Haley B., McLennan M., Koslowski M., Murthy J. and Strachan A., PUQ: A Code for Non-Intrusive Uncertainty Propagation in Computer Simulations, *Computer Physics Communications*, Vol. 194, 97–107, 2015.

Hurtado J.E., *Structural Reliability: Statistical Learning Perspectives*, Berlin: Springer, 2004.

Hurtado J.E. and Barbat A.H., Monte Carlo Techniques in Computational Stochastic Mechanics, *Archives of Computational Methods in Engineering*, Vol. 5(1), 3–30, 1998.

Hurtado J.E. and Barbat A.H., Improved Stochastic Linearization Method Using Mixed Distributions, *Structural Safety*, Vol. 18, 49–62, 1996.

Hutchinson J.W. and Thompson J.M.T., Imperfections and Energy Barriers in Shell Buckling, *International, Journal of Solids and Structures*, Vol. 148, 157–168, 2018.

Huynh V.N., Nakamori Y., Ono H., Lawry J., Kreinovich V. and Nguyen H.T., (eds.), *Interval/Probabilistic Uncertainty and Non-Classical Logics*, Berlin: Springer. 2008.

Ibrahim R.A., Structural Dynamics with Parameter Uncertainties, *Applied Mechanics Reviews*, Vol. 40, 309–328, 1987.

Ibrahim R.A., *Parametric Random Vibration*, Second edition, Mineola, NY:Dover Publications, 2008.

Ibrahim R.A., *Handbook of Structural Life Assessment*, Hoboken, NJ: Wiley, 2017.

Ichihashi H., Miyoshi T., Nagasaka K. and Satoh Y., Neuro-Fuzzy FEM and a Posteriori Error Estimation, *International Joint Conference, 4th IEEE International Conference for Fuzzy Systems and 2nd International Fuzzy Engineering Symposium, IEEE International Conference on Fuzzy Systems*, IEEE Press, Piscataway Vol. 4, pp. 2335–2342, 1995.

Igusa T., Critical Configurations of Systems Subjected to Wideband Excitation, in *Structural Dynamics: Recent Advances* (M. Petyt, H.F. Wolfe and C. Mei, eds.), London: Elsevier Science, pp. 178–187, 1991.

Igusa T., Buonopane S.G. and Ellingwood B.R., Bayesian Analysis of Uncertainty for Structural Engineering Applications, *Structural Safety*, Vol. 24(2–4), 165–186, 2002.

Ikeda K., Chida T. and Yanagisawa E., Imperfection Sensitive Strength Variation of Soil Specimens, *Journal of Mechanical Physical Solids*, Vol. 45, 293–315, 1997.

Ikeda K., Maruyama K., Ishida H. and Kagawa S., Bifurcation in Compressive Behavior of Concrete, *ACI Materials Journal*, Vol. 94(6), 484–490, 1997.

Imholz M., Faes M., Vandepitte D. and Moens D., Robust Uncertainty Quantification in Structural Dynamics under Scarse Experimental Modal Data: A Bayesian-Interval Approach, *Journal of Sound and Vibration*, Vol. 467, article 114983, 2020.

Impollonia N. and Elishakoff I., Behavior of Stochastic Shear Beams under Random Loading via Stochastic Variational Principles, *International Journal of Solids Structure*, Vol. 9(12), 1983–1996, 1997.

Ioakimidis N., *Symbolic Intervals for the Unknown Quantities in Simple Applied Mechanics Problems with the Computational Method of Quantifier Elimination*, Greece: University of Patras, 2019.

Ioakimidis N., *Application of the Method of Quantifier Elimination to the Determination of Intervals When the Uncertain Parameters Satisfy an Ellipsoidal Inequality Constraint*, Greece: University of Patras, 2020.

Ioakimidis N., *Application of Quantifier Elimination to Robust Reliability under Severe Uncertainty Conditions by Using the Info-Gap Decision Theory (IGDT)*, Greece: University of Patras, 2021.

Isaksson A.J., Frequency Domain Accuracy of Identified 2D Causal AR-Models, *IEEE Transactions on Signal Processing*, Vol. 42(2), 399–408, 1994.

Isii K., On Sharpness of Tchebycheff-Type Inequalities, *Annals of the Institute of Statistical Mathematics*, Vol. 14, 185–197, 1963.

Itao K. and Crandall S.H., Wide-Band Random Vibration of Circular Plates, *Journal of Mechanical Design*, Vol. 100, 690–695, 1978.

Iwan W.D. and Mason A.B. Jr, Equivalent Linearization for Systems Subjected to Non- Stationary Random Excitation, *International Journal of Non-Linear Mechanics*, Vol. 15, 71–82, 1980.

Iyengar R.N., Chaotic Behavior in Nonlinear Oscillators, *ZAMM: Zeitschrift für angewandte mathematik und mechanik*, Vol. 73, T46–T53, 1993.

Iyengar R.N. and Dash P.K., Study of the Random Vibration of Nonlinear Systems by Gaussian Closure Technique, *Journal of Applied Mechanics*, Vol. 45(2), 393–399, 1978.

Iyengar R.N. and Roy D., Extensions of the Phase Space Linearization Method, *Journal of Sound and Vibration*, Vol. 211, 877–906, 1998.

Izumi M., Zaiming L. and Kimura M., A Stochastic Linearization Technique and Its Application to Response Analysis of Nonlinear Systems Based on Weighted Least-Square Minimization, *Journal of Structural and Construction Engineering, Transaction of AIJ*, Vol. 395, 72–81, 1989.

Jackson M., *Uncertain: The Wisdom & Wonder of Being Unsure*, Lanham, MD: Prometheus Books.

Jahani E., Muhanna R.L., Shayanfar M.A. and Barkhordari M.A., Reliability Assessment with Fuzzy Random Variables Using Interval Monte Carlo Simulation, *Computer-Aided Civil and Infrastructure Engineering*, Vol. 29(3), 208–220, 2014.

Jain N. and Ramu P., L-Moments and Chebyshev Inequality Driven Convex Model for Uncertainty Quantification, *Structural and Multidisciplinary Optimization*, Vol. 65(7), 1–14, 2022.

Jaynes E.T., Where Do We Stand on Maximum Entropy?, in *The Maximum Entropy Formalism* (R.D. Levine and M. Tribus, eds.), Cambridge: M.I.T. Press, pp. 620–630, 1979.

Janis I.L., *Victims of Groupthink*, New York: Houghton Miffin, 1972.

Janon A., Nodet M. and Prieur C., Uncertainties Assessment Ing Global Sensitivity Indices Estimation from Metamodels, *International Journal for Uncertainty Quantification*, Vol. 4(1), 21–36, 2014.

Janssen A.E.M., Application of the Wigner Distribution to Harmonic Analysis of Generalized Stochastic Processes, in *Mathematical Centre Tracts* (I I A), Amsterdam: 1979.

Jaulin L., Kieffer M., Didrit O. and Walter E., *Applied Interval Analysis*, London: Springer, 2001.

Jiang C., Bi R.G., Lu G.Y. and Han X., Structural Reliability Analysis Using Non-Probabilistic Convex Model, *Computer Methods in Applied Mechanics and Engineering*, Vol. 254, 83–98, 2013.

Jiang C., Han X., Li W.X., Liu J. and Zhang Z., A Hybrid Reliability Approach Based on Probability and Interval for Uncertain Structures, *Journal of Mechanical Design*, Vol. 134(3), article 031001, 2012.

Jiang C., Han X. and Xie H.-C., *Nonlinear Interval Optimization for Uncertain Problems*, Berlin: Springer, 2020.

Jiang C., Li W.X., Han X., Liu L.X. and Le P.H, Structural Reliability Analysis Based on Random Distributions with Interval Parameters, *Computers and Structures*, Vol. 89(23–24), 2292–2302, 2011.

Jiang C., Lu G.Y., Han X. and Liu L.X., A New Reliability Analysis Method for Uncertain Structures with Random and Interval Variables, *International Journal of Mechanics and Materials in Design*, Vol. 8(2), 169–182, 2012.

Jiang C., Zhang Z., Han X. and Liu J., A Novel Evidence-Theory-Based Reliability Analysis Method for Structures with Epistemic Uncertainty, *Computers & Structures*, Vol. 129, 1–12, 2013.

Jiang C., Zhang Q.F., Han X., Liu J. and Hu D.A., Multidimensional Parallelepiped Model – A New Type of Non-Probabilistic Convex Model for Structural Uncertainty Analysis. *International, Journal for Numerical Methods in Engineering*, Vol. 103(1), 31–59, 2015.

Jiang C., Zheng J. and Han X., Probability-Interval Hybrid Uncertainty Analysis for Structures with Both Aleatory and Epistemic Uncertainties: A Review, *Structural and Multidisciplinary Optimization*, Vol. 6(57), 2485–2502, 2018.

Jiang X. and Mahadevan S., Bayesian Risk-Based Decision Method for Model Validation under Uncertainty, *Reliability Engineering and System Safety*, Vol. 92, 707–718, 2007.

Jiminez J.C., Shoi I. and Ozaki T., Simulation of Stochastic Differential Equations through the Local Linearization Method, A Comparative Study, *Journal of Statistical Physics*, Vol. 94, 587–602, 1999.

Johnson A.I., *Strength, Safety and Economical Dimension of Structures, Bulletin No. 12*, Stockholm: Royal Institute of Technology, 1953.

Johnston B.G., Buckling Behavior above the Tangent Modulus Load, *Journal of Engineering Mechanics Division*, Vol. 87, 79–99, 1961.

Kai V.Z., Leitmann G. and Garofalo F., Robustness of Uncertain Dynamical Systems with Delay without Matching Assumptions, in *Procedure of the 8th IFAC Symposium Identification and System Parameter Estimation*, Beijing, 1988.

Kalman R.E., Randomness Reexamined, *Modeling Identification and Control*, Vol. 15(3), 141–151, 1994.

Kamath C., On the Role of Data Mining Techniques in Uncertainty Quantification, *International Journal for Uncertainty Quantification*, Vol. 2(1), 73–94, 2012.

Kameda H. and Morikawa H., Interpolating Stochastic Processes for Simulation of Conditional Random Fields, *Probabilistic Engineering Mechanics*, Vol. 7(4), 242–254, 1992.

Kamga P.H.T., Li B., McKerns M., Nguyen L.H., Ortiz M., Owhadi H. and Sullivan T.J., Optimal Uncertainty Quantification with Model Uncertainty and Legacy Data, *Journal of the Mechanics and Physics of Solids*, Vol. 72, 1–19, 2014.

Kandler G., Füssl J. and Eberhardsteiner J., Stochastic Finite Element Approaches for Wood-Based Products: Theoretical Framework and Review of Methods, *Wood Science and Technology*, Vol. 49(5), 1055–1097, 2015.

Kang J.W. and Havichandran R.S., Random Vibration of Laminated FRP Plates with Material Nonlinearity Using High-Order Shell Theory, *Journal of Engineering Mechanics*, Vol. 125, 1081–1088, 1999.

Kang Z., Luo Y.J. and Li A., On Non-Probabilistic Reliability-Based Design Optimization of Structures with Uncertain-but-Bounded Parameters, *Structural Safety*, Vol. 33, 196–205, 2011.

Kanno Y., *Nonsmooth Mechanics and Convex Optimization*, Boca Raton: CRC Press, 2011.

Kapur K.S. and Lamberson L.R., *Reliability in Engineering Design*, New York: Wiley, 1977.

Kareem A. and Li Y., Simulation of Multi-Variate Stationary and Non-Stationary Random Processes: A Recent Development, in *Computational Stochastic Mechanics* (P.D. Spanos and S.A. Brebbia, eds.), New York: Elsevier Appl Science, pp. 533–544, 1991.

Karniadakis G.E., Quantifying Uncertainty in CFD, *Journal of Fluids Engineering*, Vol. 124(1), 2–3, 2002.

Karniadakis G.E., Uncertainty Quantification (UQ), *3rd Micro and Nano Flows Conference Thessaloniki*, Greece, 22-24 August 2011.

Karniadakis G.E. and Glimm J., Uncertainty Quantification in Simulation Science, *Journal of Computational Physics*, Vol. 217(1), 1–4, 2006.

Karniadakis G.E., Kevrekidis I.G., Lu L., Perdikaris P., Wang S. and Yang L., Physics-Informed Machine Learning, *Nature Reviews Physics*, Vol. 3(6), 422–440, 2021.

Katsidoniotaki M.I., Uncertainty Quantification Techniques with Diverse Applications to Stochastic Dynamics of Structural and Nanomechanical Systems and to Modeling of Cerebral Autoregulation, Ph.D. Thesis, Columbia University, New York, 2023.

Katsidoniotaki M.I., Psaros A.F. and Kougioumtzoglou I.A., Uncertainty Quantification of Nonlinear System Stochastic Response Estimates Based on the Wiener Path Integral Technique: A Bayesian

Compressive Sampling Treatment, *Probabilistic Engineering Mechanics*, Vol. 67, article 103193, 1–16, 2022.

Kay J.A. and King M.A., *Radical Uncertainty: Decision-Making beyond the Numbers*, Bridge Street Press, 2020.

Kay S.M., New ARMA Spectral Estimator, *IEEE Transactions on Acoustics, Speech, and Signal Processing*, Vol. 28(5).

Kaufmann A. and Gupta M.M., *Introduction to Fuzzy Arithmetic: Theory and Applications*, New York: Van Nostrand, 1985.

Kendall D.G., Foundations of a Theory of Random Sets, in *Stochastic Geometry* (E.F. Harding and D.G. Kendall, eds.), New York: Wiley, 1974.

Körner R., On the Variance of Fuzzy Random Variables, *Fuzzy Sets and Systems*, Vol. 92, 83–93, 1997.

Kazakov I.E., An Approximate Method for the Statistical Investigation for Nonlinear Systems, *Trudi Voenno-Vozdushnoi Inzhenernoi Akademii imeni Prof. N.E. Zhukovskogo*, Vol. 399, 1–52, 1954 (in Russian).

Kazakov I.E., An Extension of the Method of Statistical Linearzation, *Automation and Remote Control*, Vol. 59, 1682–1685, 1998.

Kazakov I.E., Approximate Probabilistic Analysis of the Accuracy of Operation of Essentially Nonlinear Systems, *Avomatika I Telemekhanika*, Vol. 17, 423–450, 1956 (in Russian).

Kelly J.M., Leitmann G. and Soldatos A.G., Robust Control of Base Isolated Structure under Earthquake Excitation, *Journal of Optimization Theory Application*, Vol. 53, 159–180, 1987.

Keese A., A Review of Recent Developments in the Numerical Solution of Stochastic Partial Differential Equations (Stochastic Finite Elements), Institute of Scientific Computing Technical University Braunschweig Brunswick, *Germany Informatikbericht Nr.: 2003–06*, 2003.

Van Keulen F., Haftka R.T. and Kim N.H., Review of Options for Structural Design Sensitivity Analysis. Part 1: Linear Systems, *Computer Methods in Applied Mechanics and Engineering*, Vol. 194(30–33), 3213–3243, 2005.

Khabbaz G.R., Power Spectral Density of the Response of a Nonlinear System to Random Excitation, *Journal of Acoustical Society of America*, Vol. 38, 847–850, 1964.

Kharitonov V., *Interval Uncertainty Structure: Conservative but Simple, Uncertainty Models and Measures* (H.G. Natke and Y. Ben-Haim, eds.), Berlin: Akademie Verlag, pp. 267–276, 1997.

Khozialov N.F., Safety Factors, *Building Industry*, Vol. 10, 840–844, 1929 (in Russian).

Kidane A., Lashgari A., Li B., McKerns M., Ortiz M., Owhadi H., Ravichandran G., Stalzer M. and Sullivan T.J., Rigorous Model-Based Uncertainty Quantification with Application to Terminal Ballistics, Part I: Systems with Controllable Inputs and Small Scatter, *Journal of the Mechanics and Physics of Solids*, Vol. 60(5), 983–1001, 2012.

Kim Y., Ovseyevich A. and Reshtnyak Y., Comparison of Stochastic and Guaranteed Approaches to the Estimation of the State of Dynamic Systems, *Journal of Computational Systematic Science Intistute*, Vol. 31, 56–64, 1993.

Kimura K., Komada M. and Sakata M., NonGaussian Equivalent Linearzation for Estimation for Response Distribution of Nonlinear Systems, *Transitions JSME, Part C*, Vol. 61(: 583), 831–835, 1995.

King M., Private Communication, 2020.

Kiureghian A.D., Structural Response to Stationary Excitation, *Journal of Engineering Mechanics*, Vol. 106, 1195–1213, 1980.

Kiureghian A.D., Response Spectrum Method for Random Vibration Analysis of MDF Systems, *Earthquake Engineering and Structural Dynamics*, Vol. 9, 419–413, 1981.

Kiureghian A.D., *Structural and System Reliability*, Cambridge, UK: Cambridge University Press, 2021.

Kiureghian A.D. and Ditlevsen O., Aleatory or Epistemic? Does It Matter?, *Structural Safety*, Vol. 31, 105–112, 2009.

Kleiber M. and Hien T.D., *Stochastic Finite Element Method*, New York: Wiley, 1993.

Klejnem J.P.C., Sensitivity Analysis Vs Uncertainty Analysis: When to Use What?, in *Predictability and Nonlinear Modelling of in Natural Sciences and Economics* (J. Grasman and G. van Straten, eds.), Dordrecht: Kluwer, pp. 322, 1994.

Klir G.J., Japanese Advances in Fuzzy Theory and Applications, *ONR Asian Office Science Information Bulletin*, Vol. 16(3), 65–74, 1991.

Klir G.J., The Many Faces of Uncertainty, in *Uncertainty Modelling and Analysis: Theory and Applications* (B.M. Ayyub and M.M. Gupta, eds.), Amsterdam: Elsevier Science, pp.3–19, 1994.

Klir G.J., Principles of Uncertainty: What are They? Why Do We Need Them?, *Fuzzy Sets and Systems*, Vol. 74, 15–31, 1995.

Klir G.J., *Uncertainty and Information. Foundations of Generalized Information Theory*, Hoboken, NJ: Wiley, 2005.

Klir G.J. St, Clair U.H. and Yuan B., *Fuzzy Set Theory: Foundations and Applications*, New York: Prentice Hall, 1997.

Klir G.J. and Yuan B., *Fuzzy Sets and Fuzzy Logic -Theory and Applications*, Upper Saddle River, NJ: Prentice Hall, 1995.

Kloeden P.E. and Platen E., *Numerical Solution of Stochastic Differential Equations*, Berlin: Springer, 2013.

Kogan J., Personal communication, 1980.

Koiter W.T., Elastic Stability, 28th Ludwig Prandtl Memorial Lecture, *Zeitschrift für Flugwissenschaften und Weltraumforschung*, Vol. 9, 205–220, 1985.

Koiter W.T., Foundations, and Basic Equations of Shell Theory: A Survey of Recent Progress, in *Theory of Thin Shells* (F.I. Niordson, ed.), Berlin: Springer, pp.93–105, 1969.

Koiter W.T., On the stability of elastic equilibrium (in Dutch), Ph.D. thesis, Delft University of Technology, H.J. Paris, Amsterdam; English translations (a) NASA-TTF-10, pp. 833, 1967; (b) AFFDL-TR-70-20, 1970 (translated by E. Riks).

Koiter W.T., Personal Communication, March 28, 1994.

Koiter W.T., *Stark en Steifheid*, Rotterdam: Balkema Publishers, 1972 (in Dutch.

Koiter W. T., The Effect of Axisymmetric Imperfections on the Buckling of Cylindrical Shells under Axial Compression, *Procedure Kon Ned Akad Wet*, Amsterdam, Ser. B Vol. 6, 265–279, 1963. Also, Lockheed Missiles and Space Co. Report 6-90-63-86, Palo Alto, CA, 1963.

Kolkka R.W., On the Linear Beck's Problem with External Damping, *International Journal of Non-Linear Mechanics*, Vol. 14, 497–505, 1984.

Kolmogorov A.N., Grundbegriffe der Wahrscheinlichkeitsrechnung, (Foundations of Theory of Probability), Ergeb. Math. Ihrer Grenzgeb., 2 (3) (English translation: Chelsea Publishing Co., New York, 1956), 1933.

Kolovskii M.Z., Estimating the Accuracy of Solutions Obtained by the Method of Statistical Linearization, *Automation and Remote Control*, Vol. 27, 1692–1701, 1966.

Koltunov M.A., *Creep and Relaxation*, Moscow: Visshaya Shkola, 1976 (in Russian.

Kornev V.M., On the Instability Modes of Elastic Shells under Intensive Loading, *Isvestiya Akademii Nauk SSSR Mekhanika Tverdogo Tela*, Vol. 2, 129–135, 1969 (in Russian).

Kosko B., Fuzziness vs. Probability, *International Journal of General Systems*, Vol. 17(2-3), 211-240,1990.

Kosko B., The Probability Monopoly, *IEEE Transactions on Fuzzy Systems*, Vol. 2(1), 32–33, 1994.

Kosko B., *Fuzzy Thinking: The New Science of Fuzzy Logic*, Hyperion, 1994.

Kosko B., *Fuzzy Engineering*, Prentice Hall, 1997.

Kosko B., *Fuzzy Future: From Society and Science to Heaven in a Chip*, Harmony Books, 1999.

Kosheleva O., Kreinovich V. and Nguyen T.N., Why Triangular Membership Functions are Successfully Used in f-Transform Applications: A Global Explanation to Supplement the Existing Local Ones, *Axioms*, Vol. 8(3), 95, 2019.

Kotz S. and Stroup D.F., *Educated Guessing, How to Cope in an Uncertain World*, New York: Marcel Dekker, 1983.

Kozin F., Comments upon the Technique of Statistical Linearization, *Proceedings Japan Association of Automatic Control Engineers*, Annual Convention, pp. 15.1–15.10, 1965.

Kozin F., Parameter Estimation and Statistical Linearization, in *Stochastic Structural Mechanics* (Y.K. Lin and G.I. Schuëller, eds.), Berlin: Springer, pp. 259–267, 1987a.

Kozin F., The Method of Statistical Linearization for Non-Linear Stochastic Vibrations, *Nonlinear Stochastic Dynamic Engineering Systems* (F. Ziegler and G.I. Schuëller, eds.), Berlin: Springer, pp. 45-56, 1987b.

Kovalenko I.N., *Analysis of Rare Events in Evaluation of System Effectiveness and Reliability*, Moscow: Sovyetskoe Radio, 1980 (in Russian.

Kovalenko I.N. and Kuznetsov N.Y., *Methods for Evaluation of Highly Reliable Systems*, Moscow: Radio I Svyaz, 1988 (in Russian.

Kovalenko I.N., Kuznetsov N.Y. and Pegg P.A., *Mathematical Theory of Reliability of Time Dependent Systems with Practical Applications*, New York: Wiley, 1988.

Kozlov B.A. and Ushakov I.A., *Reliability Handbook*, New York: Holt, Rinehart and Winston, 1970.

Köylüoğlu H.U., Cakmak A.S. and Nielsen S.R.K., Interval Algebra to Deal with Pattern Loading and Structural Uncertainties, in *Computational Stochastic Mechanics* (P.D. Spanos, ed.), Rotterdam: Balkema, pp. 125–133, see also *Journal of Engineering Mechanics* Vol. 121, 1149–1157, 1995.

Köylüoglu H.U., Cakmak A.S. and Nielsen S.R.K., Interval Algebra to Deal with Pattern Loading and Structural Uncertainties, *Journal of Engineering Mechanics*, Vol. 121(11), 1149–1157, 1995.

Köylüoglu H.U. and Elishakoff I., A Comparison of Stochastic and Interval Finite Elements Applied to Shear Frames with Uncertain Stiffness Properties, *Computers and Structures*, Vol. 67, 91–98, 1998.

Krakovskii M.B., Determination of Structural Reliability by Methods of Statistical Simulation, *Stroitelnaya Mekhanika i Raschet Sooruzhenii (Structural Mechanics and Analysis)*, Vol. 2, 1-13, 1982, (in Russian).

Krée P. and Soize C., *Mécanique Aleatoire*, Paris: Dunod, 1983 (in French.

Krée P. and Soize C., *Mathematics of Random Phenomena: Random Vibrations of Mechanical Structures*, Dordrecht: Reidel, 2012.

Kreiner J.H. and Putcha C.S., Safety Analysis of Tension Elements Using Various Reliability Methods, *Procedure of 3rd International Symposium on Uncertainty Model Analysis, Annual Conference of North American Fuzzy Informational Process of Society*, pp.758–763, IEEE Press.

Krenk S. and Clausen J., Calibration of ARMA Processes for Simulation, in *Reliability and Optimization of Structural Systems* (P. Thoft-Christensen, ed.), Berlin: Springer, 1987.

Kreinovich V., Decision Making under Interval Uncertainty (and Beyond), in *Human-Centric Decision-Making Models for Social Sciences*, Berlin: Springer, pp.163–193, 2014.

Kreinovich V., Solving Equations (and Systems of Equations) under Uncertainty: How Different Practical Problems Lead to Different Mathematical and Computational Formulations, *Granular Computing*, Vol. 1(3), 171–179, 2016.

Kreinovich V., Beck J., Ferregut C., Sanchez A., Keller G.R., Averill M. and Starks S.A., Monte-Carlo-Type Techniques for Processing Interval Uncertainty, and Their Potential Engineering Applications, *Reliable Computing*, Vol. 13(1), 25–69, 2007.

Kreinovich V. and Longpré L., Fast Quantum Algorithms for Handling Probabilistic and Interval Uncertainty, *Mathematical Logic Quarterly: Mathematical Logic Quarterly*, Vol. 50(4-5), 405–416, 2004.

Kreinovich V., Pownuk A. and Kosheleva O., Combining Interval and Probabilistic Uncertainty: What Is Computable?, in *Advances in Stochastic and Deterministic Global Optimization*, Cham: Springer, pp.13–32, 2016.

Kreinovich V. and Xiang G., Fast Algorithms for Computing Statistics under Interval Uncertainty: An Overview, in *Interval/Probabilistic Uncertainty and Non-Classical Logics* (V.-N. Nuynh, Y. Nakamori, H. Ono, J. Lawry, V. Kreinovich and H.T. Nguyen, eds.), Berlin: Springer, pp.19–31, 2008.

Kroese D.P., Brereton T., Taimre T. and Botev Z.I., Why the Monte Carlo Method Is so Important Today, *Wiley Interdisciplinary Reviews: Computational Statistics*, Vol. 6(6), 386–392, 2014.

Kroese D.P., Taimre T. and Botev Z.I., *Handbook of Monte Carlo Methods*, Hoboken, NJ: Wiley, 2013.

Kruskal M.D., *Asymptotology: Mathematical Methods in Physical Sciences*, Englewood Cliffs: Prentice Hall, pp. 17–48, 1963.

Kruskall M.D., *Asymptotology: Mathematical Methods in Physical Sciences*, Englewood Cliffs: Prentice Hall, 1963.

Kubota Y., Sekimoto S. and Dowell E.H., High Frequency Response of a Plate Carrying a Concentrated Mass, *Journal of Sound Vibration*, Vol. 138, 321–333, 1990.

Kulpa Z., Pownuk A. and Skalna I., Analysis of Linear Mechanical Structures with Uncertainties by Means of Interval Methods, *Computer Assisted Mechanics and Engineering Sciences*, Vol. 5, 443–477, 1998.

Kurosawa A. (Director) and Akutogawa R. (Writer), Rashomon [Motion Picture], Daiei Film Corp. Ltd, Japan, 1950.

Kurzhanski A and Valyi I, *Ellipsoidal Calculus for Estimation and Control*, Boston: Birkhäuser, 1997.

Kurzhanskii A.B., Dynamic Control System Estimation under Uncertainty Conditions, *Probabilistic Continuous Information Theory*, Vol. 9, 1980.

Ladevèze P., Puel G. and Romeuf T., Lack of Knowledge in Structural Model Validation, *Computer Methods in Applied Mechanics and Engineering*, Vol. 195(37–40), 4697–4710, 2006.

Langley A.I. and Taylor P.H., Chladni Patterns in Random Vibration, *International Journal of Engineering Science*, Vol. 17, 1039–1047, 1979.

Langley R.S., Spatially Averaged Frequency Response Envelopes for One-and Two-Dimensional Structural Components, *Journal of Sound Vibration*, Vol. 178, 483–500, 1994.

Langley R.S. and Bardell N.S., A Review of Current Analysis Capabilities Applicable to the High Frequency Vibration Prediction of Aerospace Structures, *The Aeronautical Journal*, Vol. 102(1015), 287–297, 1998.

Langley R.S., Unified Approach to Probabilistic and Possibilistic Analysis of Uncertain Systems, *Journal of Engineering Mechanics*, Vol. 126(11), 1163–1172, 2000.

Laura P.A.A., Private Communication, January 31, 2000.

Laura P.A.A., Paloto J.C. and Santos R.D., A Note on the Vibration and Stability of A Circular Plate Elastically Restrained against Rotation, *Journal of Sound and Vibration*, Vol. 41(2), 177–180, 1975.

Layon R.H., *Statistical Energy Analysis of Dynamical Systems: Theory and Applications*, Cambridge: M.I.T. Press, 1975.

Lechner J.A., Leigh S.D. and Simiu E., Recent Approaches to Extreme Value Estimation with Application to Wind Speeds. Part I: The Pickands Method, *Journal of Wind Engineering and Industrial Aerodynamics*, Vol. 41(1–3), 509–519, 1992.

Le Guin U.K., *The Left Hand of Darkness*, New York: Penguin Books, 2016.

Le Maitre O.P. and Knio O.M., *Spectral Methods in Uncertainty Quantification: With Applications to CFD*, Berlin: Springer, 2010.

Lee I., Lee U., Ramu P., Yadav D., Bayrak G. and Acar E., Small Failure Probability: Principles, Progress and Perspectives, *Structural and Multidisciplinary Optimization*, Vol. 65(11), 1–34, 2022.

Lee J., Improving the Equivalent Linearization Technique for Stochastic Duffing Oscillators, *Journal of Sound and Vibration*, Vol. 186, 845–855, 1995.

Lee C.S. and Leitmann G., Continuous Feedback Guaranteeing Uniform Ultimate Boundedness for Uncertain Delay Linear Systems: An Application to River Pollution Control, in *Proceedings, International Conference Systems Science and Engineering*, Beijing, 1988.

Lee C.S. and Leitmann G., On Optimal Long-Term Management of Some Ecological Systems Subject to Uncertain Disturbances, *International Journal of Systems Science*, Vol. 14, 979–999, 1983.

Lee S.H. and Chen W., A Comparative Study of Uncertainty Propagation Methods for Black-Box-Type Problems, *Structural and Multidisciplinary Optimization*, Vol. 37, 239–253, 2009.

Lee X.X. and Chen J.Q., A Linearization Technique for Random Vibrations of Nonlinear, in *Optimal Stochastic Control Schemes within a Structural Reliability Framework* (B.J. Leira), Cham: Springer, 2013.

Leissa A.W., On a Curve Veering Aberration, *ZAMP: Zeitschrift für angewandte Mathematik und Physik*, Vol. 25, 99–111, 1974.

Leissa A.W., Private Communication, April 18, 2000.

Leissa A.W. and Narita Y., Natural Frequencies of Simply Supported Circular Plates, *Journal of Sound and Vibration*, Vol. 70(2), 221–229, 1980.

Leitmann G., Guaranteed Asymptotic Stability for Some Linear Systems with Bounded Uncertainties, *Journal of Dynamical Systems Measurement Control*, Vol. 101, 213–216, 1979.

Leitmann G., One Approach to the Control of Uncertain Systems, *Journal of Dynamical Systems Measurement and Control*, Vol. 115, 373–380, 1993.

Leitmann G., Lee C.S. and Chen Y.H., Decentralized Control for an Uncertain Multi-Reach River System, in *Optimal Control, Lecture Notes in Control and Information Systems*, Berlin: Springer, 1986.

Lemaire M., *Mechanics and Uncertainty*, London: ISTE-Wiley, 2014.

Lemaire M., Chateauneuf A. and Mitteau J.C., *Structural Reliability*, London: ISTE-Wiley, 2009.

Leporati E., *The Assessment of Structural Safety*, Forest Groves, Oregon: Research Studies Press, 1977.

Lessing G.E., *Nathan der Weise, a Drama in Five Acts* (see also Nathan the Wise, with Related Documents, (R. Schechter, ed.), New York: Bedford/St. Martin's), 2005, 1779.

Leunberger D.G., *Introduction to Linear and Nonlinear Programming*, Reading, MA: Addison-Wesley, 1984.

Levi R., Calculs Probabilistes de la Securite Des Constructions, *Ann. Ponts Et Chaussess*, Vol. 119(4), 493–539, 1949 (in French).

Levinson M., Consistent and Inconsistent Higher Order Beam and Plate Theories: Some Surprising Comparisons, in *Refined Dynamical Theories of Beams, Plates and Shells and Their Applications* (I. Elishakoff and H. Irretier, eds.), Berlin: Springer, pp. 112–130, 1987.

Lew J.S. and Horta L.G., Uncertainty Quantification Using Interval Modeling with Performance Sensitivity, *Journal of Sound and Vibration*, Vol. 308(1–2), 330–336, 2007.

Lewis J.P., Generalized Stochastic Subdivision, *ACM Transactions in Graphics*, Vol. 6(3), 167–190, 1987.

Li J., Probability Density Evolution Method: Background, Significance and Recent Developments, *Probabilistic Engineering Mechanics*, Vol. 44, 111–117, 2016.

Li J. and Chen J.-B., Probability Density Evolution Method for Dynamic Response Analysis with Uncertain Parameters, *Computational Mechanics*, Vol. 34, 400–409, 2004.

Li J. and Chen J.-B., *Stochastic Dynamics of Structures*, Singapore: Wiley, 2009.

Li X., Zhu H., Chen Z., Ming W., Cao Y., He W. and Ma J., Limit State Kriging Modeling for Reliability-Based Design Optimization through Classification Uncertainty Quantification, *Reliability Engineering and System Safety*, Vol. 224, article 108539, 2022.

Li Y., Chen J. and Feng L., Dealing with Uncertainty: A Survey of Theories and Practices, *IEEE Transactions on Knowledge and Data Engineering*, Vol. 25(11), 2463–2482, 2012.

Li Y. and Kareem A., Simulation of Multi-Variate Nonstationary Random Processes by FFT, *Journal of Engineering Mechanics*, Vol. 117, 1991.

Li Y., Anitescu M., Roderick O. and Hickernell F., Orthogonal Bases for Polynomial Regression with Derivative Information in Uncertainty Quantification, *International Journal for Uncertainty Quantification*, Vol. 1(4), 2011.

Li Y.Q. and Zhao G.F., Unified Model for Structural Reliability Analysis Based on Fuzzy Random Probabilistic Theory, *Journal of Dalian University of Technology*, Vol. 35(4), 528–531, 1995 (in Chinese).

Li W., Chang C.W. and Tseng S., The Linearization Method Based on the Equivalence of Dissipated Structural Systems, *Journal of Sound and Vibration*, Vol. 295, 797–809, 2006.

Li Y.W., Elishakoff I., Starnes J.H. Jr and Shinozuka M., Prediction of Natural Frequency, and Buckling Load Variability due to Uncertainty in Material Properties by Convex Modeling, *Fields Institute Communication*, Vol. 9, 139–154, 1996a.

Li Y.W., Elishakoff I., Starnes J.H. and Shinozuka M., Variability of Buckling Loads by Convex Modelling, *Fields Institute Communications, American Mathematical Society*, Vol. 9, 139–154, 1996b.

Liang B. and Mahadevan S., Error and Uncertainty Quantification and Sensitivity Analysis in Mechanics Computational Models, *International Journal for Uncertainty Quantification*, Vol. 1(2), 147–161, 2011.

Liang J.W. and Feeny B., Balancing Energy to Estimate Damping Parameters in Forced Oscillators, *Journal of Sound and Vibration*, Vol. 295, 988–998, 2006.

Liang Z. and Lee G.C., *Random Vibration: Mechanical, Structural and Earthquake Engineering Applications*, Boca Raton: CRC Press, 2015.

Liesecke L., Jonscher C., Grießmann T. and Rolfes R., Investigations of Mode Shapes of Closely Spaced Modes from a Lattice Tower Identified Using Stochastic Subspace Identification, In *International Conference on Experimental Vibration Analysis for Civil Engineering Structures*, pp. 529–538, Cham: Springer Nature Switzerland, 2023.

Lin G., Engel D.W. and Eslinger P.W., *Survey and Evaluate Uncertainty Quantification Methodologies*, Richland, WA: Pacific Northwest National Lab.(PNNL), 2012.

Lin Y.K., Thoughts on Randomizing Scanlan's Theory of Bridge Dynamics, *Procedure of 80th Birthday Symposium in Honor of Professor Robert H Scanlan*, Baltimore, MD: Johns Hopkins University, pp 229–238, 1994.

Lin Y.K., Nonstationary Excitation and Response in Linear Systems Treated as Sequences of Random Pulses, *Journal of Acoustical America*, Vol. 38, 453–460, 1965.

Lin Y.K., *Probabilistic Theory of Structural Dynamics*, New York: McGraw-Hill, New, 1976.

Lin Y.K. and Cai G.Q., *Probabilistic Structural Dynamics: Advanced Theory and Applications*, New York: McGraw-Hill, 1995.

Lin Y.K., Kozin F., Wen Y.K., Casciati F., Schuëller G.I., Der Kiureghian A., Ditlevsen O. and Vanmarcke E., Methods of Stochastic Structural Dynamics, *Structural Safety*, Vol. 3, 167–194, 1986.

Lind N., On the Value of Life and Limb, *Report DIALOG 1-78*, Danish Engineering Academy, Lyngby, Denmark, 1978.

Lindberg H., An Evaluation of Convex Modeling for Multimode Dynamic Buckling, *Journal of Applied Mechanics*, Vol. 59, 929–936, 1992a.

Lindberg H., Convex Models for Uncertain Imperfection Control in Multimode Dynamic Buckling, *Journal of Applied Mechanics*, Vol. 59, 937–945, 1992b.

Lindberg H.E., Impact Buckling of a Thin Bar, *Journal of Applied Mechanics*, Vol. 32, 315–322, 1965.

Lindley D.V., The Probability Approach to the Treatment of Uncertainty in Artificial Intelligence and Expert Systems, *Statistical Science*, Vol. 2, 17–24, 1987.

Liu B., *Theory and Practice of Uncertain Programming*, Berlin: Springer-Verlag, 2009.

Liu B., *Uncertain Programming*, New York: Wiley-Interscience, 1999.

Liu B. and Munson D.C., Generation of a Random Sequence Having a Jointly Specified Marginal Distribution and Auto-Covariance, *IEEE Transactions on Acoustics, Speech and Signal Processing*, Vol. 30(6), 973–983, 1982.

Liu X.T. and Rao S.S., Vibration Analysis in the Presence of Uncertainties Using Universal Grey System Theory, *Journal of Vibration and Acoustics*, Vol. 140(3), article 031009, 2018.

Liu -X.-X. and Elishakoff I., A Combined Importance Sampling and Active Learning Kriging Model for Reliability Analysis with Random and Correlated Interval Variables, *Structural Safety*, Vol. 82, article 101875, 2020.

Louchard G. and Latouche G., (eds.), *Probability Theory and Computer Science*, Academic Press, p. IX, 1983.

Lomakin V.A., *Statistical Problems of Mechanics of Solid Deformable Bodies*, Moscow: "Nauka" Publishing House, 1970 (in Russian.

Lomakin V.A. and Tunguskova Z.G., Statistical Methods in Polymer Mechanics (Current Status and Future Aspects, *Polymer Mechanics*, Vol. 13(3), 361–363, 1977.

Lombardi M. and Haftka R.T., Anti-Optimization Technique for Structural Design under Load Uncertainties, *Computer Methods in Applied Mechanics and Engineering*, Vol. 157(1–2), 19–31, 1998.

Long X.Y., Mao D.L., Jiang C., Wei F.Y. and Li G., Unified Uncertainty Analysis under Probabilistic, Evidence, Fuzzy and Interval Uncertainties, *Computer Methods in Applied Mechanics and Engineering*, Vol. 355, 1–26, 2019.

Lorkowski J., Kreinovich V. and Aliev R., Towards Decision Making under Interval, Set-Valued, Fuzzy, and Z-Number Uncertainty: A Fair Price Approach, in *IEEE International Conference on Fuzzy Systems (FUZZ-IEEE) 2014*, 2244–2253, 2014.

Lotto B., *Deviate: The Science of Seeing Differently*, New York: Hachette Books, 2017.

Lovelace A.M., Keynote Address, in *International Conference on Structural Safety and Reliability* (A.M. Freudenthal, ed.), Oxford: Pergamon Press, pp. 4, 1972.

Luo Y.J., Kang Z., Luo Z. and Alex L., Continuum Topology Optimization with Non-Probabilistic Reliability Constraints Based on Multi Ellipsoid Convex Model, *Structural and Multidisciplinary Optimization*, Vol. 39(3), 297–310, 2008.

Lutes L.D. and Sarkani S., *Random Vibration: Analysis of Structural and Mechanical Systems*, Amsterdam: Elsevier, pp. 423–424 437–438, 2004.

Lyon R.H., *Statistical Energy Analysis of Dynamical Systems: Theory and Applications*, Cambridge, MA: MIT Press, 1975.

Ma X.-F. and Li T.-J., Dynamic Analysis of Uncertain Structures Using an Interval-Wave Approach, *International Journal of Applied Mechanics*, Vol. 10(2), article 1850021, 2018.

Mäck M. and Hanss M., Uncertainty Analysis of a Car Crash Scenario Using a Possibilistic Multi-Fidelity Scheme, in *Proceedings of the 3rd International Conference on Uncertainty Quantification in Computational Sciences and Engineering (UNCECOMP)*, ECCOMAS Procedia, paper No. 18645, Crete, Greece, 2019.

Mäck M. and Hanss M., Efficient Possibilistic Uncertainty Analysis of a Car Crash Scenario Using a Multifidelity Approach, *Journal of Risk and Uncertainty in Engineering Systems, Part B: Mechanical Engineering*, Vol. 5(4), article 041015, 2019.

Madsen H.O., Krenk S. and Lind N.C., *Methods of Structural Safety*, Englewood Cliffs: Prentice Hall, 1986, (see also the reissue: Mineola, NY: Dover Publications, 2006).

Maes M.A. and Breitung K., Reliability-Based Tail Estimation, in *Probabilistic Structural Mechanics: Advances in Structural Reliability Methods 1994*, Berlin: Springer, pp. 335–346, 1994.

Maes M.A. and Faber M.H., Bayesian Framework for Managing Preferences in Decision Making, *Reliability Engineering and System Safety*, Vol. 91(5), 556–569, 2006.

Maes M.A. and Faber M.H., Issues in Utility Modeling and Rational Decision Making, in *Proceedings of 11th IFIP WG 7.5 Reliability and Optimization of Structural Systems* (M. Maes and L. Huyse, eds.), London: Taylor and Francis Group, pp. 95–104, 2004.

Maes M.A. and Huyse L., Tail Effects of Uncertainty Modeling in QRA, in *Proceedings of 3rd International Symposium on Uncertainty Modeling and Analysis and Annual Conference of the North American Fuzzy Information Processing Society*, pp. 133–138, IEEE Press, 1995.

Maglaras G., Experimental Comparison of Probabilistic Methods and Fuzzy Sets for Designing under Uncertainty, *PhD Dissertation*, Department of Aerospace Engineering VPI, Blacksburg, 1995.

Maglaras G., Nikolaidis E., Haftka R.T. and Cudney H.H., Analytical – Experimental Comparison of Probabilistic Methods and Fuzzy Set Based Methods for Designing under Uncertainty, *Structural Optimization*, Vol. 13(2–3), 69–80, 1997.

Maglaras G., Nikolaidis E., Haftka R.T., Cudney H.H. and Evink J., Analytical-Experimental Comparison of Probabilistic Methods and Fuzzy Sets for Designing under Uncertainty, *Proc of 37th AIAA/ASME/ASCE/AHS/ASC Structures, Structural Dynamics, and Materials Conference Part 1*, pp. 2530–2540, 1996.

Maglaras G., Ponslet E., Haftka R.T., Nikolaidis E., Sensharma P. and Cudney H.H., Analytical and Experimental Comparison of Probabilistic and Deterministic Optimization, *AIAA Journal*, Vol. 34(7), 1512–1518, 1996.

Mahadevan S., Uncertainty Quantification for Decision-Making in Engineered Systems, in *Proceedings of the International Symposium on Engineering under Uncertainty: Safety Assessment and Management (ISEUSAM-2012)* (pp. 97–117). Springer, India, 2013.

Majumder L. and Rao S.S., Interval-Based Multi-Objective Optimization of Aircraft Wings under Gust
    Loads, *AIAA Journal*, Vol. 47(3), 563–575, 2009.
Makhoul I., Linear Prediction: A Tutorial Review, *Proceedings, JEEE*, Vol. 63(4), 561–580, 1975.
Mandal S., Witz J.A. and Lyons G.J., Reduced Order ARMA Spectral Estimation of Ocean Waves, *Applied
    Ocean Research*, Vol. 14, 303–312, 1992.
Mandelbrot B.B., *Fractal Geometry of Nature*, New York: Freeman, 1982.
Mandelbrot B.B. and van Ness J.W., Fractional Brownian Motions, Fractional Noises and Applications, *SIAM
    Reviews*, Vol. 10(4), 722–737, 1968.
Manevich L.I., Koblik S.G. and Pavlenko A.V., *Asymptotic Method in the Theory of Elasticity of Orthotropic Body*,
    Kiev: " Naukova Dumka" Publishers, 1981, (in Russian).
Manohar C.S. and Ibrahim R.A., Progress in Structural Dynamics with Stochastic Parameter Variations:
    1987-1998, *Applied Mechanics Reviews*, Vol. 52, 177–197, 1999.
Mantoglou A. and Wilson J.L., The Turning Bands Method for Simulation of Random Fields Using Line
    Generation by a Spectral Method, *Water Resources Research*, Vol. 18(5), 1379–1394, 1982.
Marseguerra M. and Zio E., *Basics of the Monte Carlo Method with Application to System Reliability*, Germany:
    LiLoLe-Verlag GmbH Hagen, 2002.
Marchante E.M., Monte Carlo Simulation Challenges in Structural Mechanics: An Approach with
    PROMENVIR, in *Advances in Safety and Reliability* (C. Guedes Soares, ed.), New York: Pergamon,
    pp. 1287–1292, 1997.
Marek P Personal Communication, 19 May 2000.
Marek P., Monte Carlo Simulation – Powerful Tool in Designer's Hands, in *Monte Carlo Simulation*
    (G.I. Schuëller and P.D. Spanos, eds.), Lisse: Balkema, pp.613–618, 2001.
Marek P., Guštar M. and Anagnos Th., *Simulation-Based Reliability Assessment for Structural Engineers*, Boca
    Raton: CRC Press, 1995.
Marek P., Brozzetti J. and Guštar M., (eds.), *Probabilistic Assessment of Structures Mechanik: Applied
    Mathematics and Mechanicssing Monte Carlo Simulations*, Prague: Academy Science, Czech
    Republic, 2001.
Marelli S. and Sudret B., UQLab: A Framework for Uncertainty Quantification in MATLAB, in *Vulnerability,
    Uncertainty, and Risk: Quantification, Mitigation, and Management*, 2554–2563, 2014.
Marti K., *Stochastic Optimization Methods*, Berlin: Springer, 2005.
Marti K., (ed.), *Stochastic Optimization: Numerical Methods and Technical Applications*, Berlin: Springer, 2012.
Martinez J.R., Bishay P.L., Tawfik M.E. and Sadek E.A., Reliability Analysis of Smart Laminated Composite
    Plates under Static Loads Using Artificial Neural Networks, *Heliyon*, Vol. 8(12), article e11889, 2022.
Marzouk Y.M. and Najm H.N., Dimensionality Reduction and Polynomial Chaos Acceleration of Bayesian
    Inference in Inverse Problems, *Journal of Computational Physics*, Vol. 228(6), 1862–1902, 2009.
Mascagni M., Monte Carlo Methods: Early History and the Basics, *Technical Report*, Florida State University,
    2011, available at http://www.cs.fsu.edu/~mascagni/MC_RNG_Basics.pdf, accessed
    on October 3, 2022.
Matheron, Intrinsic Random Functions and Their Applications, *Advances in Applied Probability*, Vol. 5,
    439–468, 1973.
Matheron G., *Random Sets and Integral Geometry*, New York: Wiley, 1995.
Matsousek M. and Schneider J., Undersuchungen zur Struktur des Sicherheitsproblem bei Bauwerken,
    Bericht No. 59, Institut für Baustatik und Konstruktion, Zurich: Birkhäuser, 1976 (in German).
Matsumura T. and Haftka R.T., Reliability Based Design Optimization Modeling Future Redesign with
    Different Epistemic Uncertainty Treatments, *Journal of Mechanical Design*, Vol. 135(9), article
    091006, 2013.
Matthies H.G., Uncertainty Quantification with Stochastic Finite Elements, in *Encyclopedia of Computational
    Mechanics* (E. Stein, *et al* ed.), Chapter 27, Chichester: Wiley, 2004.

Matthies H.G., Stochastic Finite Elements: Computational Approaches to Stochastic Partial Differential Equations, *ZAMM-Journal of Applied Mathematics and Mechanics/Zeitschrift für Angewandte Mathematik und Mechanik: Applied Mathematics and Mechanics*, Vol. 88(11), 849–873, 2008.

Matthies H.G., Brenner C.E., Bucher C.G. and Soares C.G., Uncertainties in Probabilistic Numerical Analysis of Structures and Solids – Stochastic Finite Elements, *Structural Safety*, Vol. 19(1997), 283–336, 1997.

Maugin G.A., Sixty Years of Configurational Forces (1950–2010), *Mechanics Research Communications*, Vol. 50, 39–49, 2013.

Mayer H., *Die Sicherheit der Bauwerke and Ihre Berechnung nach Grenzkräften anstatt nach Zulässigen Spannungen*, Berlin: Springer, 1926, (in German) (see also paper by Tichy, M., Max Mayer-Begründer der Berechnungsmethode nach Grenzzuständen, Technishe Mechanik, Magdeburg, Federal Republic of Germany, 1990.

Maymon G., *Some Engineering Applications in Random Vibrations & Random Structures*, Washington, DC: AIAA Press, 1998.

Maymon G., *Structural Dynamics and Probabilistic Analysis for Engineers*, Amsterdam: Elsevier, 2008.

Maymon G., New Approach to the Reliability Verification of Aerospace Structures, in *Modern Trends in Structural and Solid Mechanics 3: Non-deterministic Mechanics* (N. Challamel, J. Kaplunov and I. Takewaki, eds.), London: ISTE-Wiley, pp.77–94, 2021.

McCracken D.D., The Monte Carlo Method, *Scientific American*, Vol. 192(5), 90–97, 1955.

McGrayne S.B., *The Theory that Would Not Die: How Bayes' Rule Cracked the Enigma Code, Hunted down Russian Submarines, and Emerged Triumphant from Two Centuries of Controversy*, New Haven: Yale University Press, 2011.

McWilliam S., Anti-Optimisation of Uncertain Structures using Interval Analysis, *Computers and Structures*, Vol. 79(4), 421–430, 2001.

Mead D.J., Vibration Response and Wave Propagation in Periodic Structures, *Journal of Engineering Industry*, Vol. 93, 783–792, 1971.

Mead D.M., Wave Propagation in Continuous Periodic Structures: Research Contributions from Southampton, 1964–1995, *Journal of Sound and Vibration*, Vol. 190(3), 495–524, 1996.

Mei C. and Wentz K.R., Large Amplitude Random Response of Angle Ply Laminated Composite Plates, *AIAA Journal*, Vol. 20, 1450–1458, 1982.

Mei C. and Wolfe H.F., On Large Deflection Analysis in Acoustic Fatigue Design, in *Random Vibration-Status and Recent Developments* (I. Elishakoff and R.H. Lyon, eds.), Amsterdam: Elsevier, pp. 279–302, 1986.

Meng Z., Zhao J., Chen G. and Yang D., Hybrid Uncertainty Propagation and Reliability Analysis using Direct Probability Integral Method and Exponential Convex Model, *Reliability Engineering & System Safety*, Vol. 228, article 108803, 2022.

Meirovich L., *Elements of Vibration Analysis*, New York: McGraw Hill, 1986.

Mejia and Rodriguez-Iturbe, Synthesis of Random Fields from the Spectrum: An Application to the Generation of Hydrologic Spatial Processes, *Water Resources Research*, Vol. 10(4), 705–711, 1974.

Melchers R.E., *Structural Reliability and Predictions*, London: Ellis Horwood, 1987.

Melchers R.E., *Structural Reliability and Predictions*, New York: Wiley, 1999.

Melchers R.E. and Beck A.T., *Structural Reliability Analysis and Prediction*, Third edition, Hoboken, NJ: Wiley, 2018.

Mesogitis T.S., Skordos A.A. and Long A.C., Uncertainty in the Manufacturing of Fibrous Thermosetting Composites: A Review, *Composites Part A: Applied Science and Manufacturing*, Vol. 57, 67–75, 2014.

Metropolis N., Monte Carlo: In the Beginning and Some Great Expectations, in *Monte-Carlo Methods and Applications in Neutronics, Photonics and Statistical Physics*, Berlin: Springer, pp. 62–70, 1985.

Metropolis N., The Beginning of the Monte Carlo Method, *Los Alamos Science*, Vol. 15, 125–130, 1987.

Metropolis N.M., Rosenbluth A.W., Rosenbluth M.N., Teller A.H. and Teller E., Equations of State Calculations by Fast Computing Machines, *Journal of Chemical Physics*, Vol. 21(6), 1087–1092, 1953.

Metropolis N.M. and Ulam S., Monte Carlo Method, *Statistical Association*, Vol. 44(247), 335–341, 1949.

Mignolet M.P. and Spanos P.D., Recursive Simulation of Stationary Multivariate Random Processes – Part I, *Journal of Applied Mechanics*, Vol. 109, 674–680, 1987.

Mignolet M.P. and Spanos P.D., Direct Determination of ARMA Algorithms for the Simulation of Stationary Random Processes, *International Journal of Non-Linear Mechanics*, Vol. 25(5), 555–568, 1990.

Mignolet M.P. and Spanos P.D., Optimality in the Estimation of a MA System from a Long Model for Simulation Studies, *Soil and Earthquake Engineering*, Vol. 14(6), 445–452, 1995.

Mignolet M.P. and Spanos P.D., Auto-Regressive Spectral Modeling: Difficulties and Remedics, *International Journal of Non-Linear Mechanics*, Vol. 26(6), 911–930, 1991.

Mignolet M.P. and Spanos P.D., Simulation of Homogeneous Two-Dimensional Random Fields – Part I: AR and ARMA Models, *Applied Mechanics*, Vol. 114, 260–269, 1992.

Mignolet M.P. and Harish M.V., Comparison of Simulation Algorithms on Basis of Distribution, *Engineering Mechanics*, Vol. 122(2), 172–176, 1996.

Millwater H.H., Wu Y.–.T., Torng Y., Thacker B., Riha D. and Leung C., Recent Development of the NESSUS Probabilistic Structural Analysis Computer Program, *Procedure of 33rd AIAA/ASME/ASCE/AHS/ASC Structures, Structural Dynamics, and Material Conference*, pp. 614–624, 1992.

Mochio T., Samaras E. and Shinozuka M., Stochastic Linearization in a Finite Element Based Reliability Analysis, *Proceedings, Fourth International Conference on Structural Safety and Reliability*, (I. Konishi, A.H.S. Ang and M. Shinozuka, eds.), pp. 1375–1384, 1985.

Modaressi H., Discussion of the Paper by Köylüoğlu H.U., Cakmak A.S. and Nielsen S.R.K., *Journal of Engineering Mechanics*, Vol. 121, p. 645, 1997.

Moens D. and Hanss M., Non-Probabilistic Finite Element Analysis for Parametric Uncertainty Treatment in Applied Mechanics: Recent Advances, *Finite Elements in Analysis and Design*, Vol. 47(1), 4–16, 2011.

Moens D. and Vandepitte D., A Survey of Non-Probabilistic Uncertainty Treatment in Finite Element Analysis, *Computer Methods in Applied Mechanics and Engineering*, Vol. 194(12), 1527–1555, 2005.

Moens D. and Vandepitte D., Recent Advances in Non-Probabilistic Approaches for Non-Deterministic Dynamic Finite Element Analysis, *Archives of Computational Methods in Engineering*, Vol. 13(3), 389–464, 2006.

Molchanov I., *Theory of Random Sets*, Dordrecht: Springer, 2005.

Möller B. and Beer M., *Fuzzy Randomness: Uncertainty in Civil and Computational Mechanics*, Berlin: Springer, 2004.

Möller B. and Beer M., Engineering Computation under Uncertainty–Capabilities of Non-Traditional Models, *Computers and Structures*, Vol. 86(10), 1024–1041, 2008.

Möller B., Beer M., Graf W. and Hoffmann A., Possibility Theory-Based Safety Assessment, *Computer-Aided Civil Infrastructure Engineering*, Vol. 14(2), 81–91, 1999.

Möller B., Hansson S.O., Holmberg J.-E. and Rollenhagen C., eds., *Handbook of Safety Principles*, New York: Wiley, 2018.

Mooney C.Z., Monte Carlo Simulation, *Sage University Paper Series on Quantitative Applications in the Social Sciences, N. 07-116*, Newbury Park, CA, 1997.

Moore R., *Interval Analysis*, Englewood Cliffs: Prentice Hall, 1966.

Moore R.E., *Methods and Applications of Interval Analysis*, Philadelphia, PA: SIAM, 1979.

Moore R.E., (ed.), *Reliability in Computing: The Role of Interval Methods in Scientific Computing*, Boston: Academic Press, 2014.

Moore R.E., Kearfott R.B. and Cloud M.J., *Introduction to Interval Analysis*, SIAM, 2009.

Morio J. and Balesdent M., *Estimation of Rare Event Probabilities in Complex Aerospace and Other Systems: A Practical Approach*, Amsterdam: Elsevier, 2015.

Muhanna R.L. and Mullen R.L., Uncertainty in Mechanics Problems – Interval–Based Approach, *Journal of Engineering Mechanics*, Vol. 127(6), 557–566, 2001.

Muhanna R.L. and Mullen R.L., Interval Based Finite Elements for Uncertainty Quantification in Engineering Mechanics, in *IFIP Working Conference on Uncertainty Quantification*, Berlin: Springer, pp. 265–279, 2011.

Muhanna R.L., Rao M.R. and Mullen R.L., Advances in Interval Finite Element Modelling of Structures, *Life Cycle Reliability and Safety Engineering*, Vol. 2(3), 15–22, 2013.

Muhanna R.L., Zhang H. and Mullen R.L., Interval Finite Elements as a Basis for Generalized Models of Uncertainty in Engineering Mechanics, *Reliable Computing*, Vol. 13(2), 173–194, 2007.

Mumford D., The Dawning of the Age of Stochasticity, in *Mathematics: Frontiers and Perspectives*, Providence, R.I: American Mathematical Society, pp. 197–217, 1999.

Muravyov A., Turner T., Robinson J. and Rizzi S., A New Stochastic Equivalent Linearization Implementation for Prediction of Geometrically Nonlinear Vibrations, *AIAA Paper 59-1376* AIAA/ASME/ASCE/AHS/ASC Structures, Structural Dynamics Conference, St. Louis, Vol. 2, 1489–1497, 1999.

Murzewski J., *Bezpieczenstwo Konstrukeji Budowlanych*, Warsaw: Arkady Publishing House, 1970, (in Polish).

Murzewski J., *Niezawodnosc Konstrukcji Inzynierskich*, Warszawa: Arkady, 1989, in Polish.

Muscolino G., Santoro R. and Sofi A., Explicit Frequency Response Functions of Discretized Structures with Uncertain Parameters, *Computers and Structures*, Vol. 133, 64–78, 2014.

Muscolino G. and Sofi A., Stochastic Analysis of Structures with Uncertain-but-Bounded Parameters via Improved Interval Analysis, *Probabilistic Engineering Mechanics*, Vol. 28, 152–163, 2012a.

Muscolino G. and Sofi A., *Explicit Solutions for the Static and Dynamic Analysis of Discretized Structures with Uncertain Parameters, Computational Methods for Engineering Science* (B.H.V. Topping, ed.), Stirlingshire (UK): Saxe-Coburg Publications, pp. 47–73, 2012b.

Muscolino G. and Sofi A., Bounds for the Stationary Stochastic Response of Truss Structures with Uncertain-but-Bounded Parameters, *Mechanical Systems and Signal Processing*, Vol. 37, 163–181, 2013.

Nafday A.M., Strategies for Managing the Consequences of Black Swan Events, *Leadership and Management in Engineering*, Vol. 9(4),191–197, 2009.

Naess A., Prediction of Extreme Response of Nonlinear Structures by Extended Stochastic Linearization, *Probabilistic Engineering Mechanics*, Vol. 10, 153–160, 1995.

Naganuma T., Deodatis G. and Shinozuka M., ARMA Model for Two-Dimensional Processes, *Engineering Mechanics*, Vol. 113(2), 234–251, 1987.

Nahin P., *Duelling Idiots and Other Probability Puzzlers*, Princeton, NJ: Princeton University Press, 2000.

Najm H.N., Uncertainty Quantification and Polynomial Chaos Techniques in Computational Fluid Dynamics, *Annual Review of Fluid Mechanics*, Vol. 41, 35–52, 2009.

Nakagiri S. and Hisada T., *Introduction to Stochastic Finite Element Method: Analysis of Uncertain Structures*, Tokyo: Bai Fu Kan, 1985 in Japanese.

Nakagiri S. and Suzuki K., Interval Estimation Based on Finite Element Sensitivity Analysis of Stiffness Equation, *Transnational Japanese Society of Mechanical Engineering Powerpoint*, A Vol. 62(603), 2435–2439, 1996.

Nakagiri S. and Suzuki K., Interval Estimation of Eigenvalue Problem Based on Finite Element Sensitivity Analysis and Convex Model, *JSME International Journal Series American*, Vol. 40(3), 228–233, 1997.

Nakagiri S. and Yoshikawa N., Finite Element Interval Estimation by Convex Model, in *Probabilistic Mechanics and Structural and Geotechnical Reliability* (D. Frangopol and M. Grigoriu, eds.), New York: ASCE Press, pp. 278–281, 1996.

Nannapaneni S., Hu Z. and Mahadevan S., Uncertainty Quantification in Reliability Estimation with Limit State Surrogates, *Structural and Multidisciplinary Optimization*, Vol. 54(6), 1509–1526, 2016.

Narita Y. and Leissa A.W., Transverse Vibration of Simply Supported Circular Plates Having Partial Elastic Constraints, *Journal of Sound and Vibration*, Vol. 70(1), 103–116, 1980.

NASA, *Buckling of Thin-Walled Circular Cylinders*. NASA SP 8007, 1968.

Nascimento N., Stochastische Schwingungen punktweize erregter Saiten und Balken, *ZAMM: Zeitschrift für angewandte Mathematik und Mechanik*, Vol. 63, T76–T78, 1983 (in German).

Nascimento N., *Stochastische Schwingugen eindimensionaler kontinuierlicher mechanischer Systeme (Stochastic vibrations of one-dimensional continuous mechanical systems)*. M Sc Dissertation, Technical Univ Darmstadt, 1984 (in German).

Nascimento N., Der Fokussierungseffekt beim stochastisch erregtem Balken, *ZAMM: Zeitschrift für angewandte Mathematik und Mechanik*, Vol. 64, T57–T77, 1989 (in German).

Nascimento N. and Wallaschek J.,.O. effeito de focagem em systemas contimos uni e bidimensionais, in *Procedure of ADUNESP*, Brasil, 1986 (in Spanish).

Natke H.G. and Ben-Haim Y., Uncertainty: A Discussion from Various Points of View, in *Uncertainty: Models and Measures* (H.G. Natke and Y. Ben-Haim, eds.), Akademie Verlag, pp. 267–276, 1997.

Natke H.G. and Ben-Haim Y., Uncertainty: A Discussion from Various Points of View, in *Uncertainty: Models and Measures* (H.G. Natke and Y. Ben-Haim, eds.), Akademie Verlag, pp. 267–276, 1997.

Nau R. F., De Finetti Was Right: Probability Does Not Exist, *Theory and Decision*, Vol. 51(2–4), 89–124, 2001.

Nayak S. and Chakraverty S., *Interval Finite Element Method with MATLAB*, London: Academic Press, 2018.

Neal M.D., Matthews W.T. and Vangel M.G., Uncertainties in Obtaining High Reliability from Stress-strength Models. *Procedure Ninth DoD/NASA/FAA Conference Fibrous Composites in Structural Design*. Lake Tahoe, NV 1991. DOT/FAA/CT 92-95 I. pp. 503–521, 1992.

NESSUS Version 6.0 Release Notes, prepared to the National Aeronautics and Space Administration, NASA Contract NAS3-24389, SWRI Project 06-3285, prepared by SWRI and Vanderbilt Univ, 1992.

Neumaier A., *Interval Methods for Systems of Equations*, New York: Cambridge University Press, 1990.

Neumaier A., Clouds, Fuzzy Sets, and Probability Intervals, *Reliable Computing*, Vol. 10(4), 249–272, 2004.

Newland D.E., *Introduction to Random Vibrations, Spectral and Wavelet Analysis*, Harlow, UK: Longman House, 1993.

Neyman J., On the Problem of Estimating the Number of Schools of Fish, in *Publications in Statistics*, Vol. 1, Berkeley: University of California, 1945.

Ni B.Y., Jiang C. and Han X., An Improved Multidimensional Parallelepiped Non-Probabilistic Model for Structural Uncertainty Analysis, *Applied Mathematical -Modelling*, Vol. 40, 4727–4745, 2016.

Nigam N.C., *Introduction to Random Vibrations*, Cambridge: MIT Press, 1983.

Nigam N.C. and Narayanan S., *Applications of Random Vibrations*, Berlin: Springer, 1994.

Nikolaidis E., Chen S., Cudney H., Haftka R.T. and Rosca R., Comparison of Probability and Possibility Design against Catastrophic Failure under Uncertainty, *Journal of Mechanical Design*, Vol. 126(3), 386–394, 2004.

Nikolaidis E., Ghiocel D.M. and Singhal S., *Engineering Design Reliability Handbook*, Boca Raton: CRC Press, 2004.

Nikolaidis E., Mouleratos Z.P. and Pandey V., *Design Decisions under Uncertainty with Limited Information*, Boca Raton: CRC Press, 2011.

Noori M. and Davoodi H., Comparison between Equivalent Linearization and Gaussian Closure for Random Vibration Analysis of Several Nonlinear Systems, *International Journal of Engineering Sciences*, Vol. 28, 897–905, 1990.

Nowak A.S. and Collins K.R., *Reliability of Structures*, New York: McGraw-Hill, 2000.

Nowak A.S. and Collins K.R., *Reliability of Structures*, 2nd edition, Boca Raton: CRC Press, 2013.

Nowak A.S. and Collins K.R., *Reliability of Structures*, Second Edition, Boca Raton: CRC Press, 2019.

Oberkampf W.L., DeLand S.M., Rutherford B.M., Diegert K.V. and Alvin K.F., Estimation of Total Uncertainty in Modeling and Simulation, Sandia report SAND2000-0824, Albuquerque, NM, 2000.

Oberkampf W.L. and Helton J.C., *Engineering Design and Reliability Handbook: Evidence Theory for Engineering Applications*, Boca Raton: CRC Press, 2005.

Oberkampf W.L., Helton J.C., Joslyn C.A., Wojtkiewicz S.F. and Ferson S., Challenge Problems: Uncertainty in System Response Given Uncertain Parameters, *Reliability Engineering and System Safety*, Vol. 85, 11–19, 2004. (draft: November 29, 2001), http://www.sandia.gov/epistemic/.

Oberkampf W.L. and Roy C.J., *Verification and Validation in Scientific Computing*, Cambridge, UK: Cambridge University Press, 2010.

Oberkampf W.L., Trucano T.G. and Hirsch C., *Verification, Validation, and Predictive Capability in Computational Engineering and Physics*, SAND REPORT 2003-3769, Albuquerque and Livermore, Sandia National Laboratories, 2003.

Obraztsov I.F., Some Perspectives of the Plates and Shells Theory from a Point of View of Today's Aircraft, in *Theory of Plates and Shells, St*, Petersburg, "Sudostroenie" Publishers, pp. 6–12, 1975 (in Russian).

Oden J.T., Babuška I., Nobile F., Feng Y. and Tempone R., Theory and Methodology for Estimation and Control of Errors Due to Modeling, Approximation, and Uncertainty, *Computer Methods in Applied Mechanics and Engineering*, Vol. 194, 195–204, 2005.

Oden J.T. and Bathe K.J., Commentary on Computational Mechanics, *Applied Mechanics Reviews*, Vol. 31, 1053–1058, 1978.

Oden J.T., Belytschko T., Babuška I. and Hughes T.Y.R., Research Directions in Computational Mechanics, *Computer Methods in Applied Mechanics and Engineering*, Vol. 192(7–8), 913–922, 2003.

Oden J.T., Moser R. and Ghattas O., Computer Predictions with Quantified Uncertainty, Part II, *SIAM News*, Vol. 43, 1–4, 2010.

Ohsaki M., Zhang J-G. and Elishakoff I., Mutiobjective Optimization-Antioptimization for Force Design of Tensegrity Structures, *Journal of Applied Mechanics*, Vol. 79(2), paper 021015, 2012.

Oishi Sh. and Inoue H., Pseudo-Random Number Generators and Chaos, *The Transactions of the IECE of Japan*, Vol. E65(9), 534–541, 1982.

Oladyshkin S. and Nowak W., Data-Driven Uncertainty Quantification using the Arbitrary Polynomial Chaos Expansion, *Reliability Engineering & System Safety*, Vol. 106, 179–190, 2012.

Olszak W., Kaufman S., Eimer C. and Bychawski Z., *Teoria Konstrukcji Sprezonych*, Warszawa: Panstwowe Wydawnictwo Naukowe, 1961 (in Polish).

Onigawa T., Subjective Analysis of System Reliability and Its Analyzer, *Fuzzy Sets and Systems*, Vol. 83, 249–269, 1996.

Onigawa T. and Kacprzyk J. (eds.), *Reliability and Safety Analyses Under Fuzziness*, Heidelberg: Springer, 1995.

Ou J. and Wang G.-Y., Fuzzy Random Dynamical Reliability Analysis of Seismic Structures, *Proceedings of 5th International Conference on Structural Safety and Reliability*, pp. 1775–1778, 1989.

Owhadi H. and Scovel C., Toward Machine Wald, in *Handbook of Uncertainty Quantification* (R. Ghanem, D. Higdon and H. Owhadi, eds.), pp. 157–191, New York: Springer, 2017.

Owhadi H., Scovel C., Sullivan T.J., McKerns M. and Ortiz M., Optimal Uncertainty Quantification, *SIAM Review*, Vol. 55(2), 271–345, 2013.

Palmov V.A., Thin Plates under the Wide-Band Random Loading, *Proceedings of Leningrad Polytechical Institute*, Vol. 252, 97–106, 1965 (in Russian).

Panovko Y.G. and Gubanova I.I., *Stability and Oscillations of Elastic Systems: Modern Concepts, Paradoxes and Errors*, Washington, DC: NASA, 1973.

Pantelides C.P., Buckling of Elastic Columns using Convex Model of Uncertain Springs, *Journal of Engineering Mechanics*, Vol. 121(7), 837–844, 1995.

Pantelides C.P., Buckling and Post Buckling of Stiffened Elements with Uncertainty, *Thin-Walled Structures*, Vol. 26(1), 1–17, 1996.

Pantelides C.P. and Ganzerli S., Comparison of Fuzzy Set and Convex Model Theories in Structural Design, *Mechanical Systems and Signal Processing*, Vol. 15(3), 499–511, 2001.

Papadopoulos V. and Giovanis D.G., *Stochastic Finite Element Methods: An Introduction*, Berlin: Springer, 2017.

Papadopoulos V., Giovanis D., Lagaros N. and Papadrakakis M., Accelerated Subset Simulation with Neural Networks for Reliability Analysis, *Computer Methods in Applied Mechanics and Engineering*, Vol. 223, 70–80, 2012.

Papadrakakis M. and Lagaros N.D., Reliability-Based Structural Optimization using Neural Networks and Monte Carlo Simulation, *Computer Methods in Applied Mechanics and Engineering*, Vol. 191, 3491–3535, 2002.

Papadrakakis M. and Stefanou G. (eds.), *Multiscale Modeling and Uncertainty Quantification of Materials and Structures*, Basel: Springer, 2014.

Papadrakakis M., Stefanou G. and Papadopoulos V., *Computational Methods in Stochastic Dynamics*, Berlin: Springer, 2011.

Papaioannou I., Breitung K. and Straub D., Reliability Sensitivity Analysis with Monte Carlo Methods, in Proceedings of the 11th International Conference on Structural Safety and Reliability (ICOSSAR) (B. Ellingwood, G. Deodatis and D.M. Frangopol, eds.), 5335–5342, New York, 2013.

Papoulis A., *Signal Analysis*, New York: McGraw-Hill, 1977.

Paranjape K.H., Learning to See the Elephant, *Resonance*, Vol. 27(2), 177–184, 2022.

Parlos A.G., Henry A.F., Schweppe F.C., Gould L.A. and Lanning D.D., Nonlinear Multivariable Control of Nuclear Power Plants Based on the Unknown-But- Bounded Disturbance Model, *IEEE Transactions on Automatic Control*, Vol. 33, 130–137, 1988.

Paté-Cornell M.E., Uncertainties in Risk Analysis: Six Levels of Treatment, *Reliability Engineering & System Safety*, Vol. 54(2–3), 95–111, 1996.

Paté-Cornell E., On "Black Swans" and "Perfect Storms": Risk Analysis and Management When Statistics are Not Enough, *Risk Analysis, An International Journal*, Vol. 32(11), 1823–1833, 2012.

Patelli E., Alvarez D., Broggi M. and De Angelis M., Uncertainty Management in Multidisciplinary Design of Critical Safety Systems, *Journal of Aerospace Information Systems*, Vol. 12(1), 140–169, 2015.

Patelli E., Ghanem R., Higdon D. and Owhadi H., COSSAN: Multidisciplinary Software Suited for Uncertainty Quantification and Risk Management, *Handbook of Uncertainty Quantification*, pp.1–69, 2016.

Pawlak Z., *Rough Sets: Theoretical Aspects of Reasoning about Data*, Dordrecht: Kluwer Academic Publishers, 1991.

Pearl J., On Probability Intervals, *International Journal of Approximate Reasoning*, Vol. 2(3), 211–216, 1988.

Pearl J., *How to Do with Probabilities What People Say You Can't*, Computer Science Department, University of California, Los Angeles, 1985.

Pearl J., Reasoning under Uncertainty, *Annual Review of Computer Science*, Vol. 4(1), 37–72.

Pearl J., Decision Making under Uncertainty, *ACM Computing Surveys (CSUR)*, Vol. 28(1), 89–92, 1996.

Penmetsa R.C. and Grandhi R.V., Efficient Estimation of Structural Reliability for Problems with Uncertain Intervals, *Computers and Structures*, Vol. 80(12):1103–1112, 2002.

Penmetsa R.C. and Grandhi R.V., Uncertainty Propagation using Possibility Theory and Function Approximations, *Mechanics Based Design of Structures and Machines*, Vol. 31(2), 257–279, 2003.

Pesonen J. and Hyvönen E., Interval Approach Challenges Monte Carlo Simulation, *Reliable Computing*, Vol. 2(2), 155–160, 1996.

Peterson I., *The Jungles of Randomness: A Mathematical Safari*, New York: Penguin, 1998.

Peterson J.B., *Beyond Order: 12 More Rules for Life*, New York: Penguin, 2021.

Petroski H., The Fall of Skyscrapers, *American Scientist*, Vol. 19(1),16–20, 2002.

Pettit C. L., Uncertainty Quantification in Aeroelasticity: Recent Results and Research Challenges, *Journal of Aircraft*, Vol. 41(5), 1217–1229, 2004.

Pflüger A, *Stabilitätsprobleme der Elastostatik*, Berlin: Springer, 1964 (in German).

Pierre C., Mode Localization and Eigenvalue Loci Veering Phenomena in Disordered Structures, *Journal of Sound Vibration*, Vol. 126, pp. 485–502, 1988.

Pikovsky A. A., *Statics of Column Systems with Compressed Elements*, Moscow: Gosudarstvennoe Izdatel'stvo Fiziko-Matematicheskoy Literatury, 1961 (in Russian).

Piszczek K. and Niziol J., *Random Vibration of Mechanical Systems*, Ellis Horwood, Chichester, pp. 173–175, 1984.

Popov E.P. and Paltov I.N., *Approximate Methods of Investigation of Nonlinear Automatic Systems*, Moscow: Fizmatgiz Publishers, 1960 (in Russian).

Popova E. and Elishakoff I., Novel Interval Model Applied to Derived Variables in Static and Structural Problems, *Archive of Applied Mechanics*, Vol. 90, 869–888, 2020.

Popper K. R., *The Open Universe*, London: Hutchinson, 1982.

Pownuk A, Kreinovich V., *Combining Interval, Probabilistic, and Other Types of Uncertainty in Engineering Applications*, Berlin: Springer, 2018.

Pradlwarter H.J. and Schuëller G.I., Accuracy and Limitations of the Method of Equivalent Linearization for Hysteretic Multi-Story Structures, in *Nonlinear Stochastic Dynamic Engineering Systems*, (N. Bellomo and F. Casciati, eds.), Berlin: Springer, pp. 427–437, 1992.

Pradlwarter H.J., Schuëller G.I., and Melnik-Melnikov P.C., Reliability of MDoF Systems, *Probabilistic Engineering Mechanics*, Vol. 9, 235–243, 1994.

Pradlwarter H. J. and Schüeller G.I, A Practical Approach to Predict the Stochastic Response of Many-DOF Systems Modeled by Finite Elements, *Nonlinear Stochastic Mechanics*, (N. Bellomo and F. Casciati, eds.), Berlin: Springer, pp. 427–437, 1992.

Pradlwarter, H.J., Non-Gaussian Linearization an Efficient Tool to Analyze Nonlinear MDOF- Systems, *Nuclear Engineering and Design*, Vol. 128, 175–192, 1991.

Pradlwarter, H.J., and Schüeller, G.I., *Accuracy and Limitations of the Method of Equivalent Linearization for Hysteretic Multi-Store Structures, Nonlinear Stochastic Dynamic Engineering Systems*, (F. Ziegler and G. I. Schüeller, eds.), Berlin: Springer, pp. 3–21, 1988.

Preumont A, *Random Vibration and Spectral Analysis*, Dordrecht: Kluwer, 1994.

Proppe, C., Pradlwarter, H.G. and Schuëller, G.I., Equivalent Linearization and Monte Carlo Simulation in Stochastic Dynamics, *Probabilistic Engineering Mechanics*, Vol. 18, 1–15, 2003.

Proppe C., Schuëller G.I. and Pradlwarter H.J., Equivalent Linearization Revisited, *Advanced in Structural Dynamics* (J.M. Ko and Y.L. Xu, eds.), Vol. II, pp. 1207–1214, 2000.

Puggelli A., Li W., Sangiovanni-Vincentelli A.L. and Seshia S.A., Polynomial-Time Verification of PCTL Properties of MDPs with Convex Uncertainties, *International Conference on Computer Aided Verification*, pp. 527–542, Berlin: Springer, Berlin, 2013.

Pugsley A., *The Safety of Structures*, London: Edwin Arnold, 1966.

Pugsley A. and Ganesan R., Stability Analysis of a Stochastic Column Subjected to Stochastically Distributed Loadings using the Finite Element Method, *Finite Elements in Analysis and Design*, Vol. 11, 105–115, 1992.

Puig V., Stancu A. and Quevedo J., Simulation of Uncertain Dynamic Systems Described by Interval Models: A Survey, *IFAC Proceedings*, Vol.38(1), 1239–1250, 2005.

Qiu Z. P. *Convex Method Based on Non-Probabilistic Set Theory and Its Application*, National Beijing: Defense Industry Press, 2005 (in Chinese).

Qiu Z.P., Chen S.H. and Elishakoff I, Natural Frequencies of Structures with Uncertain but Nonrandom Parameters, *Journal of Optimization Theory of Applications* Vol. 86, 679–683, 1995.

Qiu Z.P., Chen S.H. and Elishakoff I., Non-Probabilistic Eigenvalue Problems for Structures with Uncertain Parameters via Interval Analysis, *Chaos, Solitons and Fractals*, Vol. 7, 303–308, 1996.

Qiu Z.P., Chen S.H. and Elishakoff I, Bound of Eigenvalues for Structures with an Interval Description of Uncertain-but-Non-Random Parameters, *Chaos, Solitons and Fractals*, Vol. 7, 425–434, 1996.

Qiu Z. P., Chen S. Q. and Wang X. J., Criterion of the Non-Probabilistic Robust Reliability for Structures, *Chinese Journal of Computational Mechanics*, Vol. 21(1): 1–6, 2004 (in Chinese).

Qiu Z.P. and Elishakoff I., Antioptimization of Structures with Large Uncertain-but-Non-Random Parameters via Interval Analysis, *Computer Methods in Applied Mechanics and Engineering*, Vol. 152(3–4), 361–372, 1998.

Qiu Z.P., Huang R., Wang X. and Qi W., Structural Reliability Analysis and Reliability-Based Design Optimization: Recent Advances, *Science China Physics, Mechanics and Astronomy*, Vol. 56(9),1611–1618, 2013.

Qiu Z. P., Müller P. C. and Frommer A., The New Non-Probabilistic Criterion of Failure for Dynamical Systems Based on Convex Models, *Mathematical and Computer Modeling*, Vol. 40(1–2), 201–215, 2004.

Qiu Z.P. and Wang L., The Need for Introduction of Non-Probabilistic Interval Conceptions into Structural Analysis and Design, *Science in China: Physics, Mechanics and Astronomy*, Vol. 59(11), article 114632, 2016.

Qiu Z.P. and Wang J., The Interval Estimation of Reliability for Probabilistic and Non-Probabilistic Hybrid Structural System, *Engineering Failure Analysis*, Vol. 17(5), 1142–1154, 2010.

Qiu Z.P. and Wang X.J., Comparison of Dynamic Response of Structures with Uncertain-but-Bounded Parameters Using Non-Probabilistic Interval Analysis Method and Probabilistic Approach, *International Journal of Solids and Structures*, Vol. 40(20), 5423–5439, 2003.

Qiu Z. P. and Wang X. J., *Set-Theoretical Convex Methods for Problems in Structural Mechanics with Uncertainties*, Beijing: Science Press, 2008 (in Chinese).

Qiu Z.P., Wu H., Elishakoff I. and Liu D.L., Data-Based Polyhedron Model for Optimization of Engineering Structures Involving Uncertainties, *Data-Centric Engineering*, Vol. 2, paper e8, 2021.

Qiu Z.P., Zheng Y. and Wang L., Recent Developments in the Non-Probabilistic Finite Element Analysis. *Journal of Harbin Institute of Technology (New Series)*, Vol. 24(4), 2017.

Qiu Z. P., Yang D. and Elishakoff I., Probabilistic Interval Reliability of Structural Systems, *International Journal of Solids and Structures*, Vol. 45(10), 2850–2860, 2008a.

Qiu Z.P., Yang D. and Elishakoff I., Combination of Structured Reliability and Interval Analysis, *Acta Mechanica Sinica*, Vol. 24(1), 61–67, 2008b.

Rackwitz R., Reliability Analysis – A Review and Some Perspectives, *Structural Safety*, Vol. 23(4), 365–395, 2001.

Radebe I.S., Response Characterization of Nanostructures Subjected to Uncertain Loading and Material Conditions by Convex Modelling, *Doctoral Dissertation*, University of KwoZulu-Natal, South Africa, 2014.

Radebe I.S. and Adali S., Minimum Weight Design of Beams Against Failure under Uncertain Loading by Convex Analysis, *Journal of Mechanical Science and Technology*, Vol. 27(7), 2071–2078, 2013.

Radebe I.S. and Adali S., Buckling and Sensitivity Analysis of Nonlocal Orthotropic Nanoplates with Uncertain Material Properties, *Composites Part B: Engineering*, Vol. 56, 840–846, 2014.

Radebe I.S. and Adali S., Minimum Cost Design of Hybrid Cross-Ply Cylinders with Uncertain Material Properties Subject to External Pressure, *Ocean Engineering*, Vol. 88, 310–317, 2014.

Radebe I.S. and Adali S., Static and Sensitivity Analysis of Nonlocal Nanobeams Subject to Load and Material Uncertainties by Convex Modeling. *Journal of Theoretical and Applied Mechanics*, Vol. 53(2), 345–356, 2015.

Radi B. and El Hami A., *Uncertainty and Optimization in Structural Mechanics*, London: ISTE-Wiley, 2013.

Rahman S. and Yadav V., Orthogonal Polynomial Expansions for Solving Random Eigenvalue Problems, *International Journal for Uncertainty Quantification*, Vol. 1(2), 163–187, 2011.

Raizer V., *Reliability of Structures: Analysis and Applications*, Backbone Publishing Company, 2009.

Raizer V. and Elishakoff I., *Philosophies of Structural Safety and Reliability*, Boca Raton: Taylor & Francis, 2022.

Reuter U., *Uncertainty Forecasting in Engineering*, Berlin: Springer; 2007.

Ramadan and Novak, Simulation of Spatially Incoherent Random Ground Motions, *Engineering Mechanics*, Vol. 119(5), 997–1016, 1993.

Ramanathan J and Zeitouni, Wavelet Transform of Fractional Brownian Motion, *IEEE Transactions on Information Theory*, Vol.

Ramu S.A., Ganesan R. and Channakeshava K. V., Critical Review of Digital Simulation Strategies for Probabilistic Structural Analysis, *Journal of Structural Engineering*, Vol. 23(1), 1–7, 1996.

Rumelin W., *Simulation of Fractional Brownian Motion, in Fractals in the Fundamenal and Applied Sciences* (H-O Peitgen, J.M. Henriques and L.F. Penedo, eds.), pp. Elsevier Science Publishers.

Rand A., *Atlas Shrugged*, New York: Pinguin, 1985.

Rand R. H. and Ambuster D., *Perturbation Methods, Bifurcation Theory and Computer Algebra*, Springer, Berlin, 1987.

Rao S. S., *Reliability-Based Design*, New York: McGraw Hill, 1992.

Rao S.S. and Berke L., Analysis of Uncertain Structural Systems using Interval Analysis, *AIAA Journal*, Vol. 35 (4), 727–735, 1997.

Rao S. S., and Liu X.T., Universal Grey System Theory for Analysis of Uncertain Structural Systems, *AIAA Journal*, Vol. 55(11), 3966–3979, 2017.

Rao S.S. and Sawyer J.P., Fuzzy Finite Element Approach for Analysis of Imprecisely Defined Systems, *AIAA Journal*, Vol. 33(12), 2364–2370, 1995.

Reardon D. and Leithead W.E., Statistical Linearization: A Comparative Study, *International Journal of Control*, Vol.52, 1083–1105, 1990.

Reddy G.R., Suziki K., Watanabe T. and Mahajan S.C., Linearization Techniques for Seismic Analysis of Piping System on Friction Support, *Journal of Pressure Vessel Technology*, Vol. 121, pp. 103–108, 1999.

Ren Y. J., Elishakoff I. and Shinozuka M., Conditional Simulation of Non-Gaussian Random Fields for Earthquake Monitoring Systems, *Chaos, Solitons and Fractals*, Vol. 5(1), 91–101, 1995.

Ren Y-J., Elishakoff I. and Shinozuka M., Finite Element Method for Stochastic Beams Based on Variational Principles, *Journal of Applied Mechanics*, Vol. 64, 664–669, 1997.

Rezaei M., Shirazi K.H. and Khodaparast H., Uncertainty Quantification of Aeroelastic Wings Flutter using an Optimized Machine Learning Approach, *Proceedings of the Institution of Mechanical Engineers, Part G: Journal of Aerospace Engineering*, article 09544100221080765, 2022.

Richtmyer R.D. and von Neumann J., Statistical Methods in Neutron Diffusion, *Report LAMS-551*, 1947.

Riemer M. and Wedig W., Bauwerke unter Wellenlasten-Fokussierungseffekte bei Poissoner- regungen, *VDI-Berichte 419*, VDI Verlag, Düsseldorf, pp. 201–207, 1981.

Righi M., Düzel S., Anderegg D., Da Ronch A., Massegur Sampietro D. and Soukhmane I., ROM-Based Uncertainties Quantification of Flutter Speed Prediction of the BSCW Wing, In *AIAA SCITECH 2022 Forum* (p. 0179), 2022.

Risken H., *The Fokker-Planck Equation: Methods of Solution and Applications*, New York: Springer, 1989.

Roache PJ., Quantification of Uncertainty in Computational Fluid Dynamics, *Annual Review of Fluid Mechanics*, Vol. 29(1),123–160, 1997.

Robbins H. E., On the Measure of Random Set, I, *Annals of Mathematical Statistics*,Vol. 15, 70–74, 1994.

Robbins H. E., On the Measure of Random Set, II, *Annals of Mathematical Statistics*, Vol. 16, 342–347, 1995.

Robert C. P. and Casella G., *Monte Carlo Statistical Methods*, Berlin: Springer, 2005.

Roberts J.B., Response of Nonlinear Mechanical Systems to Random Excitation, Part 2: Equivalent Linearization and Other Methods, *The Shock and Vibration Digest*, Vol. 13(5), 15–29, 1981.

Roberts J.B., *Statistical Linearization: Multiple Solutions and Their Physical Significance*, Structural Dynamics, (W.B. Krätzig, *et al.*, eds.), Rotterdam: Balkema, pp. 671–681, 1990.

Roberts J.B. and Spanos P.D., *Random Vibration and Statistical Linearization*, Chichester, UK: Wiley, 1991.

Roberts J.B. and Spanos P.D., *Random Vibration and Statistical Linearization*, Mineola, NY: Dover Publications, 2004.

Rocchi P., *Reliability is a New Science: Gnedenko was Right*, Berlin: Springer, 2017.

Rocchetta R., Broggi M. and Patelli E., Do We Have Enough Data? Robust Reliability via Uncertainty Quantification, *Applied Mathematical Modelling*, Vol. 54, 710–721, 2018.

Römer U., Bertsch L., Mulani S. B. and Schäffer B., Uncertainty Quantification for Aircraft Noise Emission Simulation: *Methods and Limitations. AIAA Journal*, Vol. 60(5), 3020–3034, 2022.

Roorda J., Concepts in Elastic Structural Stability, in *Mechanics Today* (S. Nemat-Nasser, ed.), Vol. 1, Oxford: Pergamon Press, pp. 322–372, 1972.

Roozen N.B., Quiet by Design: Numerical Acoustic Elastic Analysis of Aircraft Structures, *PhD Thesis*, Technical University of Eindhoven, 1992.

Rosental C., Fuzzifying the World: Social Practices of Showing the Properties of Fuzzy Logic, in *Growing Explanations: Historical Perspectives on Recent Science* (M.N. Wise, ed.), Durham: Duke University Press, pp. 159–178, 2004.

Ross T., *Fuzzy Logic with Engineering Applications*, Third Edition, West Sussex, UK: Wiley, 2010.

Rosenbluth M.N., Genesis of the Monte Carlo Algorithm for Statistical Mechanics, *AIP Conference Proceeding*, Vol. 690, 22–30, 2003.

Roy C.J. and Oberkampf W.L., A Comprehensive Framework for Verification, Validation, and Uncertainty Quantification in Scientific Computing, *Computer Methods in Applied Mechanics and Engineering*, Vol. 200(25–28), 2131–2144, 2011.

Rubinstein R.Y., *Simulation and the Monte Carlo Method*, New York: Wiley, 1981.

Rubinstein R.Y. and Kroese D.P., *Simulation, and the Monte Carlo Method*, New York: Wiley, 2016.

Rumsfeld D., *Known and Unknown: A Memoir*, New York: Penguin, 2011.

Rzhanitsyn A.R., Determination of Safety Factor in Construction, *Stroitel'naya Promishlennost*, 8, 1947 (in Russian).

Rzhanitsyn A.R., *Theory of Reliability Design of Civil Engineering Structures*, Moscow: "Stroyizdat" Publishing House, 1978 (in Russian).

Rzhanitsyn A.R., *Design of Constructions with Materials' Plastic Properties Taken into Account*, Gosudarstvennoe Isdatel'stvo Lieratury Po Stroitel'stvu I Arkhitekture, Moscow, 1954, Second Edition, Chapter 14 (in Russian) (See also a French translation: A.R. Rjanitsyn: *Calcul á la rupture et plasticite des constructions*, Paris: Eyrolles, 1959.

Sainz M.A., Armengol J., Calm R., Herrero P., Jorba L. and Vehi J., *Modal Interval Analysis*, Berlin: Springer, 2014.

Saksson A.J., Analysis of Identified 2D Non-Causal AR-Models, *IEEE Transactions on Information Theory*, Vol.

Sallak M., Schon W. and Aguirre F., Extended Component Importance Measures Considering Aleatory and Epistemic Uncertainties, *IEEE Transactions on Reliability*, Vol. 62(1), 49–65, 2013.

Samaras E., Shinozuka M. and Tsurui A., ARMA Representation of Random Processes, *Engineering Mechanics*, Vol. 111(3), 449–461, 1983.

Savoia M., Structural Reliability Analysis through Fuzzy Number Approach, with Application to Stability, *Computers & Structures*, Vol. 80(12), 1087–1102, 2002.

Savoia M., Ferracuti B. and Elishakoff I., *Fuzzy Safety Factor, in Safety and Reliability of Engineering Systems and Structures* (G. Augusti, G.I. Schuëller and M. Ciampoli, eds.), Rotterdam: Millpress, pp. 1783–1791, 2005.

Sawyer J.P. and Roa S.S., Fuzzy Finite Element Approach for the Analysis of Imprecisely Defined Systems, *AIAA Journal*, Vol. 33, 2364–2370, 1995.

Sakefeller R.T., *Convex Analysis*, Princeton, N.J: Princeton Univ Press, 1970.

Salomon J., Winnewisser N., Wei P., Broggi M. and Beer M., Efficient Reliability Analysis of Complex Systems in Consideration of Imprecision, *Reliability Engineering & System Safety*, Vol. 216, article 107972, 2021.

Samuels J.C. and Eringen A.C., Response of a Simply Supported Timoshenko Beam to a Purely Random Gaussian Process, *Journal of Applied Mechanics*, Vol. 25, 496–500, 1958.

Sankararaman S., Significance, Interpretation, and Quantification of Uncertainty in Prognostics and Remaining Useful Life Prediction, *Mechanical Systems and Signal Processing*, Vol. 52, 228–247, 2015.

Santoro R., Muscolino G. and Elishakoff I., Optimization and Anti-Optimization Solution of Combined Parameterized and Improved Interval Analyses for Structures with Uncertainties, *Computers and Structures*, Vol. 149, 31–42, 2015.

Sarkar A. and Ghanem R., Mid-Frequency Structural Dynamics with Parameter Uncertainty, *Computer Methods in Applied Mechanics and Engineering*, Vol. 191(47–48), 5499–5513, 2002.

Sason K. and Reason J., Team Errors: Definition and Taxonomy, *Reliability Engineering and System Safety*, Vol. 65, 1–9, 1999.

Sato M., *The Honda Myth: The Genius & His Wake*, New York: Vertical, Inc, 2006.

Saxe J.G., *The Blind Men and the Elephant*, Hong Kong: Enrich Spot Limited, 2016.

Scarrott C. and MacDonald A., A Review of Extreme Value Threshold Estimation and Uncertainty Quantification, *REVSTAT–Statistical Journal*, Vol. 10(1), 33–60, 2012.

Scarth C., Cooper J.E., Weaver P.M. and Silva G.H., Uncertainty Quantification of Aeroelastic Stability of Composite Plate Wings using Lamination Parameters, *Composite Structures*, Vol. 116, 84–93, 2014.

Sciacchitano A., Neal D.R., Smith B.L., Warner S.O., Vlachos P.P., Wieneke B. and Scarano F., Collaborative Framework for PIV Uncertainty Quantification: Comparative Assessment of Methods, *Measurement Science and Technology*, Vol. 26(7), paper 074004, 2015.

Schenk C.A. and Schuëller G.I., *Uncertainty Assessment of Large Finite Element Systems*, Berlin: Springer, 2005.

Scheurkogel A. and Elishakoff I., On Ergodicity Assumption in an Applied Mechanics Problem, *Journal of Applied Mechanics*, Vol. 52, 133–136, 1985.

Scheurkogel A., Elishakoff I. and Kalker J., On the Error that Can Be Induced by an Ergodicity Assumption, *Journal of Applied Mechanics*, Vol. 48, 654–656, 1981.

Schiehlen W., Personal Communication, 26 May 2000.

Schiehlen W. and Bestle D., Random Loading by Large Displacement Chaotic Motions, in *Nonlinear Stochastic Dynamic Engineering Systems* (F. Ziegler and G.I. Schuëller, eds.), Berlin: Springer, pp. 205–216, 1988.

Schmitendorf W.E., Design Methodology for Robust Stabilizing Controllers, *Journal of Guidance*, Vol. 10, 250–254, 1986.

Schmitendorf W.E. and Barmish B.R., Robust Asymptotic Tracking for Linear Systems with Unknown Parameters, *Automatica*, Vol. 22, 355–359, 1986.

Schmitendorf W.E., Methods for Obtaining Robust Tracking Control Laws, *Automatica*, Vol. 23, 675–677, 1987.

Schneider J., *Introduction to Safety and Reliability of Structures*, Zurich, Switzerland: IABSE-AIPC-IVBH, 1977.

Schöbi R. and Sudret B., Structural Reliability Analysis for p-Boxes using Multi-Level Metamodels, *Probabilistic, Engineering Mechanics*, Vol. 48, 27–38, 2017.

Schöbi R. and Sudret B., Structural Reliability Analysis for p-Boxes using Multi-Level Meta-Models, *Probabilistic Engineering Mechanics*, Vol. 48, 27–38, 2017.

Schuëller G.I., *Einführung in die Sicherheit und Zuverlässigkeit von Tragwerken*, Berlin: Verlag von Wilhelm Ernst & Sohn, 1981 (in German).

Schuëller G.I. (ed.), State of the Art Report of IASSAR on Computational Stochastic Mechanics, *Probabilistic Engineering Mechanics*, Vol. 12(4), 197–313, 1997.

Schuëller G.I., On the Stochastic Response of Nonlinear FE Models, *Archive of Applied Mechanics*, Vol. 69, 765–784, 1999.

Schuëller G.I., Efficient Monte Carlo Simulation Procedures in Structural Uncertainty and Reliability Analysis – Recent Advances, *Structural Engineering and Mechanics: An International Journal*, Vol. 32, 1–20, 2009.

Schuëller G.I. and Ang A.H.-S., Advances in Structural Reliability, *Nuclear Engineering and Design*, Vol. 134(1), 121–140, 1992.

Schuëller G.I. and Jensen H.A., Computational Methods in Optimization considering Uncertainties–An Overview, Computer, *Methods in Applied Mechanics and Engineering*, Vol. 198(1), 2–13, 2008.

Schuëller G.I., Pradlwarter H.J., Vasta M. and Harnpornchai N., Benchmark Study on Non- Linear Stochastic Structural Dynamics, in *Structural Safety and Reliability* (N. Shiraishi, M. Shinozuka and Y.K. Wen, eds.), Rotterdam: Balkema, pp. 355–362, 1998.

Schuëller G.I. and Pradlwarter H.J., Advances in Stochastic Structural Dynamics under the Perspective of Reliability Estimation, in *Structural Dynamics-EURODYN'99* (L. Fryba and J. Naprstek, eds.), Rotterdam: Balkema, pp. 267–272, 1999.

Schuëller G.I., Pradlwarter H.J., Vasta M. and Harnpornchai N., Benchamark Study on Non-Linear Stochastic Structural Dynamics, in *Structural Safety and Reliability* (N. Shiraishi, M. Shinozuka and Y.K. Wen, eds.), Rotterdam: Balkema, pp. 355–362, 1998.

Schwartz B., *The Paradox of Choice: Why More Is Less*, New York: Harper Perennial, 2004.

Schweppe F.C., Recursive State Estimation: Unknown but Bounded Errors and System Inputs, *IEEE Transactions on Automatic Control*, Vol. 13(1), 22–28, 1968.

Schweppe F.C., *Uncertain Dynamic Systems*, Englewood Cliffs, N.J: Prentice Hall, 1973.

Segal L.A., The Importance of Asymptotic Analysis in Applied Mathematics, *American Mathematics Monthly*, Vol. 73, 7–14, 1966.

Seide P., Nonlinear Stresses and Deflections of Beams Subjected to Random Time-Dependent Pressure, *Israel Journal of Techology*, Vol. 13, 143–151, 1975.

Seide P., Nonlinear Stresses and Deflections of Beams Subjected to Random Time Dependent Uniform Pressure, *Journal of Engineering for Industry*, Vol. 98, 1014–1020, 1976.

Sensmeier M., Sensharma P., Haftka R.T., Griffin R.O. and Watson L., Experimental Validation of Anti-Optimization Approach for Detecting Delamination Damage, in *36th Structures, Structural Dynamics and Materials Conference*, pp. 1508–1515, 1995.

Sepahvand K.K., Deep Learning Based Uncertainty Analysis in Computational Micromechanics of Composite Materials, *Applied Mechanics*, Vol. 2(3), 559–570, 2021.

Sepahvand K.K., Marburg S. and Hardtke H.J., Uncertainty Quantification in Stochastic Systems using Polynomial Chaos Expansion, *International Journal of Applied Mechanics*, Vol. 2(02), 305–353, 2010.

Shafer G., *A Mathematical Theory of Evidence*, Princeton, NJ: Princeton University Press, 1976.

Shafer G. and Pearl J. (eds.), *Readings in Uncertain Reasoning*, Morgan Kaufmann Publishers Inc, 1990.

Shia D. and Hui C.Y., A Monte Carlo Solution Method for Linear Elasticity, *International Journal of Solids and Structures*, Vol. 37, 6085–6105, 2000.

Shih C.J. and Wangsawidjaja R.A.S., Mixed Fuzzy-Probabilistic Programming Approach for Multiobjective Engineering Optimization with Random Variables, *Computation of Structures*, Vol. 59, 283–290, 1996.

Shinozuka M., Basic Analysis of Structural Safety, *ASCE Journal of Structural Engineering*, Vol. 59(3), 721–740, 1983.

Shinozuka M., Maximum Structural Response to Seismic Excitations, *Journal of Engineering Mechanics Division*, Vol. 96, 729–738, 1970.

Shinozuka, Simulation of Multivariate and Multidimensional Random Processes, *Journal of Acoustical Society of America*, Vol. 49, 357–367, 1971.

Shinozuka M., Monte Carlo Solution in Structural Dynamics, *Computers and Structures*, Vol. 2(5–6), 855–874, 1972.

Shinozuka M., Structural Response Variability, *Journal of Engineering Mechanics*, Vol. 113, 825–842, 1987.

Shinozuka M., Developments in Structural Reliability, Freudenthal Lecture, in *ICOSSAR'89, Proceedings of 8th International Conference on Structural Safety and Reliability*, New York: ASCE Press, pp. 1–20, 1989.

Shinozuka and Deodatis G., Stochastic Process Models for Earthquake Ground Motion, *Probabilistic Engineering Mechanics*, Vol. 3(3), 115–123, 1988a.

Shinozuka M. and Deodatis G., Response Variability of Stochastic Finite Element Systems, *Journal of Engineering Mechanics*, Vol. 114, 499–519, 1988b.

Shinozuka and Deodatis, Simulation of Stochastic Processes by Spectral Representation, *Applied Mechanics Reviews*, Vol. 44(4), 191–204, 1991.

Shinozuka and Deodatis Simulation of Multidimensional Gaussian Fields by Spectral Representation, *Applied Mechanics Reviews*, Vol. 49, 29–53, 1996.

Shinozuka M. and Jan C.M., Digital Simulation of Random Processes and Its Applications, *Sound and Vibration*, Vol. 25(1), 111–128, 1972.

Shinozuka M. and Yamazaki F., Stochastic Finite Element Analysis: An Introduction, in *Stochastic Structural Dynamics* (S.T. Ariaratnam, G.I. Schuëller and I. Elishakoff, eds.), London: Elsevier, pp. 241–292, 1988.

Shiraishi N. and Furuta H., Reliability Analysis Based on Fuzzy Probability, *Journal of Engineering Mechanics*, Vol. 109(6), 1445–1459, 1983.

Shiraishi N. and Furuta H., Assessment of Structural Durability with Fuzzy Sets, in *Proceedings of the of NSF Workshop on Civil Eng Appl of Fuzzy Sets* (C.B. Brown, J.-L. Chameanu, R. Palmer and J.T.P. Yao, eds.), Purdue University, pp. 193–218, 1985.

Simiu E., *et al.*, Extreme Wind Speeds at 129 Stations in the Continental United States, in *Building Science Series MBS BS 118*, Washington DC: National Bureau of Standards, 1979.

Singer J., Experimental Studies on Shell Buckling, *38th AIAA/ASME/ASCE/AHS/ASC Structures, Structural Dynamics, and Materials Conference and Exhibit*, AIAA Paper, 1922–1932, 1997.

Singh A.K., Chu S.L. and Singh S., Influence of Closely Spaced Modes in Response Spectrum Method of Analysis, in *Report No. SAD-126*, Chicago: Sargent & Lundy Engineers, 1973.

Singh M.P., Khdeir A.A., Maldonado G.O. and Reddy J.N., Random Response of Antisymmetric Angle-Ply Laminated Plates, *Structural Safety*, Vol. 6, 115–127, 1988.

Sinitsyn I.N., Methods of Statistical Linearization (Survey), *Automation and Remote Control*, Vol. 35, 765–776, 1974.

Sireteanu T., Bellizzi S. and Ursu I., An Extension of Gaussian Equivalent Linearization for a Class of Nonlinear Oscillator, *13th ASCE Engineering Mechanics Conference*, The Johns Hopkins University, Baltimore, June 13–16, p. 336, 1999.

Skalna I., *Parametric Interval Algebraic Systems*, Berlin: Springer, 2018.

Skudrzyk E.J., The Mean Value Method of Predicting the Dynamic Response of Complex Vibrators, *Journal of Acoustic Society of Amsterdam*, Vol. 67, 1105–1135, 1980.

Skudrzyk E.J., Understanding the Dynamic Behavior of Complex Vibrations, *Acustica*, Vol. 64, 123–147, 1987.

Smith G.N., Probability and Statistics in Civil Engineering, *Collins Professional and Technical Books*, London, 1986.

Smith P.W. and Lyon R.H., *Sound and Structural Vibration, NASA CR-160*, 1976.

Smith R.C., *Uncertainty Quantification: Theory, Implementation, and Applications*, Philadelphia: SIAM, 2014.

Sniedovich M., The Art and Science of Modeling Decision-Making under Severe Uncertainty, *Decision Making in Manufacturing and Services*, Vol. 1(1/2), 111–136, 2007.

Sniedovich M., Wald's Maximin Model: A Treasure in Disguise!, *The Journal of Risk Finance*, Vol. 9(3), 287–291, 2008.

Sniedovich M., A Bird's View of Info-Gap Decision Theory, *The Journal of Risk Finance*, Vol. 11(3), 268–283, 2010.

Sniedovich M., Black Swans, New Nostradamuses, Voodoo Decision Theories, and the Science of Decision Making in the Face of Severe Uncertainty, *International Transactions in Operational Research*, Vol. 19(1–2), 253–281, 2012a.

Sniedovich M., Fooled by Local Robustness, *Risk Analysis: An International Journal*, Vol. 32(10), 1630–1637, 2012b.

Sniedovich M., *From Statistical Decision Theory to Robust Optimization: A Maximin Perspective on Robust Decision-Making, in Robustness Analysis in Decision Aiding, Optimization, and Analytics* (M. Doumpos, C. Zopounidis and E. Grigoroudis, eds.), Cham: Springer, pp. 59–87, 2016a.

Sniedovich M., Wald's Mighty Maximin: A Tutorial, *International Transactions in Operational Research*, Vol. 23(4), 625–653, 2016b.

Sobieszczanski-Sobieski J. and Haftka R.T., Multidisciplinary Aerospace Design Optimization: Survey of Recent Developments, *Structural Optimization*, Vol. 14(1), 1–23, 1997.

Sobol' I.M., *Primer for the Monte Carlo Method*, New York: CRC, 1994.

Socha L., Linearization in Analysis of Nonlinear Stochastic Systems. Part1. Theory, *Applied Mechanics Reviews*, Vol. 58, 178–205, 2005a.

Socha L., Linearization in Analysis of Nonlinear Stochastic Systems. Part II. Applications, *Applied Mechanics Reviews*, Vol. 58, 303–315, 2005b.

Socha L., *Linearization Method for Stochastic Dynamic Problems*, Berlin: Springer, 2008.

Socha L. and Pawleta M., Corrected Equivalent Linearization of Stochastic Dynamic Systems, *Machine Dynamics Problems*, Vol. 7, 149–161, 1994.

Socha L. and Pawleta M., *Some Remarks on Equivalent Linearization of Stochastic Dynamic Systems, Stochastic Structural Dynamics* (B.F. Spencer Jr. and E.A. Johnson, eds.), Rotterdam:Balkema, pp. 106–112, 1999.

Socha L. and Soong T.T., Linearization in Analysis of Nonlinear Stochastic System, *Applied Mechanics Reviews*, Vol. 44, 399–422, 1991.

Sofi A., Structural Response Variability under Spatially Dependent Uncertainty: Stochastic Versus Interval Model, *Probabilistic Engineering Mechanics*, Vol. 42, 78–86, 2015.

Sofi A., Romeo E., Barrera O. and Cocks A., An Interval Finite Element Method for the Analysis of Structures with Spatially Varying Uncertainties, *Advances in Engineering Software*, Vol. 128, 1–19, 2019.

Soize C., Probabilistic Structural Modeling in Linear Dynamic Analysis of Complex Mechanical System I: Theoretical Elements, *La Recherche Aerospatiale*, Vol. 5, 23–48, 1986 (English edition).

Soize C., A Model and Numerical Method in the Medium Frequency Range for Vibroacoustic Predictions Using the Theory of Structural Fuzzy, *Journal of Acoustic Society of Amsterdam*, Vol. 94(Part 1), 849–865, 1993.

Soize C., Stochastic Modeling of Uncertainties in Computational Structural Dynamics: Recent Theoretical Advances, *Journal of Sound and Vibration*, Vol. 332(10), 2379–2395, 2013.

Soize C., *Uncertainty Quantification: An Accelerated Course with Advanced Applications in Computational Engineering*, Berlin: Springer, 2017.

Soize C., Desanti A. and David J.M., Numerical Methods in Electroacoustics for Low and Medium Frequency Range, *La Recherche Aerospatiale (English Edition)*, Vol. 5, 25–44, 1992.

Soize C. and Ghanem R., Physical Systems with Random Uncertainties: Chaos Representations with Arbitrary Probability Measure, *SIAM Journal on Scientific Computing*, Vol. 26(2), 395–410, 2004.

Soldatos K. and Elishakoff I., Thermoelastic Vibration of Laminated Plates According to a New Transverse Shear and Normal Deformable Theory, in *Structural Dynamics* (W.B. Krätzig, *et al*, ed.), Rotterdam: Balkema, pp. 1083–1089, 1990.

Son J. and Du Y., An Efficient Polynomial Chaos Expansion Method for Uncertainty Quantification in Dynamic Systems, *Applied Mechanics*, Vol. 2(3), 460–481, 2021.

Song C. and Kawai R., Monte Carlo and Variance Reduction Methods for Structural Reliability Analysis: A Comprehensive Review, *Probabilistic Engineering Mechanics*, article 103479, 2023.

Song J., Wei P., Valdebenito M. and Beer M., Active Learning Line Sampling for Rare Event Analysis, *Mechanical Systems and Signal Processing*, Vol. 147, article 107113, 2021.

Soong T.T., Stochastic Structural Dynamics: Research *vs* Practice, in *Stochastic Structural Dynamics* (N. Sri Namachchivaya, H. Hilton and Y.K. Wen, eds.), Urbana-Champaign, ILL: University of Illinois Press, pp. 289–300, 1988.

Soong T.T. and Grigoriu M., *Random Vibration of Mechanical and Structural Systems*, Englewood Cliffs, NJ: Prentice Hall, 1993.

Soundappan P., Nikolaidis E., Haftka R.T., Grandhi R. and Canfield R., Comparison of Evidence Theory and Bayesian Theory for Uncertainty Modeling, *Reliability Engineering and System Safety*, Vol. 85(1–3), 295–311, 2004.

Spanos P.D., Formulation of Stochastic Linearization for Symmetric or Asymmetric M.D.O.F. Nonlinear Systems, *Journal of Applied Mechanics*, Vol. 47, 209–211, 1980.

Spanos P.D., Introduction, *International Journal of Nonlinear Mechanics*, Vol. 26(6), 801–817, 1991.

Spanos P.D., Stochastic Linearization in Structural Dynamics, *Applied Mechanics Reviews*, Vol. 34, 1–8, 1981.

Spanos P.D., ARMA Algorithms of Ocean Wave Modeling, *ASME Journal of Energy Recources Technology*, Vol. 105(9), 300–309, 1983.

Spanos P.D., Di Matteo A. and Pirrotta A., Steady-State Dynamic Response of Various Hysteretic Systems Endowed with Fractional Derivative Elements, *Nonlinear Dynamics*, Vol. 98(4), 3113–3124, 2019.

Spanos P.D., Di Paola M. and Failla G., A Galerkin Approach for Power Spectrum Determination of Nonlinear Oscillators, *Meccanica*, Vol. 37, 51–65, 2002.

Spanos P.D., Eberle R.R., Hamilton D. and Mushung L., Trivariate Spectral Modeling of Space Shuttle Flight Data, *Aerospace Engineering*, Vol. 8(3), 148–155, 1995.

Spanos P.D. and Evangelatos G.I., Response of a Non-Linear System with Restoring Forces Governed by Fractional Derivatives – Time Domain Simulation and Statistical Linearization Solution, *Soil Dynamics and Earthquake Engineering*, Vol. 30(9), 811–821, 2010.

Spanos P.D. and Ghanem R.G., Stochastic Finite Element Expansion for Random Media, *Engineering Mechanics*, Vol. 115(5), 1035–1053, 1989.

Spanos P.D. and Hansen J., Linear Prediction Theory for Digital Simulation of Sea Waves, *Journal of Energy Resources Echnology*, Vol. 103(3), 243 - 249, 1981.

Spanos P. and Kontsos A., A Multiscale Monte Carlo Finite Element Method for Determining Mechanical Properties of Polymer Nanocomposites, *Probabilistic Engineering Mechanics*, Vol. 23(4), 456–470, 2008.

Spanos P.D. and Malara G., Nonlinear Random Vibrations of Beams with Fractional Derivative Elements, *Journal of Engineering Mechanics*, Vol. 140(9), 04014069, 2014.

Spanos P.D. and Mignolet M.P., Z-Transform Modeling of P-M Wave Spectrum, *Engineering Mechanics*, Vol. 112(8), 745–759, 1986.

Spanos P.D. and Mignolet M.P., Recursive Simulation of Stationary Multivariate Random Processes – Part II, *Applied Mechanics*, Vol. 54, 674–680, 1987.

Spanos P.D. and Mignolet M.P., ARMA Monte Carlo Simulation in Probabilistic Structural Analysis, *Shock and Vibration Digest*, Vol. 21(11), 3–13, 1989.

Spanos P.D. and Mignolet M.P., Simulation of Stationary Random Processes Two-Stage MA to ARMA Approach, *Engineering Mechanics*, Vol. 116(3), 620–641, 1990.

Spanos P.D. and Mignolet M.P., Simulation of Homogeneous Two-Dimensional Random Fields: Part II – MA and ARMA Models, *Applied Mechanics*, Vol. 114, 269–277, 1992.

Spanos P.D. and Schula K.P., Numerical Synthesis of Trivariate Velocity Realizations of Turbulence, *International Non-Linear Mechanics*, Vol. 277, 1986.

Spanos P.D. and Zeldin B.A., Galerkin Sampling Method for Stochastic Mechanics Problems, *Journal of Engineering Mechanics*, Vol. 120(5), 1091–1106, 1994.

Spanos P.D. and Zeldin B.A., Efficient Iterative ARMA Approximation of Multi-Variate Random Processes for Structural Dynamics Applications, *Journal of Earthquake Engineering and Structural Dynamics*, Vol. 25, 497–507, 1996.

Spanos P.D. and Zeldin B.A., Monte Carlo Treatment of Random Fields: A Broad Perspective, *Applied Mechanics Reviews*, Vol. 51(3), 219–237, 1998.

Spanos P.D. and Zeldin B.A., Efficient Iterative ARMA Approximation of Multi-Variate Random Processes for Structural Dynamics Applications, *Journal of Earthquake Engineering and Structural Dynamics*, Vol. 25, 497–507, 1996.

Spanos P.D. and Zeldin B.A., Monte Carlo Treatment of Random Fields: A Broad Perspective, *Applied Mechanics Reviews*, Vol. 51, 219–237, 1998.

Sparrow V.W., Russell D.A. and Rochat J.L., Implementation of Discrete Fuzzy Structure Models in Mathematica, *International Journal for Numerical Methods in Engineering*, Vol. 37(17), 3005–3014, 1994.

Squarcio R.M. and da Silva C.R., Uncertainty Quantification via λ-Neumann Methodology of the Stochastic Bending Problem of the Levinson–Bickford Beam, *Acta Mechanica*, Vol. 233(9), 3467–3480, 2022.

Sri Namachchivaya N. and Lin Y.K., Application of Stochastic Averaging for Nonlinear Dynamic Systems with High Damping, in *Lecture Notes in Engineering*, No 31, Berlin: Springer, pp. 277–306, 1987, also *Probabilistic Engineering Mechanics*, Vol. 3, 159–167, 1988.

Shafer G., *A Mathematical Theory of Evidence*, Princeton, N.J: Princeton University Press, 1976.

Shafer G. and Pearl J. (ed.), *Readings in Uncertain Reasoning*, Morgan Kaufmann: Publishers Inc., 1990.

Stavroulakis G., Giovanis D.G., Papadrakakis M. and Papadopoulos V., A New Perspective on the Solution of Uncertainty Quantification and Reliability Analysis of Large-Scale Problems, *Computer Methods in Applied Mechanics and Engineering*, Vol. 276, 627–658, 2014.

Steel J.A. and Craik R.J.M., Statistical Energy Analysis of Structure-Borne Sound Transmission by Finite Element Methods, *Journal of Sound Vibration*, Vol. 178, 553–562, 1994.

Stefanou G., The Stochastic Finite Element Method: Past, Present and Future, *Computer Methods in Applied Mechanics and Engineering*, Vol. 198(9–12), 1031–1051, 2009.

Steinberg A. and Ryan E.P., Dynamic Output Feedback Control of a Class of Uncertain Systems, *IEEE Transactions on Automatic Control*, Vol. 31, 1163–1165, 1986.

Stenger J., Optimal Uncertainty Quantification of a Risk Measurement from a Computer Code, *PhD Dissertation*, Paul Sabatier Université Toulouse III, Toulouse, France, 2020.

Stefanou G., The Stochastic Finite Element Method: Past, Present and Future, *Computer Methods in Applied Mechanics and Engineering*, Vol. 198(9–12), 1031–1051, 2009.

Stolfi J. and De Figueiredo L.H., An Introduction to Affine Arithmetic, *TEMA Tend Mat Apl Comput*, Vol. 4, 297–312, 2003.

Stoyanov J., *Counterexamples in Probability*, Chichester: Wiley, pp. 89–91, 1987.

Strand A., Kjølaas J., Bergstrøm T.H., Steinsland I. and Hellevik L.R., Closure Law Model Uncertainty Quantification, *International Journal for Uncertainty Quantification*, Vol. 12(3), 1–23, 2022.

Stratonovich R.L., *Selected Problems of Theory of Fluctuation in Radiotechnics*, Moscow: "Sovetskoe Radio" Publishers, pp. 368 and 545, 1961 (in Russian).

Straub D., Value of Information Analysis with Structural Reliability Methods, *Structural Safety*, Vol. 49, 75–85, 2014.

Straub D. (ed.), *Reliability and Optimization of Structural Systems*, Boca Raton: CRC Press, 2010.

Straub D. and Papaioannou I., Bayesian Updating with Structural Reliability Methods, *Journal of Engineering Mechanics*, Vol. 141(3), paper 04014134, 2015.

Streletskii N.S., *Osnovy Statisticheskogo Ucheta Koeffizienta Zapasa Prochnosti Sooruzhenii (Basics of Statistical Analysis of Safety Factors of Structures)*, Moscow: Stroiizdat Publishers, 1947 (in Russian).

Stroud W., Krishnamurthy T. and Smith S., Probabilistic and Possibilistic Analyses of the Strength of a Bonded Joint, in *19th AIAA Applied Aerodynamics Conference*, AIAA 2001-1238 paper, New York: AIAA Press, 2002.

Stuart M.G., Safe Load Tables and the Human Dimension, *Steel Construction*, Vol. 24(1), 2–12, 1990.

Su Y.L., Wang Y.J. and Stekfanko R., Finite Element Analysis of Underground Stress Utilizing Stochastically Simulated Material Properties, *Procedure of 11th United States Symposium on Rock Mechanics*, CA, 1969.

Subbotin A.I. and Chentsov A.G., *Guaranteed Optimization in Control Systems*, Moscow: "Nauka" Publishing House, 1981(in Russian).

Sudret B., *Uncertainty Propagation and Sensitivity Analysis in Mechanical Models–Contributions to Structural Reliability and Stochastic Spectral Methods, Habilitationa diriger des recherches*, Clermont-Ferrand, France: Université Blaise Pascal, 2007.

Sudret B., Global Sensitivity Analysis Using Polynomial Chaos Expansions, *Reliability Engineering and System Safety*, Vol. 93(7), 964–979, 2008.

Sudret B., Meta-Models for Structural Reliability and Uncertainty Quantification, arXiv preprint arXiv:1203.2062, 2012.

Sudret B. and Der Kiureghian A., Stochastic Finite Element Methods and Reliability: A State-of-the-Art Report, in *Tech. Rep. UCB/SEMM-2000/08*, Berkeley: Department of Civil & Environmental Engineering, University of California, 2000.

Sudret B. and Kiureghian A.D., Comparison of Finite Element Reliability Methods, *Probabilistic Engineering Mechanics*, Vol. 17(4), 337–348, 2002.

Sudret B., Marelli S. and Wiart J., Surrogate Models for Uncertainty Quantification: An Overview, *in 11th European Conference on Antennas and Propagation (EUCAP)*, pp. 793–797, IEEE, 2017.

Sullivan T.J., *Introduction to Uncertainty Quantification*, Berlin: Springer, 2015.

Sunaga T., Theory of an Interval Algebra and Its Application to Numerical Analysis, *RAAG Memoirs*, Vol. 2 547–564, 1958.

Svetlitskii V.A., *Statistical Dynamics and Reliability Theory for Mechanical Systems*, Belin: Springer, 2003.

Swanson L.W., *Linear Programming – Basic Theory and Applications*, Auckland: McGraw-Hill, 1980.

Sweppe F.C., *Uncertain Dynamic Systems*, Englewood Cliffs NJ: Prentice Hall, 1973.

Świechowski M., Godlewski K., Sawicki B. and Mańdziuk J., Monte Carlo Tree Search: A Review of Recent Modifications and Applications, *Artificial Intelligence Review*, Vol. 56(3), 2497–2562, 2023.

Taleb N.N., *The Black Swan: The Impact of the Highly Improbable*, New York: Random House, 2007.

Nomoto T., Kado K., Yagawa G. and Yoshimura S., Development of User-friendly Structural Design System for Pressure Vessels, *ISME International Journal Series A-Mechanical Mathematical Engineering*, Vol. 39, 354–361, 1996.

Takewaki I., *Critical Excitation Methods in Earthquake Engineering*, Second edition, Amsterdam: Elsevier, 2013.

Takewaki I., Abbas M. and Fujita K., *Improving the earthquake Resilience of Buildings: The Worst-Case Approach*, Berlin: Springer, 2013.

Taleb N.N., *The Black Swan: The Impact of the Highly Improbable*, New York: Random House, 2007.

Taleb N.N., *Fooled by Randomness: The Hidden Role of Chance in Life and in the Markets*, New York: Random House, 2005.

Tamir D.E., Rishe N.D. and Kandel A., *Fifty Years of Fuzzy Logic and Its Applications*, Berlin: Springer, 2015.

Tang T. and Zhou T., Recent Developments in High Order Numerical Methods for Uncertainty Quantification, *Scientia Sinica Mathematica*, Vol. 45(7), 891–928, 2015.

Tappan S. and Pham T.D., Fuzzy FE Analysis of a Foundation on an Elastic Soil Medium, *International Journal of Numerical and Analytical Methods in Geomechanics*, Vol. 17, 771–789, 1993.

Tappan S. and Pham T.D., Elasto-Plastic Finite Element Analysis with Fuzzy Parameters, *International Journal of Numerical Methods Engineering*, Vol. 38, 531–548, 1995.

Taqqu M.S., Bibliographic Guide to Self-Similar Processes and Long-Range Dependence, in *Dependence in Probability and Statistics* (E. Eberlein and M.S. Taqqu, eds.), Boston: Birkhauser, 1986.

Taraga T., Theory of an Interval Algebra and Its Application to Numerical Analysis, *RAAG Memoirs*, Vol. 2, 547–565, 1958.

Teckentrup A.L., Multilevel Monte Carlo Methods and Uncertainty Quantification, *Doctoral Dissertation*, University of Bath, 2013.

Tekkens H., Karl Popper and the Accountability of Scientific Models, in *Predictability and Nonlinear Modelling in Natural Sciences and Economics* (J. Grasman and G. Van Straten, eds.), Dordrecht: Kluwer, pp. 6–10, 1994.

Tewfik A.H. and Kim Y., Correlation Structure of the Discrete Wavelet Coefficients of Fractional Brownian Motion, *IEEE Transactions on Information Theory*, Vol. 38(2), 904–909, 1992.

Theofanous T.G., Preface to the Discussion on Quantifying Reactor Safety Margins, *Nuclear Engineering and Design*, Vol. 132(3), 403, 1992.

Tezuka S., *Uniform Random Numbers: Theory and Practice*, Boston: Kluwer Academic Publishers, 1995.

Thacker B., Anderson M.C., Senseny P.E. and Rodriguez E.A., The Role of Nondeterminism in Computational Model Verification and Validation, in *46th AIAA/ASME/ASCE/AHS/ASC Structures, Structural Dynamics and Materials Conference*, paper 1902, 2005.

Thacker B.H. and Huyse L.J., Probabilistic Assessment on the Basis of Interval Data, *Structural Engineering and Mechanics*, Vol. 25(3), 331–345, 2007.

Thacker B.H. and Paez T.L., A Simple Probabilistic Validation Metric for the Comparison of Uncertain Model and Test Results, in *16th AIAA Non-Deterministic Approaches Conference*, article 0121, 2014.

Thoft-Christensen P. and Baker M.J., *Structural Reliability Theory and Its Applications*, Berlin: Springer Verlag, 1982.

Thomson W.T., Parameter Uncertainty in Dynamic Systems, *The Shock and Vibration Digest*, Vol. 7(8), 3–9, 1975.

Tichy M., *Applied Methods of Structural Reliability*, Dordrecht: Springer, 1993.

Tichy M. and Vorlicek M., *Statistical Theory of Concrete Structures*, Prague: Academia Publishing House, 1972.

Timashev S.A., *Reliability of Large Mechanical Systems*, Moscow: "Nauka" Publishing House, 1982, (in Russian).

Timoshenko S.P. and Gere J.M., *Theory of Elastic Stability*, Auckland: McGraw Hill, 1963.

To C.W.S., The Response of Nonlinear Structures to Random Excitation, *The Shock and Vibration Digest*, Vol. 16, 13–33, 1984.

To C.W.S., *Nonlinear Random Vibration: Analytical Techniques and Applications*, second edition, Boca Raton: CRC Press, 2012.

To C.W.S., *Stochastic Structural Dynamics: Application of Finite Element Methods*, New York: Wiley, 2013.

Todinov M.T., A New Reliability Measure Based on Specified Minimum Distances before the Locations of Random Variables in a Finite Interval, *Reliability Engineering & System Safety*, Vol. 86(1), 95–103, 2004.

Todinov M.T., *Risk-Based Reliability Analysis and Generic Principles for Risk Reduction*, Amsterdam: Elsevier, 2006.

Todinov M.T., Robust Design Using Variance Upper Bound Theorem, *International Journal of Performability Engineering*, Vol. 5(4), 339–356, 2009.

Todinov M., Optimal Allocation of Limited Resources among Discrete Risk-Reduction Options, *Artificial Intelligence Research*, Vol. 3(4), 15–27, 2014.

Todinov M., *Reliability and Risk Models: Setting Reliability Requirements*, New York: John Wiley, 2015.

Todinov M., *Methods for Reliability Improvement and Risk Reduction*, New York: John Wiley, 2018.

Todinov M., Reliability Improvement and Risk Reduction through Self-Reinforcement, *International Journal of Risk Assessment and Management*, Vol. 22(1), 18–43, 2019.

Todinov M., Using Algebraic Inequalities to Reduce Uncertainty and Risk, *ASCE-ASME Journal of Risk and Uncertainty in Engineering Systems, Part B: Mechanical Engineering*, Vol. 6(4), 2020.

Todinov M., Improving Reliability and Reducing Risk by Using Inequalities, *Safety and Reliability*, Vol. 38(4), 222–245, 2018.

Tonon F., On the Use of Random Set Theory to Bracket the Results of Monte Carlo Simulations, *Reliable Computing*, Vol. 10(2), 107–137, 2004a.

Tompson A.F.B., Ababou R. and Celhar L.W., Implementation of the Three-Dimensional Turming Bands Random Field Generator, *Water Resources Research*, Vol. 25(10), 2227–2243, 1989.

Tonon F., Using Random Set Theory to Propagate Epistemic Uncertainty through a Mechanical System, *Reliability Engineering & System Safety*, Vol. 85, 169–181, 2004b.

Tonon F., Some Properties of a Random Set Approximation to Upper and Lower Distribution Functions, *International Journal of Approximate Reasoning*, Vol. 48, 174–184, 2008.

Tonon F., Bae H.R. and Pettit C.L., Using Random Set Theory to Calculate Reliability Bounds for a Wing Structure, *Structures and Infrastructure Engineering*, Vol. 2(3–4), 191–200, 2006.

Tonon F. and Bernardini A., A Random Set Approach to the Optimization of Uncertain Structures, *Computers and Structures*, Vol. 68, 583–600, 1998.

Tonon F. and Bernardini A., Multi-Objective Optimization of Uncertain Structures through Fuzzy Set and Random Set Theory, *Computer-Aided Civil an Infrastructure Engineering*, Vol. 14, 119–140, 1999.

Tonon F., Using Random Set Theory to Propagate Epistemic Uncertainty through a 2004.

Tonon F. and Bernardini A., A Random Set Approach to the Optimization of Uncertain Structures, *Computers & Structures*, Vol. 68(6), 583–600, 1998.

Tonon F. and Bernardini A., Multiobjective Optimization of Uncertain Structures through Fuzzy Set and Random Set Theory, *Computer-Aided Civil and Infrastructure Engineering*, Vol. 14(2), 119–140, 1999.

Tonon F., Bernardini A. and Elishakoff I., Concept of Random Sets as Applied to the Design and Analysis of Expert Opinions for Aircraft Crash, *Chaos, Solitons & Fractals*, Vol. 10(11), 1855–1868, 1999.

Tonon F., Bernardini A. and Elishakoff I., Hybrid Analysis of Uncertainty: Probability, Fuzziness and Anti-Optimization, *Chaos, Solitons & Fractals*, Vol. 12(8), 1403–1414, 2001.

Tonon F., Bernardini A. and Mammino A., Determination of Parameters Range in Rock Engineering by Means of Random Set Theory, *Reliability Engineering & System Safety*, Vol. 70(3), 241–261, 2000.

Tonon F. and Chen S., Inclusion Properties for Random Relations under the Hypotheses of Stochastic Independence and Non-Interactivity, *International Journal of General Systems*, Vol. 34(5), 615–624, 2005.

Tonon F. and Pettit C., Structural Reliability Application of Correlation in Random-Set Valued Variables, in *Proceedings, 9th ASCE Joint Specialty Conference on Probabilistic Mechanics and Structural Reliability*, PMC04, Albuquerque, NM, ASCE, 2004.

Tonon F. and Pettit C., Toward a Definition and Understanding of Correlation for Variables Constrained by Random Relations, in *Proceedings of the 46th AIAA/ASME/ASCE/AHS/ASC Structures, Structural Dynamics & Materials Conference, and 7th AIAA Non-Deterministic Approaches Forum*, Austin, TX, 2005.

Tonon F. and Pettit C., Toward a Definition and Understanding of Correlation for Variables Constrained by Random Relations, in *Procedings of ICOSSAR'05*, Rome, Italy: Universita' La Sapienza, 2005.

Traishi N. and Furuta M., Reliability Analysis Based on Fuzzy Probability, *Journal of Engineering Mechanics*, Vol. 109, 1445–1459, 1983.

Tsompanakis Y., Lagaros N.D. and Papadrakakis M., eds., *Structural Design Optimization Considering Uncertainties: Structures & Infrastructures Book*, Boca Raton: CRC Press, 2008.

Tung C.C., Random Response of Highway Bridges to Vehicle Loads, *Journal of Engineering Mechanics*, Vol. 93(5), 79–94, 1966.

Tye W., Factors of Safety- or of Habit?, *Journal of Royal Aeronautical Society*, Vol. 48, 487–494, 1944.

Ulam S.M., *Adventures of a Mathematician*, New York: Charles Scribner's Sons, pp. 196–198, 1976.

Ushakov I., Reliability: Past, Present, Future, in *Recent Advances in Reliability Theory: Methodology, Practice, and Inference*, pp.3–21, 2000.

Ushakov I., Reliability: Past, Present, Future, *Reliability: Theory & Applications*, Vol. 1, 10–16, 2006.

Usoro P.B., Schweppe F.C., Wormley D.N. and Gould L.A., Ellipsoidal Set-Theoretic Control Synthesis, *Journal of Dynamic Systems of Measurement Continuity*, Vol. 104, 331–336, 1982.

Utkin L.V. and Gurov S.V., A General Formal Approach for Fuzzy Reliability Analysis in the Possibility Context, *Fuzzy Sets and Systems*, Vol. 83, 203–213, 1996.

Vaicaitis, Shinozuka and Takeno, Response Analysis of Tall Buildings to Wind Loading, *Journal of Structural Division*, Vol. 101(3), 585–600, 1975.

Valdebenito M.A. and Schuëller G.I., A Survey on Approaches for Reliability-Based Optimization, *Structural and Multidisciplinary Optimization*, Vol. 42(5), 645–663, 2010.

Vallée T., Kaufmann M., Adams R.D., Albiez M., Correia J.R. and Tannert T., Are Probabilistic Methods a Way to Get Rid of Fudge Factors? Part I: Background and Theory, *International Journal of Adhesion and Adhesives*, Vol. 120, article 103255, 2022.

Vallejo L.E. and Zhou Y., Fractal Dimension of Granular Materials and Their Engineering Properties, *Proceedings of 10th ASCE Engineering Mechanics Conference*, Boulder, CO, pp. 469–472, 1995.

Valliappan S., and Pham T.D., Fuzzy Finite Element Analysis of a Foundation on an Elastic Soil Medium, *International Journal for Numerical and Analytical Methods in Geomechanics*, Vol. 17(11), 771–789, 1993.

Valliappan S. and Pham T.D., Elasto-Plastic Finite Element Analysis with Fuzzy Parameters, *International Journal for Numerical Methods in Engineering*, Vol. 38(4), 531–548, 1995.

van Danzig D., Economic Decision Problems for Flood Prevention, *Econometrica*, Vol. 24, 276–287, 1956.

Van Mierlo C., Burmberger L., Daub M., Duddeck F., Faes M.G. and Moens D., Interval Methods for Lack-of-Knowledge Uncertainty in Crash Analysis, *Mechanical Systems and Signal Processing*, Vol. 168, article 108574, 2022.

Vanmarcke E.H., Some Recent Developments in Random Vibration, *Applied Mechanics Reviews*, Vol. 32(10), 1197–1202, 1979.

Vanmarcke E.H., *Random Fields: Analysis and Synthesis*, London: MIT Press, 1983.

Vanmarke E.H. and Grigoriu, Stochastic Finite Element Analysis of Simple Beams, *Journal of Engineering Mechanics*, Vol. 109(5), 1203–1214, 1983.

Vieira H.L., Beck A.T. and da Silva M.M., Combined Interval Analysis-Monte Carlo Simulation Approach for the Analysis of Uncertainties in Parallel Manipulators, *Meccanica*, Vol. 56(7), 1867–1881, 2021.

Venter G. and Haftka R., Using Response Surface Methodology in Fuzzy Set-Based Design Optimization, in *Proceedings, 39th AIAA/ASME/ASCE/AHS/ASC Structures, Structural Dynamics, and Materials Conference and Exhibit*, pp. 177–1785, 1998.

Venter G. and Haftka R.T., Using Response Surface Approximations in Fuzzy Set-Based Design Optimization, *Structural Optimization*, Vol. 18(4), 218–227, 1999.

Verhaeghe W., Elishakoff I., Desmet W., Vandepitte D. and Moens D., Uncertain Initial Imperfections via Probabilistic and Convex Modeling: Axial Impact Buckling of a Clamped Column, *Computers and Structures*, Vol. 121, 1–9, 2013.

Verhaeghe W. and Elishakoff I., Reliability-Based Bridging of the Gap between System's Safety Factors Associated with Different Failure Modes, *Engineering Structures*, Vol. 49, 606–614, 2013.

Verhulst F., Perturbation Theory from Lagrange to Van der Pol, *Nieuw Archief voor Wiskunde*, Vol. 2, 428–438, 1984.

Verma A.K., Ajit S. and Karanki D.R., *Reliability and Safety Engineering*, London: Springer, 2010.

Villaggio P., Sixty Years of Solid Mechanics, *Meccanica*, Vol. 46, 1171–1189, 2011.

Virkler D.A., Hillberg B.M. and Goel P.K., Statistical Nature of Fatigue Crack Propagation, *Journal of Engineering Mathematics Technology*, Vol. 101, 148–153, 1979.

Volodina V. and Challenor P., The Importnace of Uncertainty Quantification in Model Reproducibility, *Philosophical Transactions A*, Vol. 379, article 20200071, 2021.

von Neumann J., Various Techniques Used in Connection with Random Digits, *Journal of Research of the National Bureau of Standards, Applied Mathematics Series*, Vol. 3(3), 36–38, 1951.

von Neumann J., *Collected Works*, Vol. 5, 768, 1963.

Voss R.F., Random Fractal Forgeries, in *Fundamental Algorithms for Computer Graphics* (R.A. Earnshaw, ed.), Berlin: Springer, pp. 805–834, 1991.

Wald A., Statistical Decision Functions Which Minimize the Maximum Risk, *The Annals of Mathematics*, Vol. 46(2), 265–280, 1945.

Wald A., *Statistical Decision Functions*, New York: Wiley, 1950.

Walker A.C., Study and Analysis of the First 120 Failure Cases, in *Symposium Structural Failures in Buildings*, London: Institute of Structures, pp. 15–40, 1980.

Wallaschek J., On the Correlation between Velocity and Displacement in Continuous Mechanical Systems, *Conference on Mechanical Vibrations*, Oberwolfach., 1986.

Wallaschek J., *Spektraldichtenmethode und Integral-kovarianzanalyse bei stochastischen Schwingungen mechanischer Kontinua (Methods of spectral analysis and integral covariance analysis of stochastic vibration of mechanical continua*, Forschritt-Berichte VDI, Reihe 11: Schwingungstechnik, Nr 95 VDI-Verlag, Düsseldorf, 1987 (in German).

Walier G.G., *Wavelets and Other Orthogonal Systems with Applications*, Boca Raton: CRC Press, 1994.

Walley P., *Statistical Reasoning with Imprecise Probabilities*, London: Chapman and Hall, 1991.

Walley P., Towards a Unified Theory of Imprecise Probabilities, *International Journal of Approximate Reasoning*, Vol. 24, 125–148, 2000.

Walz N.-P., Fuzzy Arithmetical Methods for Possibilistic Uncertainty Analysis, *Ph.D. Thesis*, University of Stuttgart, Stuttgart, Germany, 2016.

Wang G. and Wang W., Fuzzy Reliability Analysis of Aseismic Structures, *Acta Mechanica Sinica*, Vol. 2, 322–332, 1986.

Wang C., Qiang X., Xu M. and Wu T., Recent Advances in Surrogate Modeling Methods for Uncertainty Quantification and Propagation, *Symmetry*, Vol. 14(6), article 1219, 2022.

Wang C. and Zhang X.T., An Improved Equivalent Linearization Technique in Nonlinear Random Vibration, *Proceedings, International Conference on Nonlinear Mechanics*, pp. 959–964, 1985.

Wang C., Qiu Z., Xu M. and Li Y., Novel Reliability-Based Optimization Method for Thermal Structure with Hybrid Random, Interval and Fuzzy Parameters, *Applied Mathematical Modelling*, Vol. 47, 573–586, 2017.

Wang G.Y., On the Development of Uncertain Structural Mechanics, *Advances in Applied Mechanics*, Vol. 32(2), 205–211, 2002, (in Chinese).

Wang J.H., Qian J.Z., Li R.Z. and Chen T.H., Improvement and Application of Fuzzy Probabilistic Method, *Systems Engineering-Theory & Practice*, Vol. 27(5), 173–176, 2007.

Wang X.J., Robust Reliability of Structural Vibration, *Journal of Beijing University of Aeronautics and Astronautics*, Vol. 29(11), 1006–1010, 2003, (in Chinese).

Wang X.-J., Elishakoff I. and Qiu Z.-P., Experimental Data has to Decide Which of the Non-Probabilistic Uncertainty Descriptions – Convex Modeling or Interval Analysis – to Utilize, *Journal of Applied Mechanics*, Vol. 75(4), Paper 041018, 2008, 2008.

Wang X.-J., Elishakoff I., Qiu Z.-P. and Lihong Ma L.-H., Comparison of Probabilistic and Two Non-Probabilistic Methods for Uncertain Imperfection Sensitivity of a Column on a Nonlinear Mixed Quadratic-Cubic Foundation, *Journal of Applied Mechanics*, Vol. 76(1), paper 011007, 2009.

Wang X.-J., Elishakoff I., Qiu Z.-P. and Kou C.-H., Hybrid Theoretical, Experimental and Numerical Study of Vibration and Bucking of Composite Shells with Scatter in Elastic Modules, *International Journal of Solids and Structures*, Vol. 46(13), 2539–2546, 2009.

Wang X.-J., Elishakoff I., Qiu Z.-P. and Kou C.-H., Non-Probabilistic Methods for Natural Frequency and Buckling Load of Composite Plate Based on the Experimental Data, *Mechanics Based Design of Structures and Machines*, Vol. 39(1), 83–99, 2011.

Wang X.-J., Elishakoff I., Qiu Z.-P. and Ma L.-H., Comparison of Probabilistic and Two Non-Probabilistic Methods for Uncertain Imperfection Sensitivity of a Column on a Nonlinear Mixed Quadratic-Cubic Foundation, *Journal of Applied Mechanics*, Vol. 76(1), article 011007, 2009.

Wang X.-J., Qiu Z.-P. and Elishakoff I., Non-Probabilistic Set-Theoretic Model for Structural Safety Measure, *Acta Mechanica*, Vol. 198(1–2), 51–64, 2008.

Wang X.-J., Wang L., Elishakoff I. and Qiu Z.-P., Probability and Convexity are Not Antagonistic, *Acta Mechanica*, Vol. 219, 45–64, 2011.

Wang X.J., Shi Q., Fan W., Wang R. and Wang L., Comparison of the Reliability-Based and Safety Factor Methods for Structural Design, *Applied Mathematical Modelling*, Vol. 72, 68–84, 2019.

Wang X.-J., Wang L. and Qiu Z.P., A Feasible Implementation Procedure for Interval Analysis Method from Measurement Data, *Applied Mathematical Modelling*, Vol. 38(9–10), 2377–2397, 2014.

Wang X.-J., Xia Y., Zhou X. and Yang C., Structural Damage Measure Index Based on Non-Probabilistic Reliability Model, *Journal of Sound and Vibration*, Vol. 333(5), 1344–1355, 2014.

Wang Y. and McDowell D.L., (eds.), *Uncertainty Quantification in Multiscale Materials Modeling*, Duxford, U.K.: Elsevier, 2020.

Wang Y. and McDowell D.L., Uncertainty Quantification in Materials Modeling, in *Uncertainty Quantification in Multiscale Materials Modeling*, Woodhead Publishing, pp. 1–40, 2020.

Washizu K., *Variational Methods in Elasticity and Plasticity*, Oxford: Pergamon Press, 1968.

Wedig W., Zufallsschwingungen von querangestroemten Saiten, *Ingenieur-Archiv*, Vol. 48, 325–335, 1979, (in German).

Wedig W., Stationäre Zuffallsschwingungen von Balken-eineneue Methode Zur Kovarianzanalyse, *ZAMM: Zeitschrift für angewandte Mathematik und Mechanik*, Vol. 60, T89–T91, 1980, (in German).

Wedig W., Bettungs- und Dämpfungsanalyze homogene verteilter Systeme, *VDI-Berichte*, 456, 201–207, 1982, (in German).

Wedig W., Covariance Analysis of Distributed Systems under Stochastic Point Forces, in *Proceedings EQUADIFF'82* (H.W. Knobloch and K. Schmidt, eds.), Berlin: Springer, 1992.

Wei T., Li F., Meng G. and Zuo W., Static Response Analysis of Uncertain Structures with Large-Scale Unknown-but-Bounded Parameters, *International Journal of Applied Mechanics*, Vol. 13, article 2150004, 2021.

Weibull W., A Statistical Theory of the Strength of Materials, *Proceedings of Royal Swedish Institute for Engineering Research*, Stockholm, No. 151, 1939.

Weichselberger K., The Theory of Interval-Probability as a Unifying Concept for Uncertainty, *International Journal of Approximate Reasoning*, Vol. 24(2–3), 149–170, 2000.

Weintraub P., Fuzzy Finite Element Analysis, *MSME Thesis*, Department of Mechanical Engineering, Purdue University, 1997.

Weirs V.G., Fabian N., Potter K., McNamara L. and Otahal T., Uncertainty in the Development and Use of Equation of State Models, *International Journal for Uncertainty Quantification*, Vol. 3(3), 255–270, 2013.

Wemelsfelder P.J., Wetmatighden in het Optreden van Stormvlveden, *De Ingenieur*, Nov 9, 1939 (in Dutch).

Wen Y.K., Methods of Random Vibration for Inelastic Structures, *Applied Mechanics Reviews*, Vol. 42(2), 39–52, 1989.

Wen Y.K., Corotis R.B., Turkstra C.J., Reed J. and Mohammadi J., Definition of Risk, in *Guidelines for Design of Low-Rise Buildings Subjected to Lateral Forces*, Boca Raton: CRC Press, pp. 11–42, 2020.

Wentz K.R., Paul D.B. and Mei C., Large Deflections Random Response of Symmetric Laminated Composite Plates, *Shock and Vibration Bulletin*, Vol. 52, 99–111, 1982.

Wentz K.R., Paul D.B. and Mei C., Large Deflection Random Response of Symmetric Laminated Composite Plates, *Shock and Vibration Bulletin*, Vol. 52, 99–111, 1982.

Wentzel E.C., *Operations Research: Problems, Principles, Methodology*, Moscow: Nauka, 1980 (in Russian).

Wentzel E.C., *Investigation of Operations: Problems, Principles, Methodology*, Moscow: Nauka, 1980 (in Russian).

Wijker J.J., *Random Vibrations in Spacecraft Structures Design: Theory and Applications*, Dordrecht: Springer, 2009.

Wikipedia, Uncertainty Quantification, https://en.wikipedia.org/wiki/Uncertainty_quantification, accessed 1.15, 2020.

Wikipedia, the Free Encyclopedia, Entropy (Information Theory), available at https://en.wikipedia.org/wiki/Entropy, accessed on 29 December 2022.

Wills A.G. and Schön T.B., Sequential Monte Carlo: A Unified Review, *Annual Review of Control, Robotics, and Autonomous Systems*, Vol. 6, 159–182, 2023.

Wilson E.L., Der Kiureghain A. and Bayo E., A Replacement for the SRSS Method in Seismic Analysis, *Earthquake Engineering Structues Dynamics*, Vol. 9, 187–192, 1981.

Wilson G.E., Boyack B.E., Catton I., Duffey R.B., Katsma K.R. and Sli G., Quantifying Reactor Safety Margins Part 2: Characterization of Important Contributors to Uncertainty, *Nuclear Engineering and Design*, Vol. 119(1), 17–31, 1990.

Wilson K., Problems in Physics with Many Scales of Length, *Science of America*, Vol. 8, 140–157, 1979.

Wilson K.N., Renormgroup and Critical Values, *Advances in Physical Sciences*, Vol. 141(2), 193–220, 1983.

Wilson G.E., Boyack B.E., Catton I., Duffey R.B., Katsma K.R. and Sli G., Quantifying Reactor Safety Margins Part 2: Characterization of Important Contributors to Uncertainty, *Nuclear Engineering and Design*, Vol. 119(1), 17–31, 1990.

Wirshing P.H., Paez T.L. and Ortiz K., *Random Vibrations*, New York: Wiley, 1995.

Witt M. and Sobczyk K., Dynamic Response of Laminated Plates to Random Loading, *International Journal of Solids Structures*, Vol. 16, 231–238, 1980.

Wittig L.E. and Sinha A.K., Simulation of Multicorrelated Random Processes Using the FFT Algorithm, *Acoustical*.

Wolfram S., *The Mathematica Book*, Third Edition, New York: Cambridge University Press, p. 745, 1996.

Wood K.L., Antonsson E.K. and Beck J.L., Representing Imprecision on Engineering Design: Comparing Fuzzy and Probabilistic Calculus, *Research in Engineering Design*, Vol. 1, 187–203, 1990.

Wong E., *Stochastic Processes in Information and Dynamical Systems*, New York: McGraw-Hill, 1971.

Womell G.W., Karhunen-Loeve-Like Expansion for Processes via Wavelets, *IEEE Transactions of Information Theory*, Vol. 36(4), 859–861, 1990.

Wu H.-C., Fuzzy Reliability Analysis Based on Closed Fuzzy Numbers, *Information Science*, Vol. 103, 135–159, 1997.

Wu J., Zhang Y., Chen L. and Luo Z., A Chebyshev Interval Method for Nonlinear Dynamic Systems under Uncertainty, *Applied Mathematical Modeling*, Vol. 37(4), 578–4591, 2013.

Wu W.F., Comparison of Gaussian Closure Technique and Equivalent Linearization Method, *Probabilistic Engineering Mechanics*, Vol. 2(1), 2–8, 1987.

Wu X., Luo Y.X., Wen H.J. and Li M., Interval Analysis Method of Uncertain Structural Systems Using Universal Grey Number, *Chinese Journal of Computational Mechanics*, Vol. 20(3), 329–334, 2003 (in Chinese).

Wu Y.-T., Computational Methods for Efficient Structural Reliability and Reliability Sensitivity Analysis, *34th AIAA-ASME-ASCE-AHS-ASC Structures, Structural Dynamics, and Mathematics Conference*, La Jolla, CA, 1993.

Xiu D.B., *Numerical Methods for Stochastic Computations: A Spectral Method Approach*, Princeton University Press, 2010.

Xu J., Du J., Chen C., Wang Y. and Li Y., An Iterative Dimension-Wise Approach to the Structural Analysis with Interval Uncertainties, *International Journal of Computational Methods*, Vol. 15(6), article 1850044, 2018.

Xu X.F., Generalized Variational Principles for Uncertainty Quantification of Boundary Value Problems of Random Heterogeneous Materials, *Journal of Engineering Mechanics*, Vol. 135(10), 1180–1188, 2009.

Xu X.F. and Stefanou G., Convolved Orthogonal Expansions for Uncertainty Propagation: Application to Random Vibration Problems, *International Journal for Uncertainty Quantification*, Vol. 2(4), 383–395, 2012.

Yagawa G., Yoshimura S. and Nakao K., Automatic Mesh Generation of Computer Geometries Based on Fuzzy Knowledge Processing and Computational Geometry, *Integrated Computer- Aided Engineering*, Vol. 2, 265–280, 1995.

Yagawa G., Yoshimura S., Soneda N. and Nakao K., Automatic 2D and 3D Mesh Generation Based on Fuzzy Knowledge Processing, *Computational Mechanic, S*, Vol. 9, 333–346, 1992.

Yager R.R. (ed.), *Fuzzy Sets and Possibility Theory: Recent Developments*, New York: Pergamon Press, 1982.

Yager R.R., The Entailment Principle for Dempster-Shafer Granules, *International Journal of Intelligent Systems*, Vol. 1, 247–262, 1986.

Yager R.R., On the Dempster-Shafer Framework and New Combination Rules, *Information Sciences*, Vol. 41, 93–137, 1987.

Yager R.R., On Probabilities Induced by Multi-Valued Mappings, *Fuzzy Sets and Systems*, Vol. 42, 301–314, 1991.

Yager R.R., Aggregating Fuzzy Sets Represented by Belief Structures, *Journal of Intelligent and Fuzzy Systems*, Vol. 1, 215–224, 1993.

Yager R.R. and Liu L.-P. (eds.), *Classic Works of the Dempster-Shafer Theory of Belief Functions*, Berlin: Springer, 2008.

Yamazaki F. and Shinozuka M., Digital Generation of Non-Gaussian Stochastic Fields, *Journal of Engineering Mechanics*, Vol. 114(7), 1183–1197, 1988.

Yamazaki F. and Shinozuka M., Simulation of Stochastic Fields by Statistical Preconditioning, *Engineering Mechanics*, Vol. 116(2), 268–287, 1990.

Yamazaki F., Shinozuka M. and Dasgupta G., Neumann Expansion for Stochastic Finite Element Analysis, in *Stochastic Mechanics*, Vol. 1, Columbia University, 1986.

Yang J.-N., Simulation of Random Envelope Processes, *Sound and*.

Yang J.-N., On Normality and Accuracy of Simulated Random Processes, *Sound and*, Vol. 26(3), 417–428, 1973.

Yang C.Y., *Random Vibration of Structures*, New York: Wiley-Interscience, 1986.

Yang H., Gorodetsky A., Fujii Y. and Wang K.W., A Polynomial-Chaos-Based Multifidelity Approach to the Efficient Uncertainty Quantification of Online Simulations of Automotive Propulsion Systems, *Journal of Computational and Nonlinear Dynamics*, Vol. 17(5), article 051012, 2022.

Yang X., Liu J., Chen X., Qing Q. and Wen G., Hybrid Structural Reliability Analysis under Multisource Uncertainties Based on Universal Grey Numbers, *Shock and Vibration*, Vol. 2018(article 3529479), 2018.

Yao J.T.P., Damage Assessment and Reliability Evaluation of Existing Structures, *Journal of Engineering Structures*, Vol. 1, 245–251, 1979.

Yao J.T.P., Damage Assessment of Existing Structures, *Journal of Engineering Mechanics*, Vol. 106, 785–799, 1980.

Yao W., Chen X., Luo W., Van Tooren M. and Guo J., Review of Uncertainty-Based Multidisciplinary Design Optimization Methods for Aerospace Vehicles, *Progress, Aerospace Sciences*, Vol. 47(6), 450–479, 2011.

Ye D., Zun P., Krzhizhanovskaya V. and Hoekstra A.G., Uncertainty Quantification of a Three-Dimensional in-Stent Restenosis Model with Surrogate Modelling, *Journal of the Royal Society Interface, Vol*, Vol. 19(187), article 20210864, 2022.

Young D.M., Stresses in Eccentrically Loaded Steel Columns, *Publication of the International Association of Bridge Structural Engineers*, Vol. 1, 1932.

Young R.C., Algebra of Many-valued Quantities, *Mathematische Annalen*, Vol. 104, 260–290, 1931.

Young W.H., Sull due funzioni a piu valori constituite dai limiti d'una funzione di variabile reale a destra ed a sinistra di ciascun punto, *Rendiconti Accademia di Licei, Classe di Scienza Fiziche*, Vol. 5, 582–587, 1908.

Yoshikawa N., Elishakoff I. and Nakagiri S., Worst-Case Estimation of Homology Design by Convex Analysis, *Computers and Structures*, Vol. 67, 191–196, 1998.

Yu Z.D. and Gao G.A., *Theory of Random Vibration with Applications*, Tongji University Press, 1988 (in Chinese).

Yuan X., Faes M.G., Liu S., Valdebenito M.A. and Beer M., Efficient Imprecise Reliability Analysis Using the Augmented Space Integral, *Reliability Engineering & System Safety*, Vol. 210, article 107477, 2021.

Yuan X., Liu S., Valdebenito M., Gu J. and Beer M., Efficient Procedure for Failure Probability Function Estimation in Augmented Space, *Structural Safety*, Vol. 92, article 102104, 2021.

Yubin L., Zhong Q. and Guangyuan W., Fuzzy Random Reliability of Structures Based on Fuzzy Random Variables, *Fuzzy Sets Systems*, Vol. 86, 345–355, 1997.

Zadeh L.A., Fuzzy Sets, *Information Control*, Vol. 8, 338–353, 1965.

Zadeh L.A., Outline of a New Approach to the Analysis of Complex Systems and Decision Processes, *IEEE Transnational Systems Management Cybernetics SMC-3*, 28–44, 1973.

Zadeh L.A., The Concept of a Linguistic Variable, and Its Applications to Approximate Reasoning, *Information Sciences*, Vol. 8, 199–249, 1975a.

Zadeh L.A., *Fuzzy Sets and Applications: Selected Papers* (R.R. Yager, ed.), New York: Wiley, 1975b.

Zadeh L.A., The Concept of a Linguistic Variable and Its Application to Approximate Reasoning – Part 1, *Information Science*, Vol. 8(3), 199–249, 1975c.

Zadeh L.A., The Concept of a Linguistic Variable and Its Application to Approximate Reasoning – Part 2, *Information Science*, Vol. 8(4), 301–357, 1975d.

Zadeh L., Fuzzy Sets as a Basis for a Theory of Possibility, *Fuzzy Sets and Systems*, Vol. 1(1), 3–28, 1978.

Zadeh L.A., A Mathematical Theory of Evidence (book review), *AI Magazine*, Vol. 5, 81–83, 1984.

Zadeh L.A., Why the Success of Fuzzy Logic is Not Paradoxical, *IEEE Expert Intelligent Systems and Their Applications*, Vol. 9(4), 43–46, 1994.

Zadeh L.A., The Birth and Evolution of Fuzzy Logic, *International Journal of General Systems*, Vol. 17(2–3), 1990, 2008a.

Zadeh L.A., Is There a Need for Fuzzy Logic?, *Information Sciences*, Vol. 178, 2751–2779, 2008b.

Zadeh L.A., Fuzzy Logic—A Personal Perspective, *Fuzzy Sets and Systems*, Vol. 281, 4–20, 2015.

Zao Y.-G. and Lu Z.-H., *Structural Reliability Approaches from Perspectives of Statistical Moments*, Hoboken, NJ: Wiley, 2021.

Zeldin B.A., Representation and Synthesis of Random Fields: ARMA, Galerkin, and Wavelet Procedures, *PhD Thesis*, Rice Univ, Houston, TX, 1996.

Zeldin B.A. and Spanos P.D., Random Field Simulation Using Wavelet Bases, *Proceedings of 7th International Conference on Applications of Statistics and Probability in Civil Engineering*, Paris: France.

Zeldin B.A. and Spanos P.D., Random Field Representation and Synthesis Using Wavelet Bases, *Journal of Applied Mechanics*, Vol. 63, 946–952, 1997.

Zhang H., Mullen R.L. and Muhanna R.L., Interval Monte Carlo Methods for Structural Reliability, *Structural Safety*, Vol. 32(3), 183–190, 2010.

Zhang J., Modern Monte Carlo Methods for Efficient Uncertainty Quantification and Propagation: A Survey, *Wiley Interdisciplinary Reviews: Computational Statistics*, Vol. 13(5), article e1539, 2021.

Zhang J. and Ellingwood B., Orthogonal Series Expansions of Random Fields in Reliability Analysis, *Engineering Mechanics*, Vol. 120(12), 2660–2677, 2004.

Zhang R.Ch., Elishakoff I. and Shinozuka M., Analysis of Nonlinear Sliding Structures by Modified Stochastic Linearization Methods, in *Probabilistic Mechanics and Structural and Geotechnical Reliability* (Y.K. Lin, ed.), New York: ASCE Press, pp. 196–199, Extended version, *International Journal of Nonlinear Dynamics*, Vol. 5, 299–312, 1992.

Zhang X.T., Equivalent Potential Technique for Deterministic and Random Response Analysis of Nonlinear Systems, *Applied Mechanics*, IAP, 1989 (in Chinese).

Zhang X., Study of Weighted Energy Technique in Analysis of Nonlinear Random Vibration, in *Nonlinear Vibration and Chaos*, Tianjin University Press, pp. 139–144, 1992.

Zhang X.T. and Zhang R.Ch., A Stochastic Response-Consistent Equivalence Technique, *13th ASCE Engineering Mechanics Conference*, The Johns Hopkins University, Baltimore, June 13–16, 1999.

Zhang X. and Zhang R., Energy-Based Stochastic Equivalent Linearization with Optimized Power, in *Stochastic Structural Dynamics* (B.F. Spencer Jr. and E.A. Johnson, eds.), Rotterdam: Balkema, pp. 113–117, 1999.

Zhang X.T., Elishakoff I. and Zhang R.C., A Stochastic Linearization Technique Based on Minimum Mean Square Deviation of Potential Energies, in *Stochastic Structural Dynamics - New Theoretical Developments* (Y.K. Lin and I. Elishakoff, eds.), Berlin: Springer, pp. 327–338, 1991.

Zhang Z. and Jiang C., Evidence-Theory-Based Structural Reliability Analysis with Epistemic Uncertainty: A Review, *Structural and Multidisciplinary Optimization*, Vol. 63(6), 2935–2953, 2021.

Zhang Y., Liu Y. and Guo Q., A Global Sensitivity Analysis Approach for Multiple Failure Modes Based on Convex-Probability Hybrid Uncertainty, *Engineering Computations*, Vol. 38(3), 1263–1286, 2021.

Zhang Y., Kim N.H., Palliyaguru U.R., Schutte J.F. and Haftka R.T., Reduced Allowable Strength of Composite Laminate for Unknown Distribution Due to Limited Tests, *Journal of Composite Materials*, Vol. 54(21), 2823–2836, 2020.

Zhang Y., Comerford L., Kougioumtzoglou A.I., Patelli E. and Beer M., Uncertainty Quantification of Power Spectrum and Spectral Moments Estimates Subject to Missing Data, *Journal of Risk and Uncertainty of Engineering Systems, Part A: Civil Engineering*, Vol. 3(4), article 04017020, 2017.

Zhao M.Y., Yan W.J., Yuen K.V. and Beer M., Non-Probabilistic Uncertainty Quantification for Dynamic Characterization Functions Using Complex Ratio Interval Arithmetic Operation of Multidimensional Parallelepiped Model, *Mechanical Systems and Signal Processing*, Vol. 156, article 107559, 2021.

Zhao R. and Goving R., Defuzzification of Fuzzy Intervals, *Fuzzy Sets and Systems*, Vol. 43, 45–55, 1991.

Zhao R. and Govind R., Solutions of Algebraic Equations Involving Generalized Fuzzy Numbers, *Information Science*, Vol. 56, 199–243, 1991.

Zhou B., Zi B. and Zhu W., Static Response Analysis of a Dual Crane System Using Fuzzy Parameters, *Journal of Computing and Information Science in Engineering*, Vol. 21(6), 1–43, 2021.

Zhou K., Ni Z., Huang X. and Hua H., Stationary/Nonstationary Stochastic Response Analysis of Composite Laminated Plates with Aerodynamic and Thermal Loads, *International Journal of Mechanical Sciences*, Vol. 173, article 105461, 2020.

Zhou X.Y., Wang N.W., Gao K., Natarajan S., Xiong W., Jiang C., Qian S.Y. and Cai C.S., Bounds of Mechanical Properties of Fiber Reinforced Polymer Composites with Hybrid Random and Interval Uncertainties, *Thin-Walled Structures*, Vol. 182, article 110158, 2023.

Zhu E., Mandal P. and And Calladine C.R., Buckling of Thin Cylindrical Shells: An Attempt to Resolve a Paradox, *International Journal of Mechanical Sciences*, Vol. 44, 1583–1601, 2002.

Zhu L.P. and Elishakoff I., Probabilistic and Convex Modeling of Excitation and Response of Periodic Structures, *Journal of Mathematical Problems in Engineering*, Vol. 2, 143–163, 1996a.

Zhu L.P. and Elishakoff I., Hybrid Probabilistic and Convex Modeling of Excitation and Response of Periodic Structures, *Mathematics of Problems in Engineering*, Vol. 2, 143–163, 1996b.

Zhu L.P., Elishakoff I. and Starnes J.H. Jr., Derivation of Multi-Dimensional Ellipsoidal Convex Model for Experimental Data, *Mathematical Computing and Modeling*, Vol. 24, 103–114, 1996.

Zhu W.Q., *Random Vibrations*, Beijing: Science Press, 1992 (in Chinese).

Zhu W.Q. and Lei Y., Wide-Band Random Vibration of Rectangular Plates with Commensurable Aspect Ratio, *Proceedings of International Conference on Vibration Problems in Engineering*, Xian, China, pp. 890–895, 1986.

Zhu Y., Zabaras N., Koutsourelakis P.S. and Perdikaris P., Physics-Constrained Deep Learning for High-Dimensional Surrogate Modeling and Uncertainty Quantification without Labeled Data, *Journal of Computational Physics*, Vol. 394, 56–81, 2019.

Ziegler F., Private Communication to I.E., 2010.

Ziegler H., *Principles of Structural Stability*, Waltham: Blaisdell, 1968.

Zienkiewicz O.C., The Finite Element Method: From Intuition to Generality, *Applied Mechanics Reviews*, Vol. 23, 249–256, 1970.

Zimmermann H.-J., *Fuzzy Set Theory and Its Applications*, Second Edition, Dordrecht: Kluwer, 1991.

Zincik D.G. and Tennyson R.C., Stability of Circular Cylindrical Shells under Transient Axial Impulsive Loading., *AIAA Journal*, Vol. 18, 691–699, 1980.

Zio E. and Pedroni N., An Optimized Line Sampling Method for the Estimation of the Failure Probability of Nuclear Passive Systems, *Reliability Engineering and System Safety*, Vol. 95, 1300–1313, 2010.

Author Index

https://doi.org/10.1515/9783111354231-014

Subject Index

https://doi.org/10.1515/9783111354231-015

www.ingramcontent.com/pod-product-compliance
Lightning Source LLC
Chambersburg PA
CBHW082105220326
41598CB00066BA/5353